MENTAL RETARDATION: RESEARCH, EDUCATION, AND TECHNOLOGY TRANSFER

ANNALS OF THE NEW YORK ACADEMY OF SCIENCES
Volume 477

MENTAL RETARDATION: RESEARCH, EDUCATION, AND TECHNOLOGY TRANSFER

Edited by Henryk M. Wisniewski and Donald A. Snider

The New York Academy of Sciences
New York, New York
1986

Cover: Idiograms of normal X chromosome (*left*) and fragile X chromosome (*right*); drawing by Lucille Donadio.

Library of Congress Cataloging-in-Publication Data

Mental retardation.

   (Annals of the New York Academy of Sciences; v. 477)
   Proceedings of a workshop entitled Mental retardation and developmental disabilities, held 24–26 May 1985 in Staten Island and sponsored by the New York State Institute for Basic Research in Developmental Disabilities, and others.
   Includes bibliographies and index.
   1. Developmental disabilities—Congresses.   2. Mental retardation—Congresses.   3. Mentally handicapped children—Education—Congresses.   4. Technology transfer—Congresses.   I. Wisniewski, Henryk M., 1931–
II. Snider, Donald A.   III. New York Academy of Sciences.
IV. New York (State). Institute for Basic Research in Developmental Disabilities. V. Series. [DNLM: 1. Child Development Disorders—congresses.   2. Mental Retardation—congresses. W1 AN626YL v.477 1986 /
WM 300 M5535 1985]
Q11.N5 vol. 477   500 s   86-28537
[RJ135]   [618.92′8588]
ISBN 0-89766-349-7
ISBN 0-89766-350-0 (pbk.)

BiC/PCP
Printed in the United States of America
ISBN 0-89766-349-7 (cloth)
ISBN 0-89766-350-0 (paper)

ANNALS OF THE NEW YORK ACADEMY OF SCIENCES
Volume 477
December 22, 1986

MENTAL RETARDATION: RESEARCH, EDUCATION, AND TECHNOLOGY TRANSFER[a]

Editors

HENRYK M. WISNIEWSKI AND DONALD A. SNIDER

Workshop Organizing Committee

HENRYK M. WISNIEWSKI, DONALD A. SNIDER, W. TED BROWN,
RAEF K. HADDAD, ROBERT LUBIN, WAYNE SILVERMAN, AND DAVID SOIFER

CONTENTS

[a] This volume is the result of a workshop entitled Mental Retardation and Developmental Disabilities: Research, Education, and Technology Transfer, held in Staten Island, New York on May 24–26, 1985 and sponsored by the New York State Institute for Basic Research in Developmental Disabilities and cosponsored by the National Institute of Child Health and Human Development, the Institut de la Vie, and the New York Academy of Sciences.

Financial assistance was received from:
 • BRISTOL MYERS COMPANY
 • MEAD JOHNSON COMPANY
 • NEW YORK STATE OFFICE OF MENTAL RETARDATION &
   DEVELOPMENTAL DISABILITIES

Preface

HENRYK M. WISNIEWSKI AND DONALD A. SNIDER

*Institute for Basic Research
in Developmental Disabilities
New York State Office of Mental Retardation &
Developmental Disabilities
Staten Island, New York 10314*

The field of mental retardation and developmental disabilities is multidisciplinary, and it has been difficult to draw from each of the specialties to build a solid foundation of knowledge to guide practice. Today, the explosive growth of biomedical information in genetics, neurology, molecular biology, and biochemistry has provided new hope for preventing mental retardation and developmental disabilities and helping those now affected. Unfortunately, much of this basic knowledge does not readily permeate the disciplinary boundaries that subdivide the field. Clinicians are not regularly enriched by exchange with scientists generating basic research information about developmental disabilities. By the same token the basic scientists are rarely exposed to clinical perspectives or to the thinking of scientists in different disciplines working on similar problems.

For these reasons we set out to create a forum for sharing basic and clinical information across disciplinary lines in the field of mental retardation and developmental disabilities (henceforth referred to simply as mental retardation). The idea was to bring together basic scientists, clinician researchers, and educators and health policymakers to share new data, review the state of the art in key areas, and examine future directions for research, education, and technology transfer. We began by identifying a series of topical areas that we believe are critical to the developmental disabilities field and quickly found that we had to limit the papers to epidemiological and biomedical topics and some selected psychological aspects of diagnosis, if we were to have a manageable meeting.

Having agreed on this overall goal and a set of topics, we sought to attract leading researchers in each area, as well as speakers with insights on related education and policy issues. Many of the leaders in research into the causes and manifestations of mental retardation are, quite naturally, at the Mental Retardation Research Centers (better known as the Kennedy Centers) funded by the National Institute of Child Health and Human Development (NICHD) of the National Institutes of Health (NIH); therefore, we turned to Dr. Terrence Dolan, Chairperson of the Association of Mental Retardation Research Center Directors (and Director of the Waisman Center on Mental Retardation and Human Development, University of Wisconsin) to obtain papers from these centers. We looked within our own ranks at the New York State Institute for Basic Research in Developmental Disabilities (IBR), the largest research center in the country devoted exclusively to mental retardation and developmental disabilities, for new data and perspectives. We also sought out scientists who are important contributors to the respective topics, but are not affiliated with a Kennedy Center or the IBR. Dr. Sumner Yaffe, Director of the Center for Research for Mothers and Children, NICHD, one of the cosponsors, was most helpful in this regard. Dr. Maurice Marois, President of the Institut de la Vie, an organization that includes a large number of Nobel laureates and organizes conferences on important topics and issues in the life sciences, was also helpful in this task.

Our goal was to produce a unique mix of researchers, clinicians, and educators who could not only make a substantive contribution to the topics on the agenda but who would also be enriched by discussions of or different perspectives on other topics (education versus basic research versus clinical research and practice). We sought to accomplish this by providing both formal and informal opportunities for discussion.

The editors' thanks go to the other members of the IBR organizing committee: Dr. Robert Lubin, Dr. David Soifer, Dr. Raef Haddad, Dr. Wayne Silverman, and Dr. Ted Brown, and to other colleagues who helped in many other ways to make the workshop a success.

A special word of thanks goes to the New York Academy of Sciences (NYAS). We had the good fortune of having our applications for NYAS cosponsorship reviewed by Dr. Philip Siekevitz, Dr. Muriel Feigelson, and Dr. Olga Greengard of the Conference Committee. Dr. Siekevitz's subcommittee's questions and suggestions proved invaluable in focusing the program and staging the workshop.

We would also like to acknowledge the financial support of the Bristol-Myers Company and the Mead Johnson Company, which helped make the workshop possible. The involvement of Bradley Simmons, who at that time was administrative vice president of the Scientific Division of Bristol-Myers, was particularly valuable.

We would like to thank as well Paul Puccio, Director of Program Planning and Analysis, New York State Office of Mental Retardation and Developmental Disabilities, for his assistance in planning and arranging financial support for the workshop.

Before we turn to the papers in this volume we would like to say a few words to provide a public policy perspective on this workshop. Why would a state agency, the New York Office of Mental Retardation and Developmental Disabilities (NYS OMRDD) support and, in fact, encourage its research arm, the Institute for Basic Research, to sponsor and organize a meeting of national scope? For NYS OMRDD this represents an opportunity to look to the future of the field of mental retardation and to move out of the dark shadows of mental retardation services in the state's past—the Willowbrook lawsuit. Arthur Y. Webb, Commissioner of the NYS OMRDD described this situation well in his welcoming remarks with a quote from Hannah Arendt, "People in public service stand between the past and the future. The past pushes us and the future pulls us." He went on to note that it is incumbent on him to try to lay the problems of the past to rest and to develop a vision of the future of the field. His vision of the future includes an improved service delivery system enriched by education and research. To Commissioner Webb, service, education, and research are analogous to a three-legged stool. Without one of the legs, the system collapses.

Finally, as we consider the workshop in retrospect, we note that because mental retardation stems from varied causes, topics of great import were selected for which there is enough basic knowledge to share with workers in different specialties. This approach to the meeting was a success because the most frequent reaction of the participants to the workshop was that it was refreshing to hear not only new data, but also different disciplinary slants on common problems. In the first two days of the workshop, mental retardation was examined from the point of view of epidemiology, behavioral and clinical teratology, molecular biology, morphology and pathology, and developmental and clinical biochemistry. At a large conference, specialists rarely have, or at least take the opportunity, to hear more than one or two different perspectives; they listen to papers in their specialty. By holding a relatively small meeting with speakers from multiple disciplines, it was possible for every participant to hear four to five other perspectives. The corollary to this is that the participants appreciated the opportunity for some give and take, something that is largely absent in meetings with large plenary sessions. Persons who might have been inclined to defend their position in a larger, more formal setting seemed willing to entertain different points of view. This is important because it is going to take the combined and concerted efforts of epidemiologists, behavioral teratologists, clinicians, geneticists, plus basic scientists from a variety of disciplines to understand the causes of mental retardation and develop appropriate treatments for mentally retarded persons.

Mental Retardation and Developmental Disabilities: Research, Education, and Technology Transfer

An Overview

SUMNER J. YAFFE

Center for Research for Mothers and Children
National Institute of Child Health and Human Development
National Institutes of Health
Bethesda, Maryland 20205

"It has long been recognized that the prevalence of mental retardation is significantly higher in those population groups where maternal care is frequently inadequate," and "The hazards of premature birth result in a significantly higher incidence of death and damage, including mental retardation, than occurs among full-term infants": These two statements are quoted from the Prevention Section of the President's Panel on Mental Retardation, which developed a national plan in 1962 to combat mental retardation. Although we will hear from others in the program about the epidemiology of mental retardation, I wish to emphasize in my presentation and overview the importance that prenatal care and low birth weight have made and continue to make toward the problem of the mentally retarded and the developmentally disabled.

As you are aware, the mission given to the National Institute of Child Health and Human Development (NICHD) is to ensure through research the birth of healthy babies, the birth of wanted babies, and the opportunity of each infant to reach adulthood unimpaired by physical or mental handicap, and thus able to achieve his or her full potential. No mission could coincide more closely with the dreams and aspirations of the American people, and indeed the whole human family. This is what we want for all our children. The promotion of the healthy development of children, from prenatal life to adulthood, through research presents a unique but challenging opportunity to NICHD.

Although this dream is closer to realization today than 22 years ago when Congress established this institute, we still have a long way to go. The infant mortality rate has been cut by more than half in the United States during this time, due in large part to advances from research supported by this institute. Much of this research has focused on the development of neonatal intensive care and the physiologic measures employed in supporting sick newborn infants. As a consequence birth-weight-specific mortality is lowest in the United States. Nonetheless, eleven countries in the developed world have a lower overall rate; in part because their burden of low-birth-weight infants is low. The infant mortality rate among blacks in the United States is still double that among whites, and the rate of low birth weight has not changed significantly over the years since the institute was created. Although 95% of sexually active American women use some form of fertility regulation at some time, surveys indicate that more than half (52%) of the six million pregnancies each year are not wanted at all, or not wanted at the time they occur. Unwanted pregnancy warrants discussion because these women are

1

less likely to seek prenatal care, a significant factor in the rate of low birth weight. Further, unmarried mothers have consistently higher risks of delivering a low-birth-weight infant. As for opportunity, of 3.6 million infants born each year, nearly 40,000 die before their first birthday; 250,000 face an increased likelihood of disability as a consequence of low birth weight; 3% have a significant anomaly, and another 3% are mentally retarded. With this background, I would like to expand upon the subject of infant mortality and low birth weight as they relate to mental retardation and developmental disability, and finally present to you some exciting new findings in the area of prevention of low birth weight.

The United States, as I mentioned before, has achieved considerable success in reducing infant mortality. One hundred years ago, infant mortality rates were 170 per 1,000 live births. These rates have dropped so that within the last 50 years infant mortality rates have decreased from 85 per 1,000 live births in 1920 to 10.6 in 1983, a level thought almost unobtainable only a decade age. FIGURE 1 depicts these changes over the last 30 years (1955–1984). The differences between black and white infant mortality is strikingly apparent in the figure. Infant mortality rates have two main components: infant deaths within the first 28 days of life (neonatal mortality rate) and deaths from 28 days to one year of age (postneonatal mortality rate). At the turn of the century, postneonatal deaths accounted for most infant deaths, but presently, neonatal deaths are predominant. These reflect preexisting health conditions of the mother and the medical care she and her baby receive during pregnancy, at the time of delivery, and during the neonatal period. The life-style of the mother, environmental factors, and the seeking of prenatal care also influence pregnancy outcome. While there have been major improvements in the neonatal mortality rate, there is considerable opportunity for improvement. Of these early infant deaths, three-fourths occurred in babies of low birth weight; thus, in part because of the marked overall improvement in the life chances for babies, the problem of low birth weight has emerged as the single most important cause of death or subsequent handicaps in infancy. Although low-birth-weight babies (below 5.5 pounds) represent only 7% of all babies born, well

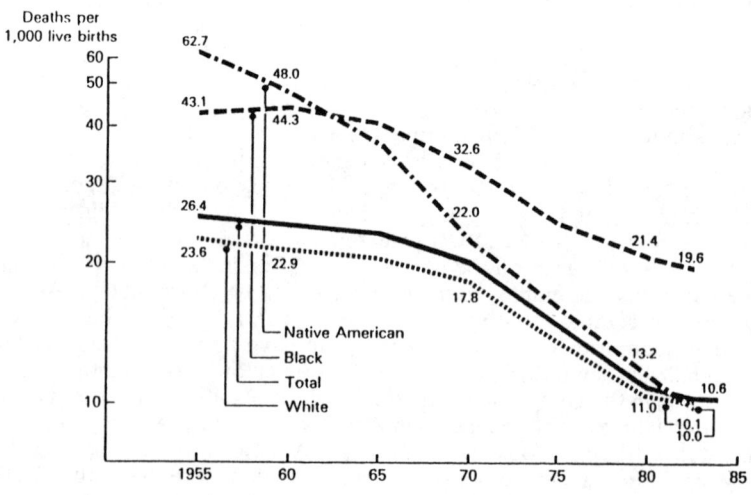

FIGURE 1. Infant mortality rates, 1955–84.

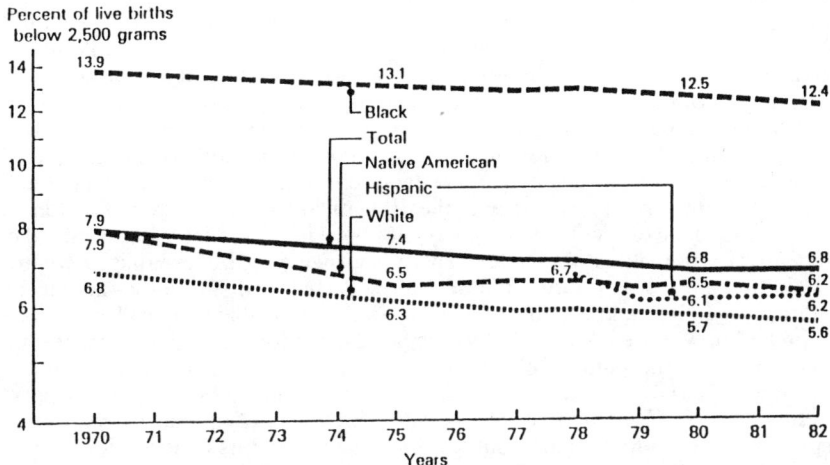

FIGURE 2. Low-birth-weight ratios, 1970–82.

over half of all present infant deaths occur among this group. Low-birth-weight babies also have many more illnesses and are much more at hazard of having serious difficulties when of school age than their normal-weight counterparts. The problem is even more serious for very tiny infants. The 1% of babies whose birth weights are very low (below 1,500 grams or 3.3 pounds) are 200 times more likely to die in the first few weeks of life, and three times more likely to suffer from congenital anomalies or developmental delay if they do survive. We have known these relationships for many years, and indeed, they were recognized in the 1962 report to President Kennedy.

Unfortunately, we have not yet been able to dramatically reduce the number of low-birth-weight babies who enter the world. The trends in low-birth-weight rates since 1970 are shown in FIGURE 2. Note the significantly higher rate of low birth weight among the black population. Low birth weight is defined as less than 2,500 grams (5.5 pounds) at birth. There are two components to the low-birth-weight population: preterm (or premature) birth and intrauterine growth retardation. Trends among these two components of the low-birth-weight population have varied over the time frame depicted in the figure. The incidence of preterm low birth weight for all races declined 7.1% while the term low-birth-weight incidence declined almost three times as much (20.9%). The incidence of preterm low birth weight among white infants during this time period declined 9.0% with the black preterm low birth weight declining by 5.8%. The term low-birth-weight incidence, however, was 24.6% lower among whites and 14.9% lower among blacks. Thus the reduction in the overall low-birth-weight incidence for both populations is principally caused by reduction in the incidence of term low birth weight or the infants with reduced intrauterine growth. With today's improved care, most infants of 1,500–2,500 grams survive with little disability. Very low-birth-weight infants (below 1,500 grams or 3.5 pounds) have significant mortality and morbidity. This group includes 1.5% of all births in the United States (50,000 per year). The impact of low birth weight upon society is profound. These infants require long periods of care in intensive care nurseries with enormous economic and social costs. It is encouraging to note that the increased survival of low-birth-weight infants has not been associated with an increase in the number with disabling conditions. The very low-birth-weight infant continues to contribute to the incidence of life-long disabilities including mental retardation, cerebral palsy, seizures, learning problems, blindness, and deafness.

Our medical advances to date have been in our abilities to keep tiny babies alive. The development of regional systems of care for mothers and the management of tiny newborns in special neonatal intensive care units have done much to improve infant survival. But despite the dramatic overall changes in infant mortality, the incidence of low-birth-weight babies born in this country has declined only very slightly in the past few years. There is growing awareness that, if we are to reach the goal that is the mission of the Child Health Institute that all surviving infants are healthy and capable of growing into responsible and productive adults, we must know more about the physiologic reasons for, and develop better methods to reduce, the number of infants born too small to function adequately. In fact, the NICHD has recently launched a multicenter clinical trial to prevent premature labor. I would like to conclude with a note of optimism. We would all agree from what I have said about low birth weight and its contribution to mental retardation that a significant solution would occur if we could prevent preterm birth. I would like to share with you some data from France that will be published in the August issue of *Pediatrics*. These data from Haguenau in eastern France represent a large study of over 16,000 live births conducted by Professor Emile Papiernik. They clearly demonstrate that prevention of preterm (premature) birth is possible. In a 12-year period (1971–1982) Papiernik and his group achieved a 31% reduction (5.4% to 3.7%) in premature births. The rate of prematurity before the institution of the prevention program was almost eight percent, similar to the rate in the United States. Thus the reduction to 3.7%, if we use the prestudy rate, is more than 50%. Of special importance to our consideration of mental retardation is that Papiernik demonstrated a 67% reduction in the rate of very low birth weight and very premature infants during the 12-year study period. These favorable results were replicated by Papiernik both at the Maternity Hospital of Clamart, a suburb of Paris where prematurity was decreased by 42% during a 10-year period (18,815 pregnancies) and on the French island of Martinique. The overall impact of the prevention of prematurity program in France has resulted in a national decrease of preterm births from 8.2% in 1972 to 5.3% in 1982, a reduction of 35%. It is heartening to think of what might be possible were this approach to be instituted in the United States.

Key to Papiernik's approach has been the development of methods that predict and then prevent preterm labor. Previously established risk factors such as maternal age, height, weight, education, and previous history of preterm birth cannot be changed. They are useful in identifying the population at risk and this is indeed done in the United States. Papiernik has attempted to define risk factors that can be modified, especially during the course of pregnancy. These include unhealthy life-style and physical exertion or stress that may enhance uterine contractility. He has incorporated these factors into a risk assessment or screening system that is widely used in France. While risk assessment is a necessary first step, it is of no value unless the pregnant woman herself is educated to understand the role that risks play in the development of preterm labor. Recognition of early labor and modification of behavior through patient education and support are essential ingredients of Papiernik's program. Prenatal care providers have been a key factor in the French approach. Their own education has enabled them to provide the support so essential to the success of the early intervention and prevention approach.

Clinical trials of predicting and preventing prematurity within the United States based upon the Papiernik program have revealed exciting and highly successful preliminary results. Creasey at the University of California in San Francisco has found a reduction in preterm deliveries from 6.7% to 2.4% over a two-year period in a small number of pregnancies. The trial has been expanded as the March of Dimes Multicenter Preterm Birth Prevention Trial. Similarly, the Los Angeles Prematurity Prevention Program has been initiated by Hobel. Once these results are reported and assuming that the programs are successful, we should be able to launch a regional or national program in the United States. Based upon the French experience, this must be a cooperative effort among all concerned parties in order to make prevention a reality. Thus the global social cost of long-term disabilities and mental retardation related to perinatal events will be reduced.

REFERENCES

1. KOOPS, B. L., L. J. MORGAN & F. C. BATTAGLIA. 1982. Neonatal mortality risk in relation to birth weight and gestational age: Update. J. Pediatr. 101: 969–977.
2. VILLAR, J & J. M. BELIZAN. 1982. The relative contribution of prematurity and fetal growth retardation to low birth weight in developing and developed countries. Am. J. Obstet. Gynecol. 143: 793–798.
3. KESSEL, S. S., J. VILLAR, H. W. BERENDES & R. P. NUGENT. 1984. The changing pattern of low birth weight in the United States: 1970–1980. J. Am. Med. Assoc. 251: 1978–1982.
4. HACK, M., B. CARON, A. RIVERS & A. A. FANAROFF. 1983. The very low birth weight infant: The broader spectrum of morbidity during infancy and early childhood. J. Behav. Dev. Pediatr. 4: 243–249.
5. SHAPIRO, S., M. C. MCCORMICK, B. H. STARFIELD & B. CRAWLEY. 1983. Changes in morbidity associated with decreases in neonatal mortality. Pediatrics 72: 408–415.
6. PAPIERNIK, E., J. BOUYER, J. DREYFUS, D. COLLIN, G. WINISDORFFER, S. GUEGEN, M. LECOMTE & P. LAZAR. 1985. Prevention of preterm births: A perinatal study in Haguenau, France. Pediatrics 76: 154–158.

DISCUSSION

ZENA STEIN: I am going to say something just a bit critical to Dr. Yaffe. I did not go to the conference in Evian, but I have studied that work. With due respect to Dr. Papeirnik, who is a great friend and whose work I admire, I would like to say we have not yet seen a controlled trial of prenatal care of any form. Whether it is love or whether it is good obstetrics or any of those wonderful things we would like to do in the prenatal period, the data are not in that tell us which program and what components of the program will reduce preterm birth. I think it is an exciting time for experiment because we have several leads.

We are doing one of these trials in Harlem, supported largely by New York State, but we will not really know the value of these programs before a couple of years.

SUMNER YAFFE: I agree with you that there was no control group and no randomized study. On the other hand, there is no question that the proof of the pudding is in the eating. According to the historical record—not rigorously controlled data—Hagueneau (the community in Eastern France where Papeirnik conducted his study) had a low-birth-weight rate of 4.6% in 1971–1974, which went down to 3.8% in 1979–1982. I don't care what the mechanism is. Although we do stand for scientific rigor, I was making these points from a societal point of view. It really doesn't make any difference how they did it; they did it. Nevertheless, I agree that in the United States you cannot convince the people who provide money for services, unless you have a controlled study.

IRVIN EMANUEL: I think it *does* matter how they did it. There is a lot of mythology about prenatal care, and until we know specifically what was done, I don't think we can duplicate it. You said that the French cannot say exactly what they have done.

SUMNER YAFFE: There is a manuscript describing the French study; it has undergone editorial review and they do have some quantifiable indices of what they did. The paper will be published in *Pediatrics* this summer.

IRVIN EMANUEL: There is so much self-selection of women who come for care at different times. I am convinced that the women who come for care early are much different, in many ways, biologically and behaviorally, from the women who don't come or come later. Until that problem of self-selection is dealt with, it is an open question.

THOMAS SHEPARD: A similar program exists in the United States that has not been mentioned. That is the program run by Robert Creasey (then at the University of California at San Francisco). He does not provide a whole lot of education. All that he does is have his people demonstrate to these women what a Braxton-Hicks contraction is and what premature labor is. The results have been excellent, so there has to be value in paying attention to these women.

IRVIN EMANUEL: The data are not in.

THOMAS SHEPARD: I know that the data are not in, but this problem has to be studied.

Introduction

DONALD A. SNIDER

Institute for Basic Research in Developmental Disabilities
New York State Office of Mental Retardation &
Developmental Disabilities
Staten Island, New York 10314

Epidemiological research was selected as the initial topic of this workshop because it serves three important functions: First, it can provide an overview of the major causes of developmental disabilities. Second, it identifies those areas that may prove most amenable to reducing the incidence and prevalence of developmental disabilities. Finally, it can help in monitoring the success of treatment and prevention efforts.

Dr. Karin Nelson of the Cerebral Palsy Section, National Institute of Neurological and Communicative Disorders and Stroke, describes what is known regarding the causes of cerebral palsy, the most common developmental disability after mental retardation. Her paper reviews the principal findings of the National Collaborative Perinatal Project.

Dr. Zena Stein of the New York State Psychiatric Institute and Columbia University describes a sample survey model for estimating the proportion of cases of serious (IQ < 55) mental retardation in developing countries in a methodological paper. The model, based on door-to-door surveys, is a tool for planning, implementing, and evaluating prevention strategies for childhood disabilities in developing countries.

Dr. Irwin Emanuel of the University of Washington (and formerly the Director of the Child Developmental and Mental Retardation Research Center there) summarizes the evidence about the factors and conditions antedating pregnancy that are important determinants of pregnancy outcome. His paper focuses on the influences of intergenerational factors, that is, the actions of one generation on the health of the next.

"Serious" Mental Retardation in Developing Countries: An Epidemiologic Approach

ZENA STEIN,[a,b] MAUREEN DURKIN,[b] AND
LILLIAN BELMONT[a,b]

[a] Epidemiology of Brain Disorders Research Department
New York State Psychiatric Institute
New York, New York 10032

[b] G. H. Sergievsky Center
Columbia University
New York, New York 10032

INTRODUCTION: THE NEED FOR FACTS AND SOME PILOT SURVEYS TO PROVIDE THEM

Although the burden of childhood disability is recognized to be heavy in developing countries where over 80% of the world's children live, we cannot realistically assess that burden lacking knowledge not only of frequencies, but also of the nature, distribution, and causes of its component parts. Yet for the dual purposes of prevention and of rehabilitation, some information is essential.

As a first step in building such a knowledge base on disability, a collaborative group planned and executed pilot surveys of childhood disability in ten communities of the developing world, using a common protocol.[1,2] A two-stage procedure was adopted, the first stage being a door-to-door survey using simple standard questionnaires, the second a clinical examination. The first stage was planned to reach all children between the ages of three to nine years (about 1,000 children) in each community, while the clinical examination was to evaluate children judged "positive" from the questionnaire, as well as a random number of "negative" children. One of the questionnaires used in the first stage, the Ten Questions, appears in the APPENDIX.

Since experience with these surveys has now been gained in ten communities of the developing world, it is time that we assess their usefulness in prevention. The findings reported here deal with "serious" mental retardation (SMR, defined here as IQ ≤ 55). Later reports will cover other disabilities also sought in the surveys, including cerebral palsy, epilepsy, and behavioral, visual, hearing, and other disorders, the focus in each case being on the more severe. An evaluation has been made of the screening procedure devised for the study, specifically in connection with its efficiency in identifying young children (three to nine years) with serious mental retardation. The method was judged to be sensitive (meaning that most affected children were probably identified); it was less efficient in terms of specificity, in that many children with other diagnoses, or lesser degrees of mental retardation, were also caught by the screening process. In those communities where the screening procedure was tested against a "key informant" approach, the screening instrument was the more sensitive approach.[3]

8

PRELIMINARY RESULTS: PREVALENCE OF SMR

A preliminary epidemiologic analysis of the findings related to serious mental retardation is presented here. TABLE 1 shows the results for SMR from eight of the participating centers. Except for India, the investigators kept quite closely to the target sample size of 1,000 (Column 1). The percent of children for whom at least one positive response was reported ranged from 7 to 30 (Column 2). In Column 3, the rate per 1,000 for children diagnosed as SMR is given, with the actual numbers in parentheses. In seven of the eight communities, the prevalence ranges from 5 to 16 per thousand. The results from India are puzzling, for while the number of children with problems indicated on the Ten Questions is very low in this community, 70 per thousand, the rate of serious mental retardation is far higher there than in any of the other sites, 40 per 1,000. Given the possibility of an artifact, we shall not comment further on the findings for this community.

The range in rates of serious mental retardation across these communities is considerable, even if the one outlier from India is omitted. Over the last two decades in the developed world, surveys of severe mental retardation have, by contrast, yielded rather constant rates, around 3.5 per thousand. We considered four reasons (not mutually exclusive) for the wide variation found in the pilot surveys of childhood disability.

First, and not very likely because the range is wide, the differences could be due to random fluctuations; however, the denominators of 1,000 are not large, given the rarity of the condition.

Second, in countries with low rates more children might have been missed in the first stage, which was the questionnaire procedure. We think this is also an unlikely explanation because our evaluation of the first-stage procedure suggested that in most sites the Ten Questions was a sensitive screening instrument, at least for "serious" mental retardation.[4]

TABLE 1. SMR ("Serious" Mental Retardation) in Children Three to Nine Years Old in Eight Countries[a,b]

Country	(1) Number of Children Surveyed	(2) Percent of Children with Any Reported Problems	(3) SMR Rate Per Thousand Children[c] (No.)
Philippines	1000	12	5.0 (5*)
Bangladesh	987	20	16.2 (16)
Sri Lanka	966	10	5.2 (5)
Malaysia	981	15	11.2 (11)
Pakistan	995	12	15.1 (15*)
India	1439	7	40.3 (58†)
Brazil	1050	30	6.7 (7*)
Zambia	1139	16	5.3 (6)
Total	8,557		

[a] A two-stage survey was conducted in which house-to-house interviews based on standard questionnaires were followed by clinical evaluations. Children with any reported problems and a presumed random sample of those without any reported problems were assessed clinically by a psychologist and/or physician as to whether or not they have SMR (IQ ≤55).

[b] Extracted from Belmont.[3]

[c] False negatives: *One child had no problems reported on the Ten Questions. †Nine children had no problems reported on the Ten Questions.

Third, there could be differences between professionals in their assessment of "serious" mental retardation (IQ ≤55) as compared to "mild." The assessments were based on "disability": this criterion and some problems of definition are discussed below.

Fourth, there could be real differences in the prevalence between communities.

The third and fourth possibilities are interesting and important. Regarding the differences in diagnostic practices between professionals, for children in developing countries we have no standardized protocols that could guide the pediatrician, psychologist, or psychiatrist in differentiating *levels* of mental handicap. Regarding the possibility that the prevalence of SMR could really be different across communities, this is also a problem needing careful study. We discuss below, first some issues concerning case-definition in SMR, and next, some interpretations of differences in prevalence.

WHAT IS A CASE?

A special problem in studying the epidemiology of mental retardation is the definition of the case. Cases to be counted must be distinguished from noncases, but the lines of demarcation are blurred by confused definition. In particular, we have to distinguish between *impairment, disability,* and *handicap.*[5-7] The manifestation of a handicapping condition as a social attribute encompasses at least three components: organic, functional, and social. The organic component, which we term *impairment,* can be any persisting physical or psychological defect that stems from molecular, cellular, physiological, or structural disorder. The functional component, which we term *disability,* refers to persisting physical or psychological dysfunction; it stems from the limitations imposed both by the impairment, and by the individual's psychological reaction to it. The social component, which we term *handicap,* refers to persisting social dysfunction, a social role that stems not from the individual but from the social expectations of others and the individual's interaction with society. Handicap describes the manner and degree in which the expected performance of social roles is altered in the presence of primary impairment and functional disability.

Organic, psychological, and social criteria yield different frequencies and make different contributions to our understanding of mental retardation. Components of this disorder, measured by each criterion, do not necessarily have a one-to-one relationship with each other, and they are made apparent by different circumstances. In fact, impairments that can be recognized at birth and those for which a one-to-one relationship with functional disability and social handicap can be predicted, as is to some degree the case in Down's syndrome, are not common. Thus, cerebral palsy is an impairment recognized by the signs of brain damage, yet not all cases of cerebral palsy suffer the functional disability of intellectual deficit, nor are they assigned the special social role of the handicapped person. Hydrocephalus is an anatomical deformity that may or may not be accompanied by cerebral impairment or functional disability or social handicap. Phenylketonuria is an inherited impairment of metabolism that, even when untreated, does not always lead to functional disability; when appropriate and timely treatment is available, both disability and handicap are averted.

Conversely, recognized functional disability cannot always be related to definitive organic lesions. In a large proportion of cases of mental retardation even with severe intellectual deficits, a specific clinical diagnosis cannot be made. In these cases, the presence of organic impairment is merely assumed. Severe mental retardation of unspecified diagnosis thus describes a residual class of cases that is heterogeneous in terms of organic impairment. Yet it is a homogeneous class in terms of functional disability and social handicap.

In mild mental retardation, on the other hand, the intellectual deficit and functional disability of the "cultural-familial" syndrome is not preceded by detectable organic impairment and is not always accompanied by the social role of mental handicap. Where, as in the communities surveyed, heavy intellectual demands of schoolwork are not the rule, the condition may not be evident at all. For this reason, mild mental retardation was not explicitly sought in the pilot surveys.

Prevalence, in our surveys, specifically indicates the frequency of disability. Assessments of disability were based on clinical judgment, since a satisfactory standard method to measure this component across cultures (for instance, with IQ tests) has not been devised. Observations relevant to impairment (hence, to the diagnosis in a clinical or pathologic sense) were also entered by the examining clinician and add credibility to the "caseness" assessment as well as to the usefulness of the survey.

INTERPRETING PREVALENCE

Incidence describes the frequency with which disorders arise in a population during a defined period of time. The search for causes of the trends and distributions of health disorders is best pursued by studies of incidence. Time order is an essential criterion in establishing causal relations, and incidence relates disorders to circumstances that exist at the time of onset of a disorder or that are antecedent to it. In the more developed world, incidence rates usually take data from service agencies; although seldom completely representative of all those in the population who suffer the disorders, they do give reasonable approximations, especially when the disorder is severe and the services adequate. Lacking adequate services, this strategy is unlikely to be useful in less developed countries.

Prevalence differs from incidence in that it describes the amount of disorder existing in a population at a particular time, regardless of time of onset. It affords a useful measure of the existing load of disorder to be provided for. Prevalence studies have the advantage that investigators can go out to discover and enumerate existing disorders in representative samples of defined populations; they need not wait for what turns up. This was the rationale for our pilot studies. But prevalence is less useful than incidence in the search for the causes of existing disabilities; it gives a cross-sectional view of a population's experience at one point in time, and cannot establish time order nor the precise circumstances in which disorders of long and variable duration arise.

Moreover, a problem of sampling inheres in a cross-sectional view of chronic health disorders. In particular, congenital anomalies that cause fetal and perinatal deaths will be missed, whereas disorders of long duration have a good chance of appearing. With severe congenital impairments there is no recovery, and duration is synonymous with survival. As a result, some who contribute to incidence do not live long enough to enter a count of prevalence. The long-lived swell and bias the numbers that contribute to prevalence; hence prevalence and incidence are not interchangeable terms for measuring frequencies.

The divergence between incidence and prevalence is exaggerated where duration varies widely across communities and where it changes through time. Both these circumstances could hold for developmental disorders. Thus in comparing prevalence rates of severe disabilities across communities, the relationship between incidence, duration and prevalence needs careful study. If incidence is assumed to be similar, then variation in prevalence will depend on differences in duration (or, with disabilities that are life-long, survival). If incidence is dissimilar, but duration is stable across communities, then dissimilarities in prevalence reflect differences in incidence. When *either* incidence *or* duration is known, inferences about the other term can be drawn from prevalence studies; and surveys may often provide *duration* because with disabilities present at birth, the age of an affected child indicates duration. But it is an uncertain assumption that impaired survivors at ages three to nine years fairly represent impairments among all births: not all those impaired at birth will live until the ages when they might be surveyed. Nor is all impairment immediately apparent at birth, and knowledge of its presence must often be inferred from functional disability and social handicap at later ages. Epidemiologic inference, however, has enabled the age of inception of impairment to be determined within reasonable limits for several conditions. In Down's syndrome and phenylketonuria, in neural tube defects and in mental retardation due to intrauterine irradiation and rubella infection an approximate age of inception can be given.

In serious mental retardation, then, the major obstacle to inferring incidence from knowledge of prevalence is the absence of data about the occurrence of mental retardation among individuals who died before they could enter a prevalence count. Since we are dealing with survivors at every juncture, we have no options here but to use prevalence advisedly.

EPIDEMIOLOGICAL ANALYSIS: WHAT IS PREVENTIBLE?

In what follows, and with due caution, we shall define a case primarily in terms of *disability*. *Impairment* is seldom known with precision but we shall use what we have. Handicap is not considered. The frequencies generated refer perforce to *prevalence* in young children; no assumption is made about the age of onset, though we will explore it where possible. There is no knowledge of infant deaths of affected children; hence there is no secure knowledge of incidence.

Epidemiological analysis brings us now to consider three uses of the data, which will be exemplified in FIGURE 1 and TABLES 2, 3, and 4 below for various estimates of risk. FIGURE 1 and TABLE 2 show proportional morbidity by factor (PMF); TABLE 3 shows the odds ratio (OR) for serious mental retardation for a selection of putative risk factors. In TABLE 4 we select likely risk factors and associations for a discussion of a final risk estimate, the attributable fraction for the population studied (AF_{ps}).[8]

Proportional Morbidity by Factor (PMF)

A table such as TABLE 2 can be derived from records of the children who are "cases." The table shows the distribution of cases in two of the developing country sites (in Malaysia and Pakistan; the use of the pilot material is only for illustration, the numbers being much too small to provide secure percentage points) and in two communities studied by Hagberg and Gustavson and colleagues in Sweden.[9,10] If the cases studied are representative of the cases in that city, region, or country, they give an indication of the kind of cause or accompanying

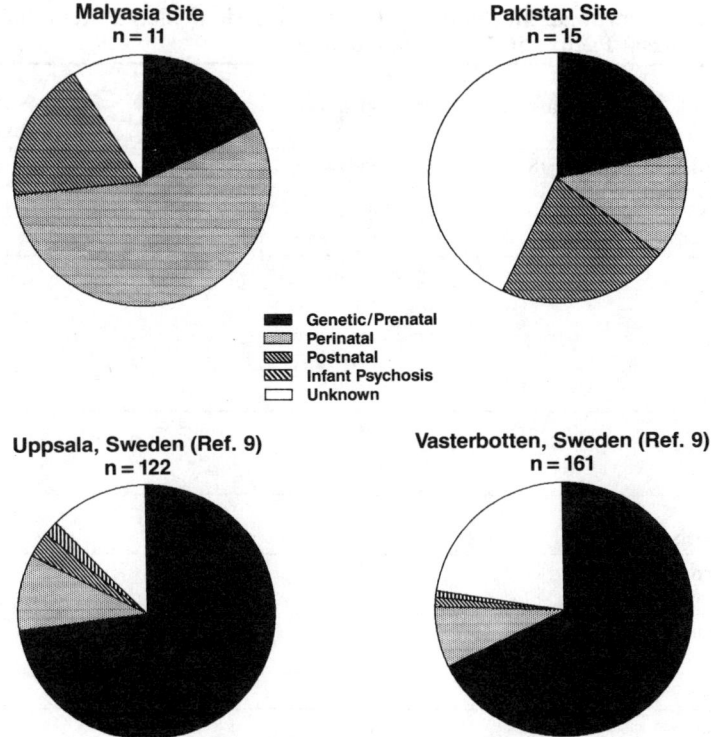

Malyasia Site
n = 11

Pakistan Site
n = 15

Genetic/Prenatal
Perinatal
Postnatal
Infant Psychosis
Unknown

Uppsala, Sweden (Ref. 9)
n = 122

Vasterbotten, Sweden (Ref. 9)
n = 161

FIGURE 1. Proportional morbidity by factor (PMF) for SMR in four sites.

problem to be met with in affected children. As TABLE 2 shows, one may also compare PMF across regions and countries. Thus a first impression (FIG. 1) from the pilot surveys is that causes other than those operative before birth comprise a larger proportion of SMR in Malaysia and Karachi than they do in Sweden—but these are only impressions.

Note that in TABLE 2, and in the tables that follow, a distinction could be made between a factor that is a supposed "cause" and factors that may be described as "associations." The distinction, when it can be made, is a vital one, implying directionality and time order—a cause leads to an effect. The directionality between an association and the outcome may not be well established; however, the information about associations still has its uses, suggesting lines of inquiry, and adding descriptive material relevant to prevention and rehabilitation.

A shortcoming of the PMF is its failure to take account of the frequency of the supposed "cause" or "associated condition" among *unaffected* persons in that population. Hence the PMF is inherently likely to produce an overestimate of the risk attributed to a factor, because the value does not offset for the frequency with which the factor occurs without contributing to the occurrence of cases. Thus, the PMF does not actually indicate the presence of an association between factor and outcome. An example would be the relationship of measles, malnutrition, or

TABLE 2. Proportional Morbidity by Factor for Serious Mental Retardation and Various Causal Factors in Four Communities

Assigned Cause	Malaysia No. = 11	Karachi, Pakistan No. = 15	Uppsala, Sweden[a] No. = 122	Vasterbotten, Sweden[b] No. = 161
Before birth	18%	20%	73%	68%
Genetic	18%	7%	43%	52%
Unknown	—	13%	20%	16%
Acquired	—		10%	
Perinatal (breech, twin, preterm birth, dystocia)	55%	13%	10%	8%
Postnatal (infection, trauma)	18%	20%	3%	1%
Infantile psychosis	—	—	2%	1%
Unknown, CNS signs	—	27%	7%	22%
Unknown, aclinical	9%	13%	5%	
All causes	100%	100%	100%	100%
Estimated prevalence (per 1000)	11.2	15.1	2.8	3.5

[a] Taken from Hagberg.[9]
[b] Taken from Hagberg[9] and Gustavson et al.[10]

prematurity to mental retardation; cause could not be inferred from PMF alone because all these conditions may be common among unaffected children. Despite its shortcomings, if the only data available are those based on hospital or clinic patients, then the PMF may often be the only statement that can be made. It can be useful in flagging strong associations, especially with exposures known to be rare in the general population, or with rare diseases occurring in special contexts, such as cretinism.

Relative Risk and Odds Ratio

We need to advance beyond PMF (and hence, to reach out beyond clinic populations) if the purpose is to investigate associations and discover causes. Let us first consider "relative risk." "Relative risk" (RR) is the incidence in the exposed population divided by the incidence in the unexposed population; it provides a direct measure of the relationship between exposure to a given factor (cause) and an outcome. An RR of one indicates no association between a factor and a disorder, while an RR greater than one indicates a positive association.

Suppose prematurity was suspected as a cause of SMR; then, if retardation rates were to be compared in all premature and term infants, finding a doubling of the SMR rate among premature infants would give a relative risk of two. However, the relative risk of a factor for childhood disability can only rarely be assessed because it requires incidence data; studies comparing incidence rates in exposed and unexposed groups (for example, in premature and term infants) would require a long, complex, and expensive procedure. If one is starting from a weak knowledge base, such studies would be inappropriate because they typically focus on one suspected factor at a time.

Instead of RR then, epidemiologists turn, as we do here, to odds ratios (OR) based on comparisons between "cases" (affected children) and "controls" (not affected) in respect of the study factor. To generate this most important set of estimates, we have to select, interview, and examine a number of unaffected as well as affected children, as we did in the pilot surveys. Odds ratios, unlike PMF, cannot be based on study of the affected only. They require case-control studies.

The hardest aspects of carrying out case-control studies and those requiring most thought and exploratory work are the basis on which the controls are selected and the blinded (or neutral) way in which observations relating to the factor of interest are elicited. If this is not done carefully, biases and misclassification may lead to incorrect inferences.

In matched case-control analyses, the results are tabulated in fourfold tables such as those shown in TABLE 3. TABLES 3(A) and 3(B) illustrate calculation of OR's to assess the presence of an association between maternal age over 35 and SMR in the communities surveyed in Malaysia and Pakistan, respectively. TABLES 3(C) and 3(D) show the same for the OR's relating consanguinity to SMR in the two sites. One would expect OR's for a given risk factor and a given disorder to be similar across sites. The disparities between sites shown in TABLE 3 are most likely due to the small numbers in the pilot study. Thus, the confidence limits are very wide; they overlap between sites and most include one (one indicates no association between the factor and SMR). This means that the intersite differences as well as the differences between the observed OR values and one may be due to chance alone.

The approximation of relative risk by the odds ratio is valid with two provisos: (1) Cases and controls are drawn from the same population; and (2) the disorder is uncommon (say, less than 10% in the population).

Because the approach described here involves surveying an entire community (as the population from which cases and controls are selected or identified) and because SMR is relatively rare, both of these provisos are met. In order to produce results capable of achieving statistical significance, the approach will have to be applied to larger communities (i.e., containing several thousand three- to nine-year-old children).

In calculating OR's for SMR we would consider, as in TABLE 2, not only factors we feel intuitively are "causes" but also factors for which neither time order nor directionality is inferred. A further task of the analyst, after establishing a raised risk, is to attempt to determine the causal connections between the factors and SMR.

Attributable Fraction in the Population Studied

Although the odds ratio calculated from case-control studies does indicate the strength of association between a risk factor and a disorder, it provides no infor-

TABLE 3. Calculation of Odds Ratios from Matched Case-Control Analyses: Estimating the Association between Selected Risk Factors and Serious Mental Retardation in Two Communities[a]

A. Malaysia		Control Maternal Age		B. Pakistan		Control Maternal Age	
		>35	≤35			>35	≤35
		(a)	(b)			(a)	(b)
	>35	1.5	2.5		>35	0	1
Case Maternal Age				Case Maternal Age			
		(c)	(d)			(c)	(d)
	≤35	0.5	7.5		≤35	2	12

OR = b/c = 2.5/0.5 = 5[b]   OR = b/c = 1/2 = 0.5
95% Confidence limits (1.5, 17.3)[c]   95% Confidence limits (0.2, 1.2)

C. Malaysia		Control Consanguinity		D. Pakistan		Control Consanguinity	
		+	−			+	−
		(a)	(b)			(a)	(b)
	+	2	1		+	5	5
Case Consanguinity				Case Consanguinity			
		(c)	(d)			(c)	(d)
	−	1	6		−	3	2

OR = b/c = 1/1 = 1   OR = b/c = 5/3 = 1.7
95% Confidence limits (0.4, 2.5)   95% Confidence limits (0.7, 3.7)

[a] Each entry in the tables represents one pair of a case and a control of the same age and sex. Only pairs discordant with respect to exposure to the risk factor (cells b and c) contribute to calculation of OR.

[b] This table was adjusted by adding 0.5 to the value in each cell. The adjustment was necessary in order to calculate the odds ratio, because cell (c) contained the value zero.

[c] The formulas for confidence limits for ORs derived from matched pair case-control studies are provided by Fleiss.[13]

mation about the frequency of the supposed causal factor in the population; hence it does not indicate the public health importance of the factor. This can be obtained from another epidemiologic measure, the attributable fraction (AF), which indicates the fraction of all cases that may be attributed to the factor in question. In the present context, attributable fractions can be estimated for various risk factors for the total number of cases occurring in the population under study. Here the population under study (ps) is the particular community surveyed. The AF in the population under study can be expressed as:

$$AF_{ps} = I_{ps} - I_u/I_{ps}$$

where I_{ps} is the incidence in the population studied and I_u is the incidence in the unexposed segment of the same population. An algebraically equivalent formula when OR can be substituted for RR is:

$$AF_{ps} = [Pe \ (OR - 1)]/[Pe \ (OR - 1) + 1]$$

where Pe is the prevalence of the exposure in the population studied.[13] Thus, AF_{ps} can be estimated from case-control studies, provided that Pe for the factor under study is known or can be estimated. In the surveys under discussion, information on exposure to a number of factors was obtained for all children in the communities by means of the house-to-house interviews. Therefore, the survey method provided stable Pe values for those factors inquired about in the interviews. Pe may also be estimated in case-control studies from the prevalence of exposure in the controls, if the disorder is rare and the controls are representative of all noncases in the population studied.

In studying a single factor, for instance, prematurity, and a single disorder, for instance, SMR, we may find that the attributable fractions diverge widely across study populations. Since AF_{ps} depends on both OR and the frequency of the risk factor in the population under study, the divergence could be due to differences in either or both of these terms across communities.

TABLE 4 illustrates variations in AF_{ps} for the same two risk factors and communities presented in TABLE 3. Although these estimates are necessarily very unstable due to the unstable OR estimates (a result of the small numbers of matched case-control pairs in each site), still for the two factors, maternal age and consanguinity, analyzed in TABLE 3, and for the same two communities, they illustrate the relative contributions of the odds ratio and the prevalence of the exposure factor to the AF_{ps}. In the Malaysian community, up to 27% of the cases of SMR may be associated with maternal age over 35 years, while the same risk factor is not associated with any of the cases in the Pakistani community. On the other hand, in the Malaysian community consanguinity does not appear to be associated with SMR at all, while in the Pakistani community it is associated with 27% of the cases. Note that although the estimated OR for maternal age over 35 in Malaysia (5) is much higher than that for consanguinity in Pakistan (1.7), the AF_{ps} in both cases is estimated to be 27%. This underscores the contribution of the respective Pe values, the frequency of exposure to the factor in that community.

Furthermore, in studying a single risk factor (for instance prematurity) and different disorders (for instance SMR and seizures) the fraction of cases associated with the risk factor (i.e., the AF_{ps}) will vary for the two disorders, even within the same population. This is because the OR for the association between prematurity and SMR is likely to differ from that for the association between prematurity and seizures.

A further caution is in order. We noted the implication of the term "attributable" is that elimination of the factor would reduce the incidence of the disease by the amount or fraction indicated by the estimated attributable fraction; however,

TABLE 4. The Contributions of Odds Ratios (OR) and Prevalence of Exposure to the Risk Factor (Pe) to the Attributable Fraction in the Population Under Study $(AF_{ps})^a$

Risk Factor	Malaysia			Pakistan		
	OR	Pe	AF_{ps}	OR	Pe	AF_{ps}
Maternal age >35 years	5	9.2%	27%	0.5	8.3%	0
Consanguinity	1	15.8%	0	1.7	55%	27%

$^a AF_{ps} = Pe (OR - 1)/Pe (OR - 1) + 1.$

attributable fractions are usually calculated from data generated from observational studies rather than from experimental intervention studies. It cannot, therefore, simply be assumed that all differences in disease frequency between the exposed and unexposed are in fact causally related to the factor under study. Other uncontrolled or unmeasured factors may be involved and not accounted for.

In addition, several factors may overlap in their effects to produce less than their sum, while others may interact to produce more than their sum. For this reason some authors have developed attributable risk estimates adjusted for potential confounding variables.[14,15] The usual attributable fractions calculated for single factors are perhaps best interpreted as upper limits of the proportion of disease that could be prevented by eliminating the factor.

In summary, the following conditions must be met in order for an AF_{ps} calculated from a case-control study to be a valid estimate of the proportion of cases of severe mental retardation (or some other condition) in a population that is due to a given risk factor:

(1) The cases and controls are representative of cases and noncases in the population studied;
(2) The disorder is rare (say, prevalence less than 10%) in the population studied;
(3) The risk factor is causally associated with severe mental retardation;
(4) For both cases and controls, exposure to the factor under study does not affect the probability of survival to a given age;
(5) Being severely retarded does not affect one's risk for exposure;
(6) Important confounding and interacting variables have been adequately measured and controlled; and
(7) Biases due to misclassification and measurement error are minimized.

CONCLUSION

The sample survey method is here proposed as a tool in the planning, implementation and evaluation of preventive strategies for childhood disabilities in developing countries.

Among the three estimates, PMF, OR, and AF_{ps}, the last is obviously the one of most use in public health. It enables the planner to address questions and then to set realistic targets. For example:

What can be predicted about the prevalence of SMR from knowledge of the frequency of a particular risk factor (e.g., proportion of "unattended" births, prevalence of goiter, consanguinity)? How much might be prevented by the reduction of a risk factor?

A population-based survey, and not a clinic-based analysis, is needed to provide the answers to this type of question.

Tasks for the Future

In order to estimate rates and risks that are stable enough to guide a preventive policy, one would want to repeat the door-to-door survey in several representative sites. This should not prove too difficult, for the pilot procedures have proved to be inexpensive, noninvasive, and acceptable. We have learned in the pilot work that interviewers need no particular background apart from knowledge of the community, motivation, and specific training in the survey procedures. Also, house-to-house surveys should always carry the promise of some kind of remediation or care for the disabled members.

For any particular community, it may be useful to add locally needed information to the survey instrument, for example, place of birth (home or hospital), preconceptional iodine supplementation of the mother, immunization status, and the number of children who had died. Such items may be added to the questionnaires, others may be deleted. In successive surveys in different locations and times, such items may be evaluated as risk factors for disability.

Some factors may "cause" not one, but a range of disabilities, and planners may be as interested in the fact that "unattended birth," for instance, relates to a grouping of serious disabilities, as that it relates to a particular category of disorder. The methods described here are easily adapted to that perspective, with the emphasis being on degree of dysfunction rather than on category of impairment.

Two methodological issues are identified as priorities for further development. One, which could follow naturally from the type of extensions envisioned above, is to improve the fit between the factors identified in the screening procedure, and the fuller understanding of these factors that could be derived at the professional examination. For example, the screening questionnaire might be limited to a single question on the circumstances at birth of the child—who was in attendance or complications—whereas, for "cases" and "controls," the positive answers would be explored in detail at the professional examination. Therefore, the questionnaire would provide an indicator for the "factor" in the population studied.

The second issue is case definition and description, an essential part of the epidemiological study, and with serious mental retardation, the area in most need of development. First, a consensus on criteria needs to be developed in a workshop setting; second, standard protocols must be created that are in conformity with these criteria; third, investigators have to be trained in their usage; fourth, field studies need to test the reliability and validity of the assessments.

SUMMARY

In this paper we first present methods and preliminary results of pilot surveys of "serious" mental retardation (IQ ≤ 55); the surveys included screening and diagnostic components and were carried out in the less-developed world. Next we discuss two problems raised by these surveys: one is the diagnosis of a case and its clinical dimensions, and the other is the interpretation of prevalence. In the next section we illustrate epidemiological approaches to the analysis of such data, in particular their relevance to prevention. Lastly, we propose that the two-stage survey approach developed in the course of the pilot work can provide a valuable basis for planning and prevention, if certain key conditions can be met.

NOTES AND REFERENCES

1. The International Pilot Study of Severe Childhood Disability was completed at 10 sites in nine developing countries: Bangladesh, Brazil, India, Malaysia, Nepal, Pakistan (Karachi and Lahore), the Philippines, Sri Lanka, and Zambia. The Pilot Study was a collaborative effort between principal investigators and their teams at study sites and a number of other bodies: the Bishop Bekkers Foundation Workshops, the New

York State Psychiatric Institute, Rehabilitation International/UNICEF Technical Support Program, and the Sergievsky Center, Columbia University.
2. BELMONT, L., Ed. 1981. Int. J. Ment. Health. **10:** 1.
3. BELMONT, L. 1984. The international pilot study of severe childhood disability. Final report: Screening for severe mental retardation in developing countries. Bishop Bekkers Foundation and Institute Research Series, No. 1. Utrecht, The Netherlands. (Mimeo)
4. BELMONT, L. Screening for severe mental retardation in developing countries. The international pilot study of severe childhood disability. *In* Science and Service in Mental Retardation. J. M. Berg, Ed. Methuen. London. In press.
5. STEIN, Z., & M. SUSSER. 1971. Changes over time in the incidence and prevalence of mental retardation. *In* Exceptional Infant Vol. 2: Studies in Abnormalities. J. Hellmuth, Ed. Brunner-Mazel. New York. pp. 305–340.
6. SUSSER, M., & W. WATSON. 1971. Sociology in Medicine, 2nd Edition. Oxford University Press. London.
7. World Health Organization. 1980. International Classification of Impairment, Disabilities, and Handicaps. WHO. Geneva.
8. SUSSER, M., A. HAUSER, J. KIELY, N. PANETH, & Z. STEIN. 1985. Prenatal and perinatal risk factors in perinatal mortality, mental retardation, and epilepsy. *In* Prenatal and Perinatal Factors Associated with Brain Disorders. J. Freeman, Ed. USDHHS. Bethesda, MD. pp. 359–440.
9. HAGBERG, B. 1978. Severe mental retardation in Swedish children born 1959–1970: Epidemiological panorama and causative factors. *In* Major Mental Handicap: Methods and Costs of Prevention. Ciba Foundation Symposium 59. Elsevier–North Holland. Amsterdam. pp. 29–51.
10. GUSTAVSON, K., G. HOLMGREN, R. JONSELL, K. JONSELL, & H. BLOMQUEST. 1977. Severe mental retardation in a northern Swedish county. J. Mental Defic. Res. **21:** 161–180.
11. SUSSER, M. 1973. Causal Thinking in the Health Sciences. Oxford University Press. New York.
12. FLEISS, J. 1981. Statistical Methods for Rates and Proportions, Second Edition. John Wiley & Sons. New York.
13. LEVIN, M. 1953. The occurrence of lung cancer in man. Acta Unio Int. Contra Cancrum. **9:** 531–541.
14. MIETTINEN, O. 1974. Proportion of disease caused or prevented by a given exposure, trait, or intervention. Am. J. Epidemiol. **99**(5): 325–332.
15. WALTER, S. 1983. Effects of interaction, confounding, and observational error on attributable risk estimation. Am. J. Epidemiol. **117**(5): 598–604.

APPENDIX

Ten Questions

1. Compared with other children, did the child have any serious delay in sitting, standing, or walking?   YES   NO
2. Does the child have difficulty seeing?   YES   NO
3. Does the child appear to have difficulty hearing?   YES   NO
4. When you tell the child to do something, does he seem to understand what you are saying?   YES   NO
5. Does the child have weakness and/or stiffness in the limbs and/or difficulty in walking or moving his arms?   YES   NO
6. Does the child sometimes have fits, become rigid, or lose consciousness?   YES   NO

7. Does the child learn to do things like other children his age?   YES   NO
8. Does the child speak at all (can he make himself understood in words; can he say any recognizable words)?   YES   NO
9. Is the child's speech in any way different from normal (clear enough to be understood by people other than his immediate family)?   YES   NO
10. Compared with other children his age, does the child appear in any way backward, dull, or slow?   YES   NO

Cerebral Palsy: What Is Known Regarding Cause?

KARIN B. NELSON

Developmental Neurology Branch
National Institute of Neurological and Communicative
Disorders and Stroke
National Institutes of Health
Bethesda, Maryland 20892

Of the several categories of problems called "developmental disabilities," only mental retardation has been the subject of as much etiologic investigation as cerebral palsy (CP). CP is also the disability linked most frequently and most convincingly to perinatal events. Despite this, much remains unknown about the causes of CP, and it must be asked whether much that we "know" can be satisfactorily documented.

One reason for the scarcity and uncertainty of knowledge relating maternal and pregnancy factors and events of birth to neurologic outcome is the difficulty and expense of long-term studies of children whose early histories are well recorded. The largest study designed to investigate factors that might be important in the cause of CP was the Collaborative Perinatal Project of the National Institute of Neurologic and Communicative Disorders and Stroke (the NCPP). Analysis of data from that project is finally being completed, and this is a favorable opportunity to review what that experience has taught us.

The medical literature on the cause of CP is generally taken to begin with the writings of John Little[1] about 140 years ago. With the addition of clinical series from Eastman and DeLeon,[2] Lilienfeld and Parkhurst,[3] Steer and Bonney,[4] Mayer and Wingate,[5] Dale and Stanley,[6] and many others, this literature has stressed the importance of obstetric factors and of premature delivery. I shall refer to the pathogenetic factors recognized on the basis of these studies as the "classical risk factors."

NCPP experience confirms that a number of the classic risk factors (but not, as will be seen, all) were associated with heightened risk of CP; however, the absolute magnitude of risk associated with these risk factors was not enormous. For example, one of the birth factors most commonly cited is breech delivery. Dale and Stanley[6] found that infants presenting for delivery in the breech had CP four times more frequently than controls; NCPP experience was similar.[7] The absolute level of risk was not extremely high, however: only 1% of term infants delivered in the breech had CP. In general, most of the known risk factors for CP are uncommon, and most are not immensely powerful as predictors; most children with the risk factors do not develop CP.

A number of classic risk factors could *not* be confirmed. For example, we did not find that maternal diabetes was a risk factor for CP. Nor was either prolonged or precipitate labor. While it appears to be generally assumed that duration of labor is associated with the risk of CP, documentation of this statement appears very slender, and neither Dale and Stanley[6] nor we in the NCPP found it to be true.

Eastman and DeLeon[2] commented in 1955 that "lists of obstetrical conditions [said to cause CP] are imposing, but it should be noted that they have been prepared largely on a presumptive basis, and that they lack any specific documentation. In other words, these are lists of conditions which, on the basis of theoretical reasoning, might be *presumed* to cause cerebral palsy."

There may be only meager basis in empiric data to document the association of certain risk factors with increased incidence of CP.

In NCPP experience, for most quantifiable risk factors, it was only unusual and extreme *degrees* of the factor that carried the majority of the risk. Perhaps the most dramatic example is that of the Apgar score. Mild depression of the infant, with Apgar scores in the range four to six at one or at five minutes or later, were not associated with much increase in risk of CP.[8] Only in the presence of severe depression for a relatively long time did risk rise impressively. For the small proportion of babies over 2,500 g whose Apgar scores remained below four for 20 minutes despite active resuscitative efforts, the observed rate of CP exceeded 50%. Only severe depression for a substantial period of time was associated with a high rate of CP. (Since mild, brief depression of Apgar score is common, and CP is not common, this result is perhaps not surprising: We would expect much more CP than there is if mild depression of Apgar score were a good predictor of risk.)

All in all, obstetric factors were not associated with increased risk of CP unless children who experienced them were symptomatic in the first minutes, hours, and days of life. The infant whose brain is irreversibly injured during the process of delivery is *sick* as a newborn; he is symptomatic neurologically,[8,9] and often is symptomatic with respect to other organ systems as well, with oliguria and hematuria, with cardiac, pulmonary, and other system involvement. The first measure of neonatal neurologic status applied to most infants is Apgar scoring, and infants who experience an obstetric complication that is associated with long-term handicap usually show their involvement on that early measure. In the NCPP, children whose births were complicated and who had a five-minute Apgar score of five or less contributed 13 times the number of cases of CP that the size of this group would predict; this group of children with complicated births and immediate neonatal signs of depression was indeed a high-risk group. On the other hand, it was a small group: only 1.5% of 52,000 births had both birth complications and low Apgar score, and the excess of cases they contributed accounted for only 13% of the cases of CP in the NCPP population. In contrast, children whose births were complicated but who had five-minute Apgar scores of six or better constituted 62% of the population, and this group contributed a smaller proportion of CP cases than their prevalence as a group predicted. Children with birth complications and good Apgar scores were not at heightened risk of CP (TABLE 1).

Thus the excess of risk of CP associated with birth complications was restricted to the small group of babies who had both complicated births and low Apgar scores. Most cases of CP came from the large group not at increased risk by virtue of obstetric or immediate neonatal condition.

In some infants, brain-injuring events may occur late in pregnancy, but long enough before delivery to allow the infant to stabilize and pass his first neurologic examination, the Apgar, with a good score. A recent report notes pathologic evidence of intrauterine injury in the brains of some stillborn infants.[11] We have no estimate of the frequency of such occurrences in surviving children.

The tragedy of Minamata Bay proved that intrauterine exposure to high levels of ingested mercury could cause fetal damage that included CP among its manifestations; however, while we must remain alert for new evidence, there is little other evidence that environmental agents or nutritional factors are important

TABLE 1. Obstetric Complications, Mortality, and Cerebral Palsy[a]

Obstetric Complication	Five-Minute Apgar Score								
	0–3			4–6			7–10		
	No.	Death[b]	CP	No.	Death	CP	No.	Death	CP
Any	293	18.4[c]	5.9	818	6.4[c]	0.7	26,197	0.9	0.2
None	77	7.8	1.7	200	2.5	1.1	15,119	1.0	0.3

[a] Birth weights ≥ 2,500 g.
[b] Values given in percent of total.
[c] Statistical significance.

contributors to CP. Congenital infection, hyperbilirubinemia, and some genetic disorders can cause motor handicap. A few syndromes of unknown cause include chronic nonprogressive motor handicap. Overall, however, it is sobering to consider that in general the list of factors we know to be risk factors for CP is not a great deal longer than the list known to John Little 140 years ago.

It is sometimes assumed that all the developmental disabilities have similar causal bases; however, in the NCPP we have observed that seizure disorders in children who did not also have CP were not associated with very low birth weight, with low Apgar scores, or with birth complications.[12] Broman[13] has found that low Apgar scores and other measures of birth asphyxia are not important antecedents of mental retardation in children without other neurologic abnormality, and we have observed that children who survived very low Apgar scores without developing CP did not demonstrate an excess of mental retardation.[8] Except when they occur in the same individual, the developmental disabilities may differ in their major antecedents.

In the NCPP, we did find indications of a few novel risk factors that, if other studies confirm the associations, may increase the present set of hypotheses.

In general, we probably know less than we may have thought about the cause of CP, and of the other developmental disabilities. That was the conclusion of a recent publication entitled *Prenatal and Perinatal Factors Associated with Brain Disorders*,[14] sponsored by the National Institute of Child Health and Human Development and the National Institute of Neurologic and Communicative Disorders and Stroke. We know something, although far from enough, about prenatal and perinatal conditions that are associated with increased risk; we know far less about what proportion of CP in the population is accounted for by each of these factors, or by all the known factors combined. For example, we know that a sufficient dose of asphyxia, however one may elect to define that term, can destroy the brain of a neonate. We do not know, however, under the conditions in which humans are actually born, how much of the CP in that human population is contributed by asphyxia. We know that very low-birth-weight infants and the term infants who are cared for in newborn intensive care nurseries are individually at heightened risk, and often substantially heightened risk; but we do not know, for any population I am aware of, what proportion of cases of CP in that population arise from identified high-risk nurseries, or from community hospitals, or from other sectors of the newborn population.

We do not know whether the incidence of CP is rising, falling, or remaining constant. Since mild and moderate early motor signs resolve with considerable frequency, so that CP prevalence differs by age[15] and age of ascertainment is not comparable in existing studies, and since in general the definition of CP is so vague, except for the few studies with only a single examiner, some studies relating to changes in frequency of CP may not be meaningful.

We know that congenital malformations are more frequent outside the nervous system in children who have CP, but we know very little about the role of early and subtle maldevelopment within the brain, such as has been reported to date in scattered and heterogeneous reports in focal epilepsy,[16] in infantile spasms,[17] in dyslexia,[18] and in Down's syndrome.[19,20] Floyd Gillis[21] has stressed that the neuropathology of the first half of gestation is still a nearly uncharted area. At this meeting, held at the New York State Institute for Basic Research in Developmental Disabilities, it is appropriate to underline again the importance of careful histopathologic investigation in the developmental disabilities. For CP, it would be especially valuable to have detailed histologic study of brains that did not show gross evidence of destructive perinatal injury.

It is common in medicine for pathology to point the way for etiologic studies, and careful neuropathologic investigation may suggest new hypotheses as to mechanism. Despite the difficulty and expense, and the uncharted waters, there is clearly a need for effort in this direction in CP and the other developmental disabilities.

Our analyses[7] suggest that a smaller proportion of CP is attributable to known factors of pregnancy, labor, and delivery than was earlier supposed. If this is correct, then a substantial proportion of cases of cerebral palsy are still candidates for explanation. I would guess that the now-dominant model in CP, that of the fetus who enters labor with an intact brain and experiences grave and irreversible harm during the process of birth, describes only the minority of cases. I would guess that in many and perhaps in a majority of cases, disorders in the *formation* of that immensely complex instrument, the human brain, are at the root of the problem in CP, and perhaps play a role in some proportion of the other developmental disabilities. The exciting and rapidly expanding field of the developmental neurosciences will undoubtedly have a great deal of explanatory power for the developmental disabilities. A major challenge ahead of us lies in finding the links between the emerging knowledge of the basic neurobiology of development, and the clinical problems that we work to understand, to ameliorate, and in the future, to prevent.

REFERENCES

1. LITTLE, W. J. 1843. Lectures on the deformity of the human frame. Lancet **1:** 318–320.
2. EASTMAN, N. J. & M. DELEON. 1955. The etiology of cerebral palsy. Am. J. Obstet. Gynecol. **69:** 950–961.
3. LILIENFELD, A. M. & E. PARKHURST. 1951. A study of the association of factors of pregnancy and parturition with the development of cerebral palsy. Am. J. Hygiene **53:** 262–282.
4. STEER, C. M. & W. BONNEY. 1962. Obstetric factors in cerebral palsy. Am. J. Obstet. Gynecol. **83:** 526–531.
5. MAYER, P. S. & M. B. WINGATE. 1978. Obstetric factors in cerebral palsy. Obstet. Gynecol. **51:** 399–406.
6. DALE, A. & F. J. STANLEY. 1980. An epidemiological study of cerebral palsy in western Australia, 1956–1975. II. Spastic cerebral palsy and perinatal factors. Dev. Med. Child Neurol. **22:** 13–25.
7. NELSON, K. B. & J. H. ELLENBERG. 1984. Obstetric complications as risk factors for cerebral palsy or seizure disorder. J. Am. Med. Assoc. **251:** 1843–1848.
8. NELSON, K. B. & J. H. ELLENBERG. 1981. Apgar scores as predictors of chronic neurologic disability. Pediatrics **68:** 34–44.

9. CHRISTENSEN, E. & J. MELCHIOR. 1967. Cerebral Palsy—a Clinical and Neuropathological Study. William Heinemann Medical Books Ltd. London.
10. BRANN, A. 1985. Factors during neonatal life that influence brain disorders. *In* Prenatal and Perinatal Factors Associated With Brain Disorders. J. M. Freeman Ed.: 263–358. NIH Publication No. 85-1149. Bethesda, MD.
11. SIMS, M. E., S. B. TURKEL, G. HALTERMAN & R. H. PAUL. 1985. Brain injury and intrauterine death. Am. J. Obstet. Gynecol. **151:** 721–3.
12. NELSON, K. B. & J. H. ELLENBERG. 1984. Perinatal risk factors for nonfebrile seizure disorders in children free of epilepsy. *In* Advances in Epileptology: XVth International Symposium. R. J. Porter, Ed.: 385–389. Raven Press. New York.
13. BROMAN, S. H. 1979. Perinatal anoxia and cognitive development in early childhood, *In* Infants Born at Risk. T. F. Field, Ed.: 29–52. S. P. Medical and Scientific Books. New York.
14. FREEMAN, J. M, Ed. 1985. Prenatal and Perinatal Factors Associated with Brain Disorders. NIH Publication No. 85-1149. Bethesda, MD.
15. NELSON, K. B. & J. H. ELLENBERG. 1952. Children who "outgrew" cerebral palsy. Pediatrics **69:** 529–536.
16. MENCKE, H-J & D. JANZ. Neuropathological findings in primary generalized epilepsy: A study of eight cases. Epilepsia **25:** 8–21.
17. HUTTENLOCHER, P. R. 1974. Dendritic development in neocortex of children with mental defect and infantile spasms. Neurology **24:** 203–210.
18. GALABURDA, A. M. & T. L. KEMPER. 1979. Cytoarchitectonic abnormalities in developmental diplexia: A case study. Ann. Neurol. **6:** 94–100.
19. ROSS, M. H., A. M. GALABURDA & T. L. KEMPER. 1984. Down's syndrome: Is there a decreased population of neurons? Neurology **34:** 909–916.
20. WISNIEWSKI, K. E., M. LAURE-KAMIONOWSKA & H. M. WISNIEWSKI. 1984. Evidence of arrest of neurogenesis and synaptogenesis in brains of patients with Down's syndrome (letter). N. Engl. J. Med. **311:** 1187–1188.
21. GILLES, F. H. 1985. Neuropathologic indicators of abnormal development, *In* Prenatal and Perinatal Factors Associated with Brain Disorders. J. M. Freeman, Ed.:53–108. NIH Publication No. 85-1149. Bethesda, MD.

Maternal Health during Childhood and Later Reproductive Performance[a]

IRVIN EMANUEL

Department of Epidemiology
Department of Pediatrics
University of Washington
Seattle, Washington 98195

In seeking to understand problems of pregnancy outcome, the dominant concept seems to be that pregnancy is, in effect, a relatively acute condition lasting an average of about ten lunar months. In such a conceptualization, the important factors to be studied relate to the course of pregnancy and to the perinatal period. A corollary is that adequate prenatal and intrapartum care will largely eliminate complications of pregnancy and delivery.

Recently, however, there has been a reexamination of the relative importance of perinatal events and of prenatal and intrapartum care.[1-4]

Because this emphasis on perinatal events and complications of pregnancy has not led to adequate understanding and prediction of abnormal pregnancy outcomes, I would like to summarize the evidence that has accumulated over a number of years, which strongly suggests that factors and conditions that antedate pregnancy are important determinants of the quality of pregnancy outcome, at least with respect to certain problems.

INTERGENERATIONAL FACTORS

These factors have been referred to as intergenerational factors, signifying that the experiences of one generation influence the health of the next generation. Intergenerational factors may be defined as those factors, conditions, exposures, and environments experienced by one generation that relate to the health, growth, and development of the next generation. I will attempt to delineate certain risk factors that are preconceptional in timing, and are largely those that can be considered to influence the growth and development of mothers during the childhood period. Much of the evidence is indirect, but the accumulation of studies from a number of countries points in the direction of the importance of maternal childhood growth and development as an important influence on the quality of later reproduction.

[a] This study was supported in part by Grant No. MCJ-9043 from the Division of Maternal and Child Health, Bureau of Health Care Delivery and Assistance, U.S. Public Health Service.

27

CHILD HEALTH AND ADULT HEALTH

Women with various chronic diseases are at high risk for certain abnormal pregnancy outcomes. The most thoroughly studied condition is diabetes mellitus, which is associated with a number of adverse outcomes.[5] Further evidence comes from the National Natality Survey of 1980, which found an increased risk of low-birth-weight births to women who had underlying medical conditions.[6] If overt maternal disease can be important in pregnancy outcome, then it would seem biologically plausible that more subtle evidence of suboptimal health status of mothers might also be important, although it is much more difficult to document; however, the sum total evidence does point in the direction that child health is related to adult health, of which reproductive health is but one facet. With respect to problems of human reproduction, intergenerational factors have been associated with those problems for which the poor are at higher risk than the well-off.

The concept that child health is related to adult health can be traced at least as far back as 1934. Kermack et al.,[7] using a cohort analysis of mortality for England and Wales and for Scotland, calculated age-specific standard mortality ratios, with the standard rates being those of the first year for which mortality rates were available (England and Wales, 1841–1850; Scotland, 1860–1862). The mortality ratios for the adult years related rather consistently to the years of birth of each cohort. The fall in infant mortality lagged behind the decline at other age groups. The authors drew several conclusions: First, that adult health is determined to a great extent by childhood environment; second, improvement of infant mortality rates had to await improvement of maternal health. They attributed the improvement in mortality during this period to improvement in social and environmental conditions. These investigators stated that, with respect to adult health, their data "implied that the care of children during their first ten to fifteen years of life is of supreme importance."

CHILDHOOD ENVIRONMENTAL AND REPRODUCTIVE OUTCOME

The concept that maternal health during childhood is important for later pregnancy outcome has been largely developed by Baird and others in Scotland, beginning in the 1940s.[8-26]

In British studies extending from the 1950s through the mid-1970s, the low-birth-weight rate and fetal death or perinatal mortality rates were related to the occupational class both of the gravida's father and her husband.[10,12,21,23] A reasonable interpretation of such associations is that the conditions under which a mother grew up and the conditions under which the pregnancy occurred are both important in determining pregnancy outcome.

On the other hand, Drillien[20] concluded that the social class into which a woman marries is of minor importance compared to the social class of her childhood. Similarly, in marital mobility studies of women who lived with their parents after marriage, the pattern of low birth weight appeared to be more influenced by the circumstances of the mother's family of upbringing than by the conditions experienced during pregnancy.[26] In any event, the evidence is that the social and environmental conditions under which the mother spent her childhood relate to the quality of her reproductive experience.

Between 1948 and 1972 in Aberdeen, mothers who had babies with unexplained low birth weight tended to be born during the years of the Great Depression.[14] Baird[17] has also presented some preliminary evidence that there is a multi-generational effect in the production of unexplained low birth weight. Many studies of the Aberdeen investigators can be summarized as follows: Women who married upward in social class, in comparison with women who did not, had the following characteristics: they were taller and were subjectively judged to be healthier, they came from smaller families of upbringing, they showed upward occupational mobility compared to their fathers, they had higher educational attainment, and they had higher IQs. Tall maternal stature and upward mobility were associated with lower rates of low birth weight, fetal death, and perinatal mortality. The Aberdeen workers hypothesized that the biological mechanisms involved related to growth disturbance during childhood that may predispose to later problems with reproduction.

PREGNANCY OUTCOME OF BLACK WOMEN

It is a cause of great concern that black women in the United States have about a two fold risk of delivering low-birth-weight babies and of infant mortality compared to white women. In the Harlem study of nutritional supplementation during pregnancy, foreign-born black women who had been selected as being at high risk for delivering low birth weight babies unexpectedly had low birth weight rates of 3.8%, much lower than that of the American white population. It was determined that these foreign-born black women had a higher childhood socioeconomic background than the comparable group of U.S.-born black women, which the authors offered as a partial explanation for their more favorable pregnancy outcome.[27] National data also showed that at all educational levels, foreign-born black women have lower rates of low birth weight babies than do American-born black women.[28] Put together, these two studies suggest the possibility that the higher risk for poor pregnancy outcome among American black women may be related in some fashion to their childhood environments.

MATERNAL STATURE AND PREGNANCY OUTCOME

Several British studies found that in all paternal occupational class groups, perinatal mortality and low-birth-weight rates were inversely related to maternal stature (Baird).[9-16,18] The 1958 British Perinatal Mortality Survey found that several cause-specific perinatal mortality rates were also related to maternal stature.[19] These included low birth weight, antepartum hemorrhage, central nervous system malformations, and death due to mechanical factors. It had previously been determined that short women more commonly have abnormal pelvic shapes[29] and that short women are at elevated risk for cesarean section.[12] An Irish study confirmed the relationship between pelvic measurements and maternal stature, and also found positive correlations between year of birth and stature, and between year of birth and most pelvic dimensions and indices.[30]

MOTHER'S BIRTH WEIGHT AND LATER REPRODUCTION

Several studies place the possible growth disturbance in the mother's own intrauterine existence. Ounstead & Ounstead[31] found that mother's birth weight was related to intrauterine growth of their babies. Mothers of small-for-dates babies had the lowest mean maternal birth weight, and mothers of large-for-dates babies had the highest mean maternal birth weight. In an Aberdeen study, sisters of mothers of small-for-dates babies had lighter babies than their sisters-in-law, the general population, or the sisters of women who delivered prematurely.[24] The authors interpreted the results as supporting the hypothesis that the mother's own intrauterine experience influences her later reproductive outcomes. Recently, my colleagues and I more intensively studied maternal birth weight as it relates to future pregnancy outcome.[32] In our study, mother's birth weight and other data were obtained from the mother's own birth certificate. We found that mother's birth weight was related significantly to the following: maternal stature, prepregnancy weight and pregnancy weight gain; babies' birth weights, gestational duration and intrauterine growth; the need for neonatal intensive care, and respiratory distress syndrome and transient tachypnea of the newborn; and marginally to miscarriage. Using all the preconceptional variables available, mother's birth weights were still significantly related to babies' birth weights after adjustment for all other factors. Of special interest was the finding that 64% of babies of mothers who themselves weighed ≤2,000 gm at birth had abnormal outcomes, compared to 26% of babies of mothers with higher birth weights.

Recently, data from one center in the Collaborative Perinatal Project was similarly studied.[33] They also found a significant relationship between mothers' and babies' birth weights. They did not find mothers' birth weights related to gestational age, nor did the English study.[31] These two studies depended on maternal knowledge of their own birth weights, and the Collaborative Perinatal Project cohort was described as middle class. Our series was from a tertiary center, and the presumably accurate maternal birth weights were obtained from the mothers' own birth certificates. Clearly, studies from representative populations are needed to further investigate this potentially important relationship between mother's birth weight and future reproductive performance.

HYPOTHETICAL MECHANISM AND TIMING

The evidence that suggests that maternal growth disturbance—prenatal or postnatal—may predispose to later adverse reproductive outcome is largely indirect, and such measurements as maternal birth weight and maternal adult stature are crude and need to be translated into more direct mechanisms. We have suggested that deficiencies in these crude measures may be related to growth disturbance of organ systems, including the reproductive and/or endocrine systems.[34] Autopsies on perinatally dead, growth-retarded infants found histological abnormalities in the adrenals[35] and the pancreas.[36] A similar approach that includes other ages and other organs might be fruitful. A Guatemalan study of a chronically malnourished population yielded data that also suggests a critical period for growth disturbance. In this series, babies' birth weight was related to maternal head circumference, even after controlling for maternal stature and weight.[37] The authors suggested that since head circumference is largely determined by the first few years of growth, the very early environment of the mother may be reflected in the fetal growth of her infants.

Perhaps a caveat is in order relative to some of the measures used in the studies that I have reported. Studies in the United States, including our own, usually, but not always, find that the relationship between maternal stature and the baby's birth weight disappears when adjusted for prepregnant weight. Since the "catch-up" phenomenon is strong, adult stature may not be the most appropriate variable to measure, although it is a convenient one. There is also evidence that maternal stature, weight, and weight gain relate to babies' birth weight differently depending on maternal body habitus.[38,39] In studies in this country, birth size, and family socioeconomic circumstances are strongly related to postnatal growth.[40,41] In less developed countries where growth and adult stature are also strongly related to family circumstances during childhood, maternal height is independently related to newborn size[42] and infant mortality.[43]

MENARCHEAL AGE AND REPRODUCTION

There is additional evidence that family circumstances during the mother's childhood may relate to the development of the reproductive system. Median menarcheal age is directly related to the number of siblings in a girl's family of upbringing.[44] Early menarche is evidence of physiologic maturity, and one might expect that it should relate to better reproductive outcome. However, several studies have related "gynecologic age"—the number of years between menarche and first birth—to pregnancy outcome with conflicting results relative to birth weight.[45,46] In Malaysia, mothers with menarcheal age greater than 18 years had smaller babies.[47] Paradoxically, two studies found that women with early menarche—less than 13 years[48] or less than 11 years[49]—were at high risk for having spontaneous abortions. The latter study also found that the early menarche group had significantly more women with no reported pregnancies and with no live births. An autopsy study of correlates of perinatal central nervous system problems found that late menarche—greater than fifteen years—was associated with an increased risk of ganglionic eminence hemorrhage and subarachnoid hemorrhage.[50] Such studies suggest that women at the extremes of menarcheal age may be at elevated risk for some kinds of perinatal problems. In any event, it is difficult to imagine that the dramatic downward secular trend in menarcheal age does not in some way relate to the trends in some problems of pregnancy outcome.

CHILDHOOD FAMILY STRUCTURE AND LATER REPRODUCTION

In our study of maternal birth weight, it was possible to identify mothers who had been in three types of families of orientation: natural nuclear families, adoptive nuclear families, and single-parent families. The mean birth weight of babies of mothers from the two types of nuclear families was identical, and this was 1.5 pounds greater than the babies of mothers who themselves had been born to single mothers.[51] While the numbers are small, these data do suggest that girls who are raised in two-parent nuclear families for some reason are destined for better reproductive health upon reaching adulthood, possibly related to better care during childhood. In support of this concept is a New Zealand study that showed that adoptive parents were more conscientious than natural parents in providing their children with what is considered good preventive pediatric care, and the natural nuclear parents had a better record than single parents. The same study showed that early childhood morbidity rates showed the same gradient in the three kinds of families.[52]

There is additional indirect evidence that family conditions during childhood may be related to future reproductive problems. In the 1958 British Perinatal Mortality Survey, stillbirth rates were directly related to the number of siblings in the mother's family of upbringing. Of three maternal grandfather occupational classes, the gradient was most pronounced in the unskilled worker group.[25] This suggests that in all social classes, the more people who must share family resources, the more pronounced the effect on child health. The effects in the poorest families with the most meager resources were most pronounced.

MORE EVIDENCE FOR PRECONCEPTIONAL DETERMINANTS

The quality of reproductive outcome probably is related to a number of factors including the following: genetic constitution of the parents and embryo/fetus; state of health and nutrition of the mother at conception and during the course of pregnancy; and a possible multitude of extrinsic environmental factors. Health of the mother at conception depends on a sum total of a lifetime's experience as well as genetic factors. Several studies suggest that by the time reproduction begins, significant risks for adverse outcome are already present. A longitudinal study of the Norwegian population of births found that at all pregnancy orders, perinatal mortality, and low-birth-weight rates were directly related to attained sibship size.[53] A study of all pregnancies of British women physicians reported a similar pregnancy order—gravidity level relationship for fetal deaths.[54] In both these studies, the relationship held even for the first pregnancy. For instance, the women physicians with the most pregnancies had the highest fetal death rates at all pregnancy orders, and those with only two pregnancies had the lowest fetal death rates. The investigator interpreted these findings as due largely to the interaction of two variables—risk hetereogeneity among the women, and reproductive compensation. She suggested that women and not pregnancies should be studied.

An American study found that women tended to produce babies of similar birth weight and gestational duration characteristics in subsequent pregnancies.[55] More recently a longitudinal study of the total Norwegian population of births also found a marked tendency for women to produce babies of similar birth weight and gestational age combinations in consecutive pregnancies, regardless of maternal age, birth order, or complications of pregnancy, labor, or delivery.[56] These authors hypothesized that women were "programmed" to produce babies of certain characteristics. In other words, this "programming" is present at the time reproduction begins.

NEURAL TUBE DEFECTS

I would now like to turn to the question of the neural tube defects (NTD), anencephaly and spina bifida. These important congenital malformations of the nervous system show may epidemiologic similarities to those of low birth weight/ prematurity, stillbirth, and infant mortality. These include relationships to socioeconomic status, ethnic group differences, marked secular trends, maternal age, and birth order.[34,57] I believe there is also substantial support for the importance of intergenerational/preconceptional factors with respect to these important malformations.

The two most striking epidemiologic features of NTDs are the relationship to socioeconomic factors and the marked secular trends. An early study from Scotland found that the rate of anencephaly was inversely related to socioeconomic status, and that within two of the three occupational class groups, rates were inversely related to maternal stature.[58] The relationship of NTDs to social class and maternal stature was confirmed in the 1958 British Perinatal Mortality Survey,[19] and the social class effect has been duplicated in studies from the United States,[59] Canada,[60] and Taiwan.[61] Because of the higher rates in shorter women, the Scottish workers hypothesized that disturbance of growth during a critical period of development may predispose women to have babies with these malformations.

In a London study, recurrence risk for NTDs was found to be unrelated to the occupation of the father of the baby, but recurrence was significantly less if the baby's maternal grandfather was a white collar worker.[62] In our Northwest study, we also found an association between women's social class of origin and NTDs that was independent of the effects of the husband's social class.[63]

Another striking epidemiologic feature of neural tube defects is the epidemicity. Infant mortality data from England and Wales has shown two epidemic peaks for spina bifida, one in the 1930s and 1940s, and another in the 1950s.[64] Similar epidemic patterns have been found for anencephaly in Scotland, Birmingham, England, and Dublin.[65] Other Western European countries have also had epidemics.[66,67] Some Canadian provinces have also had a downward trend, but the data do not extend back far enough to determine when the epidemic peaks have occurred.[68] In a hospital study from Boston and Providence, a single epidemic peak was seen with a peak year of 1932, after which the rates began to decline while unemployment rates continued to rise.[69]

The epidemics discussed above were spread out and do not resemble what might be considered a "point source" epidemic. On the other hand, apparently two more sudden epidemics have been described. In the Hunger Winter of 1944–45 in the Netherlands, there was an increase in the occurrence of spina bifida.[70] In different parts of Germany during and shortly after World War II, there were epidemics of NTDs.[71] There appear therefore to be perhaps two types of NTD epidemics, each possibly requiring different explanations. The sudden epidemic in the Netherlands and in Germany were associated with acute food shortages and other economic and personal stresses. More will be said about the slowly developing epidemics later.

More recently, rates of NTDs have been declining in most countries for which data are available, including the United States,[72] South Wales,[73] England,[74] Scotland,[16] Northern Ireland,[75] the Irish Republic,[76] Australia,[77] the Netherlands,[78] and Germany.[71] In some areas there have been dramatic declines, and it appears that no single factor can account for the falling rates. Specifically, prenatal detection and termination cannot fully explain the trends, and the role of periconceptional vitamin intake has been questioned.[79]

NEURAL TUBE DEFECTS AND GEOGRAPHY

Interesting geographic variations have also been documented. There are similar east-west gradients in the United States[80] and Canada.[68] These trends remain unexplained, although some of the high rate areas in both countries are in chronically economically depressed regions such as Appalachia and the Maritime Provinces. The geographic variation of NTDs in England and Wales also corresponds to the distribution of births among the different social classes and of maternal stature[81] in a fashion that might be at least a partial explanation.

NEURAL TUBE DEFECTS AND MIGRATION

Large birth prevalence rate differences of NTD have been reported among different ethnic groups. The highest rates have been reported among the Irish and Welsh in the British Isles. High rates have also been reported in Scotland, parts of England, Egypt, and among the Sikhs. Low rates have been reported among black groups in several countries, in spite of their poor socioeconomic circumstances in some situations.[57] Ethnic group differences are usually considered to be indicative of genetic causative factors, but a small number of studies of migrants suggest that nongenetic factors may be operating as well. For example, in Boston, Irish-born women had rates less than women who resided in Dublin or Belfast at about the same time, and American-born women of Irish descent had still lower rates.[59] Anthropometric studies of immigrants to the United States from the early part of this century and later revealed that for most population groups, American-born children of immigrants were taller and heavier then their parents.[82] Because of the reported relationship of maternal stature to NTD rates, one may hypothesize that improved childhood growth may have been a factor in the lower NTD rates among the Boston Irish women.[34] Other migrant studies suggest early environmental factors as well. Oriental Jewish women who migrated to Israel had higher NTD rates than Israeli-born women;[83] however, Israeli-born Jewish women whose fathers had migrated from Africa and Asia had NTD rates equivalent to those of other Israeli-born women. Women born in the United Kingdom who migrated to the Oxford area had anencephaly rates resembling those of their regions of origin.[84] Because of the evidence of maternal childhood factors in the causal pathways for NTDs, we suggested that the long-term epidemics could partially be explained on that basis both in the United States and the United Kingdom. Mothers delivering defective babies during the peak years were born and grew up during periods of economic difficulties. These difficulties were marked by high rates of unemployment for both countries and by high emigration for the United Kingdom and by high immigration for the United States.[34] Recently Baird[16] analyzed the secular trends of anencephalic stillbirths in Scotland and in different maternal age groups. Curves were generated when the rates were plotted either by years of occurrence of the stillbirth or by years of mothers' birth. The peak years for most maternal age groups indicated that the mothers tended to be born during the Great Depression.

A UNIFYING HYPOTHESIS FOR NEURAL TUBE DEFECTS

We suggested a unifying hypothesis for causality of neural tube defects with three categories of causal factors: genetic, immediate, and intergenerational. The evidence for genetic factors relates to ethnic group differences, increased risk in siblings and more distant relatives, and most particularly, to the findings that the risk for producing affected children appears to be similar for parents of either sex with spina bifida (See Elwood & Elwood,[57] page 234). The evidence for immediate factors—those relating to the pregnancy in question—include maternal age, birth order, and seasonal relationships. Spina bifida has been associated with maternal use of the anticonvulsant valproate,[85] and this is also evidence that immediate factors may be important. Sharper epidemics that were temporally related to extreme war-related difficulties in Germany and in the Netherlands also support the operation of immediate factors.

The evidence for the operation of intergenerational factors in NTDs include: the long-term epidemic waves, the cohort effect, the geographic variation in the United Kingdom, and the relationship to maternal stature and to social class of mother's family of origin and to migrant studies. This tripartite etiologic model may apply to other problems of pregnancy outcome that also show relationships to socioeconomic factors. In any event, it does not seem likely that the complex epidemiologic patterns of neural tube defects can be explained simply on the basis of deficiency of one or more vitamins in the periconceptional period.

CONCLUSIONS

In summary, I believe there is very substantial evidence that intergenerational factors are important in several problems of pregnancy outcome, including low birth weight, fetal mortality, perinatal mortality, and neural tube defects. This evidence is of a largely indirect nature. Nevertheless, there appears to be sufficient evidence to warrant an attempt to further investigate this hypothesis and to develop measures that are relevant, and to attempt to elucidate the biologic mechanisms. It seems likely and biologically plausible that healthy children become healthy adults with reduced risk for a variety of health problems, including reproductive problems. The evidence for this simple principle has been accumulating over the last five decades and, I believe, has been largely ignored in the search for other causal factors that presumably are more amenable to immediate interventions.

It does not seem likely that immediate interventions will solve the problems of interest. On the other hand, it appears that the roots of these problems to some extent are laid down in childhood. If this is so, then a new sense of urgency is warranted since necessary interventions will take longer. I believe it would be appropriate, based on the available evidence, to reiterate what Kermack and his colleagues concluded in 1934, that the care of children is of supreme importance.

REFERENCES

1. STANLEY, F. & E. ALBERMAN. 1984. The Epidemiology of the Cerebral Palsies. Clinics in Developmental Medicine. No. 87. J. B. Lippincott. Oxford; Philadelphia, PA.
2. ILLINGWORTH, R. S. 1985. A paediatrician asks—why is it called birth injury? Br. J. Obstet. Gynaecol. **92:** 122–130.
3. FREEMAN, J. M., Ed. 1985. Prenatal and Perinatal Factors Associated with Brain Disorders. National Institutes of Health Publication No. 85-1149. Bethesda, MD.
4. NISWANDER, K., G. HENSON, D. ELLBOURNE, I. CHALMERS, C. REDMAN, A. MACFARLANE & P. TIZARD. 1984. Adverse outcome of pregnancy and the quality of obstetric care. Lancet **2:** 828–831.
5. MILLS, J. L. 1982. Malformations in infants of diabetic mothers. Teratology **25:** 385–394.
6. HUTCHINS, V., S. S. KESSEL & P. J. PLACEK. 1984. Trends in maternal and infant health factors associated with low infant birth weight, United States, 1972 and 1980. Public Health Rep. **99:** 162–172.

7. KERMACK, W. D., A. G. MCKENDRICK & P. L. MCKINLAY. 1934. Death rates in Great Britain and Sweden. Some general regularities and their significance. Lancet I: 698–703.
8. BAIRD, D. 1947. Social class and foetal mortality. Lancet II: 531–535.
9. BAIRD, D. 1952. Preventive medicine in obstetrics. N. Engl. J. Med. 246: 561–568.
10. BAIRD, D. 1962. Environmental and obstetrical factors in prematurity, with special reference to experience in Aberdeen. Bull. WHO 26: 291–295.
11. BAIRD, D. 1964. The epidemiology of prematurity. Am. J. Obstet. Gynecol. 65: 909–924.
12. BAIRD, D. 1965. Variations in fertility associated with changes in health status. J. Chronic Dis. 18: 1109–1124.
13. BAIRD, D. 1974. Sociological considerations of maternal and infant capabilities. In Horizons in Perinatal Research. N. Kretchmer and E. G. Hasselmeyer, Eds. pp. 10–22. Wiley. New York.
14. BAIRD, D. 1974. The epidemiology of low birth weight: Changes in incidence in Aberdeen, 1948–72. J. Biosocial Sci. 6: 323–341.
15. BAIRD, D. 1975. The interplay of changes in society, reproductive habits, and obstetric practice in Scotland between 1922 and 1972. Br. J. Prev. Soc. Med. 29: 135–146.
16. BAIRD, D. 1980. Environment and reproduction. Br. J. Obstet. Gynaecol. 87: 1057–1067.
17. BAIRD, D. 1985. Changing problems and priorities in obstetrics. Br. J. Obstet. Gynaecol. 92: 115–121.
18. BAIRD, D., R. ILLSLEY. 1952. Environment and childbearing. Proc. R. Soc. Med. 46: 53–59.
19. BAIRD, D. & A. THOMSON. 1969. The effects of obstetric and environmental factors on perinatal mortality by clinico-pathological causes. In Perinatal Problems. The Second Report of the 1958 British Perinatal Mortality Survey. N. R. Butler and E. D. Alberman, Eds. E&S Livingston Ltd. Edinburgh, Scotland. pp. 211–226.
20. DRILLIEN, C. M. 1957. The social and economic factors affecting the incidence of premature birth. Part I—Premature births without complications of pregnancy. J. Obstet. Gynaecol. Br. Emp. 64: 161–184.
21. ILLSLEY, R. 1955. Social class selection and class differences in relation to stillbirths and infant deaths. Br. Med. J. II: 1523–1524.
22. ILLSLEY, R. 1980. Professional or Public Health? Sociology in Health and Medicine. Chapter 2. The Nuffield Provincial Hospitals Trust. London. pp. 11–44.
23. ILLSLEY, R., R. G. MITCHELL, Eds. 1984. Low Birth Weight: A Medical, Psychological, and Social Study. Wiley. New York.
24. JOHNSTONE, F. & L. INGLIS. 1974. Familial trends in low birth weight. Br. Med. J. II: 659–661.
25. KINCAID, J. C. 1965. Social pathology of foetal and infant loss. Br. Med. J. I: 1057–1060.
26. THOMSON, A. M. 1959. Maternal stature and reproductive efficiency. Eugen. Rev. 51: 157–162.
27. VALANIS, B. M. & D. RUSH. 1979. A partial explanation of superior birth weights among foreign-born women. Soc. Biol. 26: 198–210.
28. TAFFEL, S. 1980. Factors associated with low birth weight. United States. 1976. Vital & Health Statistics, Series 21, Number 37. National Center for Health Statistics. Washington, D.C.
29. BERNARD, R. N. 1951–52. The shape and size of the female pelvis. Trans. Edinburgh Obstet. Soc. 104: 1–15.
30. HOLLAND, E. L., G. W. CRAN, J. H. ELWOOD, J. H. M. PINKERTON & W. THOMPSON. 1982. Associations between pelvic anatomy, height, and year of birth of men and women in Belfast. Ann. Hum. Biol. 9: 113–120.
31. OUNSTED, M. & C. OUNSTED. 1968. Rate of intrauterine growth. Nature 220: 599–600.
32. HACKMAN, E., I. EMANUEL, G. VAN BELLE & J. DALING. 1983. Maternal birth weight and subsequent pregnancy outcome. J. Am. Med. Assoc. 250: 2016–2019.
33. KLEBANOFF, M. A., B. I. GRAUBARD, S. S. KESSEL & H. W. BERENDES. 1984. Low birth weight across generations. J. Am. Med. Assoc. 252: 2423–2427.

34. EMANUEL, I. & L. E. SEVER. 1973. Questions concerning the possible association of potatoes and neural-tube defects, and an alternative hypothesis relating to maternal growth and development. Teratology **8:** 325–332.
35. NAEYE, R. L., M. M. DIENER, H. T. HARCKE, JR. & W. A. BLANC. 1971. Relation of poverty and race to birth weight and organ and cell structure in the newborn. Pediatr. Res. **5:** 17–22.
36. VAN ASSCHE, F. A., F. A. DEPRINS, L. AERTS & M. VERJANS. 1977. The endocrine pancreas in small-for-dates infants. Br. J. Obstet. Gynaecol. **84:** 751–753.
37. LECHTIG, A., H. DELGADO, R. LASKY, C. YARBROUGH, R. E. KLEIN, J-P. HABICHT & M. BEHAR. 1975. Maternal nutrition and fetal growth in developing countries. Am. J. Dis. Child. **129:** 553–556.
38. WINIKOFF, B. & C. H. DEBROVNER. 1981. Anthropometric determinants of birth weight. Obstet. Gynecol. **58:** 678–684.
39. HAFER, R. F. 1985. The effect of maternal size on perinatal outcome. Master of Public Health Thesis. University of Washington. Seattle, WA.
40. GARN, S. M. & M. T. KEATING. 1980. Effect of various prenatal determinants on size and growth through seven years. Ecol. Food Nutr. **9:** 109–112.
41. GARN, S. M., H. A. SHAW & K. D. MCCABE. 1978. Effect of socioeconomic status on early growth as measured by three different indicators. Ecol. Food Nutr. **7:** 51–55.
42. HABICHT, J-P., C. YARBROUGH, A. LECHTIG & R. E. KLEIN. 1973. Relationship of birth weight, maternal nutrition, and infant mortality. Nutr. Rep. Int. **7:** 533–546.
43. MARTORELL, R., H. L. DELGADO, V. VALVERDE & R. E. KLEIN. 1981. Maternal stature, fertility, and infant mortality. Hum. Biol. **53:** 303–312.
44. ROBERTS, D. F., L. M. ROZNER & A. V. SWAN. 1971. Age at menarche, physique, and environment in industrial Northeast England. Acta Paediatr. Scand. **60:** 158–164.
45. ZLATNIK, F. J. & L. F. BURMEISTER. 1977. Low "gynecologic age:" An obstetric risk factor. Am. J. Obstet. Gynecol. **128:** 183–186.
46. HOLLINGSWORTH, D. R. & J. M. KOTCHEN. 1981. Gynecologic age and its relation to neonatal outcome. Birth Defects: Orig. Artic. Ser. **17:** 91–105.
47. DAVANZO, J., J-P. HABICHT & W. P. BUTZ. 1984. Assessing socioeconomic correlates of birth weight in peninsular Malaysia: Ethnic differences and changes over time. Soc. Sci. Med. **18:** 387–404.
48. LIESTØL, K. 1980. Menarcheal age and spontaneous abortion: A causal connection. Am. J. Epidemiol. **111:** 753–758.
49. SANDLER, D. P., A. J. WILCOX & L. F. HORNEY. 1984. Age at menarche and subsequent reproductive events. Am. J. Epidemiol. **119:** 765–774.
50. GILLES, F. H., A. LEVITON & E. C. DOOLING. 1983. The Developing Human Brain. Growth and Epidemiologic Neuropathology. Chapters 15 and 16. John Wright—PSG Inc. Boston, MA.
51. EMANUEL, I., E. HACKMAN, J. DALING & G. VAN BELLE. Unpublished data.
52. FERGUSSON, D. M., J. HORWOOD & F. T. SHANNON. 1981. Birth placement and child health. N. Z. Med. J. **93:** 38–41.
53. BAKKETEIG, L. S. & H. J. HOFFMAN. 1979. Perinatal mortality by birth order within cohorts based on sibship size. Br. Med. J. **2:** 693–696.
54. ROMAN, E. 1984. Fetal loss rates and their relation to pregnancy order. J. Epidemiol. Community Health **38:** 29–35.
55. YERUSHALMY, J. 1967. Biostatistical methods in investigations of child health. Am. J. Dis. Child. **114:** 470–476.
56. BAKKETEIG, L. S., H. J. HOFFMAN & E. E. HARLEY. 1979. The tendency to repeat gestational age and birth weight in successive births. Am. J. Obstet. Gynecol. **135:** 1086–1103.
57. ELWOOD, J. M., J. H. ELWOOD. 1980. The Epidemiology of Anencephalus and Spina Bifida. Oxford University Press. Oxford.
58. ANDERSON, W. J. R., D. BAIRD & A. M. THOMSON. 1958. Epidemiology of stillbirths and infant deaths due to congenital malformations. Lancet **1:** 1304–1306.
59. NAGGAN, L. & B. MACMAHON. 1967. Ethnic differences in the prevalence of anencephaly and spina bifida in Boston, Massachusetts. N. Engl. J. Med. **277:** 1119–1123.

60. HOROWITZ, I. & A. D. MCDONALD. 1969. Anencephaly and spina bifida in the Province of Quebec. Can. Med. Assoc. J. **100:** 748–755.
61. EMANUEL, I. 1972. Nontuberous neural-tube defects. Lancet **2:**879.
62. CARTER, C. O. & K. EVANS. 1973. Spina bifida and anencephaly in greater London. J. Med. Genet. **10:** 209–234.
63. SEVER, L. E. & I. EMANUEL. 1981. Intergenerational factors in the etiology of anencephalus and spina bifida. Dev. Med. Child. Neurol. **23:** 151–154.
64. ROGERS, S. C. & M. MORRIS. 1971. Infant mortality from spina bifida, congenital hydrocephalus, monstrosity, and congenital diseases of the cardiovascular system in England and Wales. Ann. Hum. Genet. London **34:** 295–305.
65. LECK, I. & S. C. ROGERS. 1967. Changes in the incidence of anencephalus. Br. J. Prev. Soc. Med. **21:** 177–180.
66. ROGERS, S. C. 1972. Changes in infant mortality from spina bifida and/or hydrocephalus and from congenital heart disease in Europe, 1948–67. Health Trends **4:** 52–54.
67. ROGERS, S. C. & J. WILKIN. 1976. Congenital malformations in Europe in war and at peace. Health Trends **8:** 20–21.
68. ELWOOD, J. M. 1974. Anencephalus in Canada, 1943–1970. Am. J. Epidemiol. **100:** 288–296.
69. MACMAHON, B. & S. YEN. 1971. Unrecognized epidemic of anencephaly and spina bifida. Lancet **1:** 31–33.
70. STEIN, Z., M. SUSSER, G. SAENGER & F. MAROLLA. 1975. Famine and human development. The Dutch Hunger Winter of 1944–1945. Oxford University Press. New York. pp. 224–228.
71. KOCH, M. & W. FUHRMANN. 1984. Epidemiology of neural tube defects in Germany. Hum. Genet. **68:** 97–103.
72. WINDHAM, G. C., & L. D. EDMONDS. 1982. Current trends in the incidence of neural tube defects. Pediatrics **70:** 333–337.
73. ROBERTS, C. J., B. M. HIBBARD, G. H. ELDER, K. T. EVANS, K. M. LAURENCE, A. ROBERTS, J. S. WOODWARD, I. B. ROBERTSON & M. HOOLE. 1983. The efficacy of a serum screening service for neural-tube defects: The South Wales experience. Lancet **1:** 1315–1318.
74. OWENS, J. R., F. HARRIS, E. MCALLISTER & L. WEST. 1981. 19-year incidence of neural tube defects in area under constant surveillance. Lancet **2:** 1032–1035.
75. NEVIN, N. C. 1981. Neural tube defects. Lancet **2:** 1290–1291.
76. COFFEY, V. P. 1983. Neural tube defects in Dublin 1953–1954 and 1961–1982. Ir. Med. J. **76:** 411–413.
77. DANKS, D. M. & J. L. HALLIDAY. 1983. Incidence of neural tube defects in Victoria, Australia. Lancet **1:** 65.
78. ROMIJN, J. A. & P. E. TREFFERS. 1981. Anencephaly in the Netherlands: A remarkable decline. Lancet **1:** 64–65.
79. WALD, N. J. & P. E. POLANI. 1984. Neural tube defects and vitamins: The need for a randomized clinical trial. J. Obstet. Gynecol. **91:** 516–523.
80. GREENBERG, F., L. M. JAMES & G. P. OAKLEY, JR. 1983. Estimates of birth prevalence rates of spina bifida in the United States from computergenerated maps. Am. J. Obstet. Gynecol. **145:** 570–573.
81. ILLSLEY, R. & J. C. KINCAID. 1963. Social correlations of perinatal mortality. *In* Perinatal Mortality. The First Report of the 1958 British Perinatal Mortality Survey. E&S Livingston Ltd. Edinburgh. pp. 270–286.
82. KAPLAN, B. A. 1954. Environment and human plasticity. Am. Anthropol. **56:** 780–800.
83. NAGGAN, L. 1971. Anencephaly and spina bifida in Israel. Pediatrics **47:** 577–586.
84. HOBBS, M. S. T. 1969. Risk of anencephaly in migrant and nonmigrant women in the Oxford area. Br. J. Prev. Soc. Med. **23:** 174–178.
85. ROBERT, E. & F. ROSA. 1983. Valproate and birth defects. Lancet **2:** 1142.

DISCUSSION

ROBERT GUTHRIE: I wanted to add lead exposure to the list of important causes of mental handicaps in different countries, especially the so-called developing countries. During a project in which we screened Kuwaiti children, we found a high incidence of blood lead elevations. This lead exposure is due to the universal use of kohl, a black eye cosmetic, in Kuwaiti women. Exposure begins in infant girls, and extends right through a woman's life. Often this kohl includes lead disulfide, or galena, which reflects light and causes the eyes to sparkle. This practice appears to be spread across all Arab countries and to India and Bangladesh (where it is called *surma*) and can lead to congenital lead toxicity. We have found "lead lines" present (on x-ray) in the newborn infant's skeleton at birth, symptoms such as seizures, and even death in early infancy.

PEGGY BROOKS-BERTRAM: How do the weights of babies born to black adolescents compare with those of white adolescents?

IRVIN EMANUEL: The rates of low birth weight have gone down at all maternal ages, but blacks are at higher risk at all ages.

ZENA STEIN: My impression is that among teenage mothers the birth weights are good where there are special programs, but that where there are not special programs they seem to be at a disadvantage. Perhaps the other question was whether blacks and whites differ on preterm births as well as intrauterine growth retardation.

PART II. ENVIRONMENTAL RISK FACTORS DURING
EMBRYOGENESIS, FETAL DEVELOPMENT, AND POSTNATAL
DEVELOPMENT

Introduction

DONALD A. SNIDER

*Institute for Basic Research in Developmental Disabilities
New York State Office of Mental Retardation &
Developmental Disabilities
Staten Island, New York 10314*

There are many environmental agents that have been directly linked to mental retardation and there is no way that a meeting of this scope could address them all. Instead, we decided to have two papers on alcohol because it is estimated to be the leading specific environmental cause of mental retardation. We sought to balance the program with papers on other environmental agents where a summary of the state of knowledge seemed feasible, some exciting new data have been produced, or future research objectives identified.

Dr. Patricia Rodier of the University of Rochester Medical School describes the need for data on the teratogenic potential of anesthetics, as well as their effectiveness. Dr. Rodier defines the issue this way: Because of their wide use in the delivery rooms of hospitals, inhalant anesthetics, more than any other environmental agent, present a high rate of exposure to developing humans.

Dr. Charles Vorhees of the Children's Hospital Research Foundation in Cincinnati continues the theme introduced by Dr. Rodier as he writes on the effects of fetal exposure to phenytoin and trimethadione, two widely used anticonvulsants. Dr. Vorhees' research with animal models draws parallels between the results of anticonvulsant exposure and the fetal alcohol syndrome.

Dr. Ann Streissguth of the University of Washington, one of the coinvestigators in the two studies that originally defined and described the fetal alcohol syndrome (FAS), provides a methodological perspective.

Dr. Robert Sokol of the Hutzel Hospital and Wayne State University in Detroit discusses determinants of susceptibility to alcohol teratogenicity. His findings on susceptibility have important implications for public health and clinical approaches to prevention.

Dr. Ernest Abel, who was then at the New York State Research Institute on Alcoholism (and is now also at the Hutzel Hospital, Detroit), summarizes the four preceding papers, describing how the trends in research reflect societal concerns. For this reason, he believes alcohol will continue to receive a high level of attention.

Dr. Thomas Shepard of the University of Washington notes the dramatic increase in the amount of chemicals produced each year. (There are now about 5,000,000 chemicals to which our population has significant exposure.) Despite the methodological difficulties, Dr. Shepard believes that we need to do more to establish the effects of workplace chemicals on the offspring of the persons exposed.

Dr. Jeanne Stellman of the Women's Occupational Health Research Center at Columbia sparked controversy with her remarks on hazards from Agent Orange to video display terminals (VDTs).

Rubella and cytomegalovirus (CMV) have historically been major causes of severe mental retardation. CMV causes more MR than rubella did before the immunization program for that disease. Dr. Robert Pass of the Children's Hospital of the University of Alabama at Birmingham examines the progress and difficulties made in preventing CMV in light of the immunization program for rubella.

Inhalant Anesthetics as Neuroteratogens

PATRICIA M. RODIER

Department of Obstetrics and Gynecology
University of Rochester
Rochester, New York 14642

The possibility that anesthetics, particularly inhalant anesthetics, might be terato-genic, has been a source of concern for many years. Occupational exposures, exposures to environmental contaminants, and exposures to over-the-counter or prescription drugs may occur at high rates within some subpopulations of pregnant women, but it is difficult to imagine a class of agents to which developing humans are exposed with such frequency as they are to agents commonly used during delivery. No teratogen could produce gross malformations at the time of parturition, instead, the question is whether anesthetics may permanently alter organ systems that are developing rapidly at the time of birth. The central nervous system is of special interest in this regard, not only because it exhibits basic developmental processes long after birth, but because it does not exhibit the continuous development of tissues such as blood and intestinal epithelium. Thus, the nervous system remains sensitive to teratogens over a very long period, but it does not share the regenerative potential of other late-developing systems. These characteristics put the CNS at risk at stages when most body systems are refractory to teratogens.

Investigators have measured a variety of neurobehavioral endpoints in children exposed to anesthetics versus those not exposed (reviewed by Brackbill[1]), and the majority of these comparisons suggest significant differences. It is difficult to control all the confounding variables in such studies, and especially difficult to prove that the use of different drugs at parturition is not itself indicative of differences in pregnancy or labor that might be responsible for the observed effects. A second major issue in human studies is whether the behavioral differences are long-lasting or whether they represent a transient effect of the drugs. Separation of pharmacologic effects from teratogenic effects requires long-term follow-up, which has been attempted only rarely in human studies. While the data are limited, the typical finding in prospective studies is that behavioral effects persist for at least several months,[2,3] and retrospective studies suggest that differences may still be detected at later time points.[4,5]

Laboratory investigations of the effects of anesthetics on developing animals have focused on exposures during organogenesis, with gross malformations as the endpoint. The most recent studies of this type suggest that nitrous oxide is particularly hazardous, even when compared to other inhalants, such as xenon[6] and halothane.[7] *In utero* exposure of rodents to halothane has been shown to lead to behavioral abnormalities on several tasks, when the exposure occurs around the time of implantation or neural tube closure[8] but the same measures were unaffected when exposure occurred during the fetal period. Fetal exposure to halothane or enflurane did alter performance on a maze task in the study of Chanlon, *et al.*[9] Together, these studies indicate that inhalants can be injurious to body form and brain function and the functional effects are long-lasting. They do not address the question of whether functional effects can be induced by perinatal exposure.

A completely different, but important line of investigation is that which details the effects of inhalants on proliferating cells. The property of interference with cell production was considered to be common to all inhalants as early as Andersen's 1966 review.[10] Sturrock and Nunn,[11] among others, have described the effects of halothane on the cell cycle. More recently, the mechanism of nitrous oxide inhibition of DNA synthesis via interference with methionine synthetase has been reviewed by Nunn and Chanarin.[12] This line of evidence indicates that inhalants interfere with cell proliferation in a variety of mammalian tissues, both *in vitro* and *in vivo*, but there have been no attempts to test inhalants for interference with neuron production in the developing brain.

The studies to be summarized below were based on the hypothesis that nitrous oxide and halothane injure the developing brain by interfering with cell production. Because many known brain-damaging agents exhibit this mechanism, we chose to compare the effects of inhalants to some well-established effects of known CNS teratogens, such as X-ray, azacytidine, and methylazoxymethanol (reviewed by Rodier[13]). The test hypotheses were that the inhalants would (1) cause a pattern of decreased cell production, followed by a rebound in mitotic activity; (2) cause a permanent reduction in cell number and tissue size in the brain regions actively forming during exposure; and (3) cause specific behavioral effects, depending on the brain regions injured, and therefore depending on the time of exposure.

Two treatment times—the 14th day of mouse gestation (day of finding a plug = day G1) and postnatal day two (day of birth = day PN1)—were selected because of the differences in brain regions adding neurons on those days and the differences in behavioral outcomes of antimitotic injuries on those days.

GENERAL METHODS

In all the experiments, dams or neonates were exposed to either a sham gas mixture (75% N_2 and 25% O_2), nitrous oxide (75% N_2O and 25% O_2), or halothane (0.5% halothane in the sham gas mix). Prenatal exposures lasted six hours, and postnatal exposures lasted four-hours. (In subsequent experiments, we have examined many of the same parameters with four-hour prenatal exposures and six-hour postnatal exposures, and the two-hour difference seems to have no effect on any of the measures.) Animals to be sacrificed soon after exposure, as in the studies of mitotic activity, received a single exposure in a semi-open flow chamber. Animals to be studied behaviorally, or for adult neuroanatomy, were exposed on both treatment days, to equalize handling and other effects of the exposures. They differed only in the gas mixtures they received. Thus, the prenatal nitrous oxide exposure group received 75% N_2O and 25% O_2 on G14 and the sham mix on PN2, while the postnatal halothane exposure group received a sham treatment prenatally and 0.5% halothane on PN2. Controls received the sham mix on G14 and PN2. Behavioral testing groups ranged from about 50–70 animals per group. Histological comparisons were carried out on nine animals/treatment group in the studies of mitosis and 25 animals/group in studies of adult brain structure.

BEHAVIORAL STUDIES

TABLE 1 summarizes the effects of inhalant exposure on a number of behavioral measures from PN6 to six months of age. It had been predicted that delays in

TABLE 1. Some Behavioral Effects of Early Exposure to Inhalants

Behavior	Prenatal		Postnatal	
	N_2O	Halothane	N_2O	Halothane
Surface righting	delayed[a]	delayed[a]	delayed	delayed
Air righting	delayed[a]	delayed[a]	delayed	delayed
Pivoting	persistent[a]	persistent[a]	persistent	—[b]
Walking	delayed[a]	delayed[a]	—[b]	delayed
Preweaning activity	—[b]	—[b]	low	low
Young adult activity	—[a,b]	low[a]	low	low
Mature adult activity	—[a,b]	—[a,b]	—[b]	low

[a] Unexpected significant effects or unexpected absence of effects.
[b] Not significantly different from controls.

reflex development, persistence of infantile locomotion, and delays in walking would characterize postnatally exposed animals, while prenatal treatment would have little effect on these measures. Activity was expected to be high in the prenatal treatment groups as they matured and low or normal in the postnatal groups.

While both nitrous oxide and halothane had many significant effects on pre-weaning motor development[14] and on general activity,[15] the pattern of effects was not that seen with antimitotic agents. On almost every measure, prenatally treated animals departed from the performance characteristic of animals treated to injure proliferating cells. They were never hyperactive, and the prenatal halothane group had an excess proportion of animals with low activity. They showed as much impairment in motor development as the postnatal treatment groups. Post-natally treated mice did resemble animals treated at this stage with antimitotic agents, but since all four groups behaved similarly, it seems unlikely that they could have sustained the very different injuries predicted. Of course, it is possible that the agents did act to interfere with mitosis, but that some other injury, common to all the treatment groups, is actually dictating the observed behavioral abnormalities. Not only were the early and late treatment groups similar, but the two inhalants had similar effects as well.

MORPHOLOGY AT THE TIME OF TREATMENT

The hypothesis that these agents alter cell production by changing mitotic activity was tested directly by exposing animals as for behavioral testing, then counting mitotic figures and cell numbers as well as the thickness of proliferative zones. Mitotic figures were characterized as "early" through metaphase, or as "late" for telophase and anaphase. These measures were collected from pre-sumptive cerebral cortex, hippocampus, midbrain, and cerebellum in fetal mice, and from the external germinal layer of the cerebellum in neonates, at four differ-ent sacrifice times, to follow the response of the tissue over time. No group showed evidence of mitotic arrest, as has been observed in plant tissue exposed to halothane.[16] On the other measures, there were several significant effects in fetal brain, but no strong indication of the classic depression and rebound observed with antimitotic agents (e.g. FUdR[17]). In several cases, the various measures seemed unrelated—mitosis would decrease while cell numbers increased. Such effects are difficult to explain. In neonates, halothane produced some significant but unexpected effects. The only treatment that produced the expected pattern of results was postnatal nitrous oxide. In these animals, cell production was sharply decreased for at least twelve hours after exposure, and then exceeded control levels at 24 hours and 48 hours.

In the same animals examined for fetal brain response to inhalants, the development of blood was evaluated. This tissue has been reported to be sensitive to antimitotic effects of inhalants by many investigators.[18] By counting the immature blood cells (normoblasts) versus the mature cells (erythrocytes) in the choroid plexus, it was obvious that blood development had been significantly slowed by nitrous oxide exposure. Halothane's effect was similar, but fell short of significance. From these studies we conclude that inhalants have a classic antimitotic action on some tissues, at some stages of development, but spare others, or at least alter them in unfamiliar ways.

A final measure from animals sacrificed soon after exposure was body weight. General interference with proliferation throughout the rapidly growing fetus or neonate sharply affects weight, and the effects are usually long-lasting.[19] We knew from the animals prepared for behavioral testing that inhalant exposure did not alter body weight for extended periods, but we wondered whether weight might be affected even briefly. The results are shown in TABLE 2. Exposure *in utero* led to a rapid, significant reduction in weight that only began to recover at 24 and 48 hours after sacrifice. Postnatal exposure had no detectable effect on weight; there was not even a trend toward weight reduction. The discrepancy in effects at the two treatment times raises the question of whether the fetal weight effect may be maternally mediated. That is, it may reflect some redistribution of blood or fluids from the fetus to the mother, rather than an antimitotic effect of the drug on fetal tissue. This hypothesis would help explain how the weight difference occurs so rapidly. Six hours is less than the length of a cell cycle, and thus seems too short a time to allow much growth inhibition via a block of cell production. At any rate, the weight effect *in utero* could itself interfere with development. The fact that this rather startling effect of inhalants has not been observed previously should serve to remind investigators that our knowledge of the actions of these drugs on the fetus is completely inadequate. The details of this group of studies on immediate effects of inhalants on brain, blood, and body weight are reported in Rodier, Aschner, Lewis, and Koëter.[20]

As in the behavioral studies, both halothane and nitrous oxide had many effects, but not all were consistent with the original hypothesis of the studies.

TABLE 2. Body Weight (g) after Inhalant Exposure

Sacrifice Time	Controls	N$_2$O	Halothane
		Prenatal Exposure (6 h)	
Immediate	0.217	0.194^a	0.193^a
12 h	0.285	0.268^b	0.266^a
24 h	0.333	0.280	0.266
48 h	0.551	0.500	0.471
		Postnatal Exposure (4 h)	
Immediate	1.69	1.65	1.71
12 h	1.80	1.83	1.81
24 h	2.10	1.88	2.02
48 h	2.13	2.42	2.13

^{a} $p < 0.01$
^{b} $p < 0.05$

Postnatal nitrous oxide does appear to have antimitotic action on the developing cerebellum, but the meaning of the other effects is unclear.

ADULT MORPHOLOGY AFTER EXPOSURE TO INHALANTS

This study is still in progress,[21] but analysis of several parameters of the adult cerebellum in exposed and control mice reveals substantial permanent changes in neuroanatomy, as indicated in TABLE 3. These measures were taken to contrast

TABLE 3. Significant Decreases in Brains of Adult Mice Exposed to Inhalants during Development

Parameter	Prenatal		Postnatal	
	N_2O	Halothane	N_2O	Halothane
Size of cerebellum	ns	$p < 0.05$	$p < 0.05$	$p < 0.05$
Number of Purkinje cells	$p < 0.05$	ns	ns	ns
Number of molecular layer cells	ns	$p < 0.05$	$p < 0.05$	$p < 0.05$

effects on prenatally forming cells, such as the Purkinje cells, with effects on postnatally forming cells, such as those in the molecular layer. If nitrous oxide and halothane acted on the developing brain solely by interfering with cell production, the G14 treatment would decrease cell numbers mainly in cerebral cortex, hippocampus, midbrain, and diencephalon. The Purkinje cells form predominantly on days 11, 12, and 13. The molecular layer cells are produced in great numbers during the first week after birth, and thus would be reduced by treatment on PN2. As in the behavioral studies, postnatal groups fit the pattern predicted for antimitotic agents. The size of the cerebellum was significantly reduced, compared to controls, and the number of molecular-layer cells in a unit width of vermian cerebellar cortex was significantly decreased by inhalants, while the number of Purkinje cells was normal. In light of the behavioral data, it is not surprising that the prenatally exposed halothane group had the same pattern of adult structural changes. This damage could explain the slow motor development and low activity levels observed in this group. What is mysterious is how a prenatal exposure could create an injury in cells that form after birth. Perhaps this cell loss is secondary to some other effect of the prenatal treatment, or perhaps a continued presence of halothane or its metabolites causes the damage. The one deviant group is that exposed to nitrous oxide prenatally. In these mice, exposure apparently caused some loss of the recently formed Purkinje cells. Taken together, these results suggest that antimitotic activity may account for some, but not all, of the effects of inhalants on the brain. Perhaps more importantly, they indicate that early exposure to nitrous oxide or halothane has permanent effects on brain morphology, just as it has lasting effects on function.

CONCLUSION

As a class of potential teratogens, inhalant anesthetics deserve serious attention because of their widespread use. To some readers, it may seem that evidence of teratogenicity of these agents presents an impossible problem, for the use of anesthetics may be necessary in a variety of clinical situations, but that is true only if the teratogenic action of these agents is perfectly correlated with the anesthetic action. Fortunately, both the present studies and the recent studies of Lane et al.[6,7] suggest that this is not the case. In the studies outlined above, the nitrous oxide exposure left animals not only conscious but rather active, while the halothane exposure was borderline anesthetic; that is, halothane animals appeared to sleep soon after exposure began. They could be aroused by noise or by touch, but they were clearly much more affected than the nitrous oxide animals. Yet, nitrous oxide had slightly more teratogenic effects. Lane's group found the same pattern when the two agents were compared for ability to cause malformations during organogenesis; thus, it is clear that the teratogenic effects of anesthetics are not determined by, or produced by, the degree of anesthesia obtained. This means that some potent anesthetics may have little or no teratogenic effect. Conversely, one cannot enhance the safety of the fetus by selecting an ineffective anesthetic, because light anesthesia is not necessarily less teratogenic than deep anesthesia. Now that the lack of relationship has been demonstrated, it is possible that anesthetics with no teratogenic effects already exist or could be developed.

REFERENCES

1. BRACKBILL, Y. 1979. Obstetrical medication and infant behavior. In Handbook of Infant Development. J. D. Osofsky, Ed. Wiley, New York, pp. 76–125.
2. ALEKSANDROWICZ, M. K. & D. R. ALEKSANDROWICZ. 1974. Obstetrical pain-relieving drugs as predictors of infant behavior variability. Child Dev. **45:** 935–945.
3. FRIEDMAN, S. L., Y. BRACKBILL, A. T. CARON & R. F. CARON. 1978. Obstetric medication and visual processing in four-and five-month-old infants. Merrill-Palmer Q. **24:** 111–128.
4. MULLER, P. F., H. E. CAMPBELL, W. E. GRAHAM, H. BRITTAIN, T. A. FITZGERALD, M. A. HOGAN, V. H. MULLER & A. H. RITTENHOUSE. 1971. Perinatal factors and their relationship to mental retardation and other parameters of development. Am. J. Obstet. Gynecol. **109:** 1205–1210.
5. GOLDSTEIN, K. M., D. V. CAPUTO & H. B. TAUB. 1976. The effect of prenatal and perinatal complications on development at one year of age. Child Dev. **47:** 613–621.
6. LANE G. A., M. L. NAHRWOLD, A. R. TAIT, M. TAYLOR-BUSH, P. J. COHEN & A. R. BEAUDOIN. 1980. Anesthetics as teratogens: Nitrous oxide is fetotoxic, xenon is not. Science **210:** 899–901.
7. LANE, G. A., P. M. DuBOULAY, A. R. TAIT, M. TAYLOR-BUSH & P. J. COHEN. 1981. Nitrous oxide is teratogenic: Halothane is not. Anesthesiology **55:** A252.
8. SMITH R. F., R. E. BOWMAN & J. KATZ. 1978. Behavioral effects of exposure to halothane during early development in the rat. Anesthesiology **45:** 413–420.
9. CHANLON, J., C. K. TANG, S. ROMANOTHAN, M. EISNER, R. KATZ & H. TURNDORF. 1981. Exposure to halothane and enflurane affects learning function in murine progeny. Anesth. Analg. **60:** 794–797.
10. ANDERSEN, N. D. 1966. The effect of CNS depressants on mitosis. Acta. Anaesth. Scand. **10:** Suppl. 22: 1–36.
11. STURROCK, J. & J. F. NUNN. 1976. Effects of halothane on DNA synthesis and the presynthetic phase (G1) in dividing fibroblasts. Anesthesiology **45:** 413–420.
12. NUNN, J. F. & I. CHANARIN. 1985. Nitrous oxide inactivates methionine synthetase. In Nitrous Oxide. E. I. EGER, Ed. Elsevier, New York, pp. 211–233.
13. RODIER, P. M. 1980. Chronology of neuron development: Animal studies and their clinical implications. Dev. Med. Child Neurol. **22:** 525–545.
14. KOËTER, H. B. W. M. & P. M. RODIER. 1986. Behavioral effects in mice exposed to nitrous oxide or halothane: Prenatal vs postnatal exposure. Neurobehav. Toxicol. Teratol. **8:** 189–194.

15. RODIER, P. M. & H. B. W. M. KOËTER. 1986. General activity from weaning to maturity in mice exposed to halothane or nitrous oxide. Neurobehav. Toxicol. Teratol. **8:** 195–199.

16. NUNN, J. F., J. D. LOVIS & K. L. KIMBALL. 1971. Arrest of mitosis by halothane. Br. J. Anaesth. **43:** 524–530.

17. ANDREOLI, J., P. M. RODIER & J. LANGMAN. 1973. The influence of a prenatal trauma on formation of Purkinje cells. Am. J. Anat. **137:** 87–102.

18. LASSEN, H. C. A., E. HENRIKSSEN, F. NEUKIRCH & H. S. KRISTENSEN. 1956. Treatment of tetanus: Severe bone marrow depression after prolonged nitrous-oxide anesthesia. Lancet **1:** 527–530.

19. RODIER, P. M. & S. S. REYNOLDS. 1977. Morphological correlates of behavioral abnormalities in experimental congenital brain damage. Exp. Neurol. **57:** 81–93.

20. RODIER, P. M., M. ASCHNER, L. S. LEWIS & H. B. W. M. KOËTER. 1986. Cell proliferation in developing brain after brief exposure to nitrous oxide or halothane. Anesthesiology **64:** 680–687.

21. RODIER, P. M., R. SALTZMAN & H. B. W. M. KOËTER. 1986. Morphometric evaluation of adult brain after fetal or neonatal exposure to inhalant anesthetics. In progress.

DISCUSSION

QUESTION: Since the data do raise concern about the safety of anesthetics in obstetrics, what should be our course of action?

PATRICIA RODIER: We need to collect detailed evaluation of CNS effects for a variety of anesthetics, so that the teratogenic risk to CNS can be considered in selecting anesthetics. The studies should concentrate on perinatal exposure, rather than exposure during organogenesis. Until more information on the effects of perinatal anesthetic exposure is available, we should at least revise our thinking to match the present data, which indicate that anesthetic effectiveness and teratogenic potential are not necessarily correlated.

Fetal Anticonvulsant Exposure: Effects on Behavioral and Physical Development[a]

CHARLES V. VORHEES

Institute for Developmental Research
Children's Hospital Research Foundation
and
Department of Pediatrics
University of Cincinnati College of Medicine
Cincinnati, Ohio 45229

It is estimated that from 0.33 to 0.52% of the American population has some form of epilepsy.[1,2] There are about 3.5 million births in the United States per year, and if 0.52% of the mothers have epilepsy, then up to 18,200 of these babies are born to epileptic women. Although estimates of the birth defect rate among all newborns varies, most place it at around 3%.[1] Women with epilepsy deliver children with an incidence of birth defects two to three times the population average,[1,2] thus, each year, approximately 1,100–2,200 infants with birth defects are born to women with epilepsy. A significant number of these cases might be prevented if it were shown that the increased birth defect rate among epileptics was due to the anticonvulsant medications prescribed to them or an interaction of these drugs with epileptic disease (TABLE 1).

As striking as these numbers are for birth defects, they may represent only the tip of the iceberg. Ten years ago Hanson and Smith,[3] Hanson et al.,[4] and Zackai et al.[5] described the fetal trimethadione and fetal hydantoin syndromes. Recognition of these syndromes suggested that there could be many more affected infants than previously estimated by studies that looked only for serious malformations, since subtler effects are thought to occur more frequently than severe ones. If correct, this understanding could substantially raise the number of affected infants above 2,200 per annum.

Hanson et al.[4] suggest that there may be a spectrum of fetal anticonvulsant effects ranging from severe to mild. Severe cases are those with the full syndrome and serious sequelae, such as mental retardation. Some of these may also have major malformations. Moderate cases would be those having the full syndrome, but without serious sequelae. Hanson et al. estimate that 11% of infants born to women taking phenytoin have the full syndrome. Mild cases, those with partial or incomplete expressions of the syndrome, may be the most prevalent subgroup. Hanson et al. estimated that 31% of women taking phenytoin during pregnancy have children in the incompletely affected category. An emerging appreciation of these incomplete forms has served to alter perceptions of prenatal anticonvulsants, because they represent a previously undetected and perhaps misdiagnosed group of developmentally disabled children.

[a] This work was supported by NSF Grant BNS79-24710 and NIH Grants AA06032 and HD19090.

49

TABLE 1. Prevalence of Epilepsy and Epilepsy-Related Birth Defects

- 3,500,000 births occur in the United States annually.
- Up to 0.52% of the population has epilepsy.
- Therefore, 18,200 births annually are to women with epilepsy.
- Children born to women with epilepsy have 2–3 times the birth defect rate of the rest of the population.
- Therefore, there are 1100–2200 children who have birth defects born annually to women with epilepsy.
- This represents 550–1,650 excess birth defects annually in this group.
- If these excess cases are due to anticonvulsant drugs, then a search for safer drugs could significantly alter this incidence.

Like the fetal alcohol syndrome (FAS), the fetal anticonvulsant syndromes consist of at least three symptom clusters. These are facial dysmorphogenesis, growth retardation, and mental subnormality.[3,4,5] In addition, distal phalangeal defects have been described for some of these syndromes and represent a fourth symptom cluster. Since the original fetal trimethadione and hydantoin syndromes were described, three additional fetal anticonvulsant syndromes have been suggested. These involve the drugs phenobarbital,[6] primidone,[7,8] and valproic acid.[9] A summary of the symptoms of all five of the fetal anticonvulsant syndromes is provided in TABLE 2.

An indication of the prevalence of use of different anticonvulsants among epileptic women of childbearing age has recently been provided by Kelly et al.[9] Selected aspects of their findings are shown in TABLE 3.

The pivotal issue in this area is whether the defects described above are caused by epilepsy per se, anticonvulsant drugs, or a combination of both. De-

TABLE 2. Fetal Anticonvulsant Syndromes: Tabulation of Clinical Symptoms[a]

Symptom	Hydantoin	Trimethadione	Barbital	Primidone	Valproate
Mental retardation	+	+	+	+	±
Growth retardation	+	+	+	+	±
Short nose	+		+		+
Low nasal bridge	+		+		+
Hypertelorism	+		+		
Ptosis	+		+		
Epicanthus	+	+		+	+
Cleft palate	+	+		+	
Cleft lip	+	+		+	
Abnormal ears		+	+	+	±
V-shaped eyebrows		+		+	
Hypoplastic nails	+			+	
phalanges	+		+		
Long upper lip					+
Shallow philtrum					+
Thin upper vermilion					+
Downturned corners of mouth					+
Other defects	±	±	±	±	±

[a] + indicates the symptom is usually present, ± indicates disagreement among authors as to whether the symptom is present, and a blank indicates the symptom is not or is rarely reported to be present.

TABLE 3. Anticonvulsant Use among Epileptic Women

1. Prospective study of 468 epileptic women by Kelly *et al.*, Am. J. Med. Genet. 1984. **19:** 435–443.
2. 95.7% of these women were on anticonvulsant medications.
3. Of those on anticonvulsants, 42.6% were on monodrug therapy and 57.4% on polydrug therapy.
4. Of those on anticonvulsants, 19.9% were on phenytoin alone and 42.0% on phenytoin and another anticonvulsant(s).
5. 9.4% were on phenobarbital and 28.3% were on phenobarbital and another anticonvulsant (usually phenytoin).
6. 7.4% were on carbamazepine, 4.0% on valproate, 1.1% on ethosuximide, and <1% on other drugs.

spite much research, investigations trying to resolve this point in humans beings have been largely unsuccessful.[9,11,12] This is an area where animal research may be particularly valuable in sorting out some of the confounding factors.

The research of Finnell[13] is particularly noteworthy in this regard. Finnell demonstrated that phenytoin, at therapeutically relevant blood levels, was able to produce a wide spectrum of malformations in inbred mice. He demonstrated that the malformation frequency was proportional to oral dose and plasma drug concentrations. Finnell also showed that in quaking mice with a genetic seizure disorder, malformation rate was a simple function of plasma phenytoin levels and did not interact with the genetic background of the seizure-prone mice. This research speaks to the issue of the contribution of phenytoin to the production of major malformations, but does not address the issue of the CNS features that are part of the fetal anticonvulsant syndromes. Our interest has been directed toward developing a model of the nonmalformation aspects of these syndromes.

Of the three symptom clusters most frequently identified as part of these syndromes, the two we have been most interested in have been CNS dysfunction and growth retardation. From the outset our goal has been to attempt to model the human syndromes, and to design the experiments to address points that have not been resolved in the human studies. Our working hypothesis has been that the fetal anticonvulsant syndromes are drug-induced disorders. Given this hypothesis, we have elected to develop our model in normal animals. Our hypothesis is falsifiable because if it is incorrect and the anticonvulsant syndromes require the existence of epilepsy in the organism as a precondition for occurrence, then we would be unable to develop a reasonable animal model of these disorders. As shown below, we have produced models of the fetal hydantoin and trimethadione syndromes, as have others recently for the fetal hydantoin syndrome, thereby supporting the hypothesis that these are drug-induced disorders.

METHODS

Details of the methods and results for most of the experiments described herein have been provided elsewhere.[14,15] Briefly, the data that follow have been drawn from three experiments (TABLE 4). All were done on the offspring of pregnant Sprague-Dawley CD rats administered phenytoin or trimethadione by gavage once daily throughout all or part of days 7–18 of gestation. Pregnant dams were allowed to deliver and raise their offspring normally. Shortly after birth,

TABLE 4. Experiments in Rats to Model the Fetal Hydantoin and Fetal Trimethadione Syndromes

Experiment	Group	Dose (mg/kg)	Exposure Period (days of gestation)	No. of Litters Tested
	Control	0	7–18	23
Exp. 1	Phenytoin	200	7–18	15
	Trimethadione	250	7–18	13
	Control	0	7–10	12
		0	11–14	11
		0	15–18	11
Exp. 2	Phenytoin	200	7–10	13
		200	11–14	12
		200	15–18	13
	Trimethadione	250	7–10	12
		250	11–14	11
		250	15–18	12
Exp. 3	Control	0	7–18	13
	Phenytoin	200	7–18	17
		150	7–18	12
		100	7–18	15

newborns were examined and randomly selected for subsequent assessment. We monitored growth and mortality as well as physical, neurological, and behavioral development from day three to adulthood. A summary of the assessments made on the animals is provided in TABLE 5.

In the first experiment the dose of phenytoin was 200 mg/kg given on days 7–18 of gestation. In the second experiment the dose was the same but the

TABLE 5. Cincinnati Behavioral and Developmental Test Battery for Rats

Test[a]	Postnatal Day of Testing
Surface righting	3–12
Negative geotaxis	6–12 even-numbered days
Swimming ontogeny	6–24 even-numbered days
Pivoting locomotion	7–11 odd-numbered days
Early locomotion	7–11 odd-numbered days
Auditory startle development	10–16
Incisor eruption	8 until incisors emerge
Eye opening	12 until eye are open
Figure-8 activity	15–17[b] and 40–42 (A)[c]
Open-field activity	15–17[b] and 40–42 (A)
Neonatal T-maze	15–24[b]
Vaginal patency	30 until complete patency
Biel or Cincinnati water maze	50–55 (A)
Spontaneous alternation	40 (B)[d]
M-Maze	43–44 (B)
Passive avoidance	50 until response remembered for 3 minutes

[a] All offspring in each litter tested unless otherwise specified. Not all tests are conducted in each experiment.
[b] One-half of each litter tested.
[c] One-half of each litter tested and designated subgroup A.
[d] One-half of each litter tested and designated subgroup B.

exposure period was subdivided into one of three shorter periods, that is, days 7–10, 11–14, and 15–18. In these first two experiments, groups were also prepared that were given trimethadione in the same regimen, and the dose of this drug was 250 mg/kg. In the third experiment, the exposure period was the same as in Experiment 1, but the dose was varied, and only phenytoin was administered. The doses were 100, 150, and 200 mg/kg. These are higher than human doses, in part because of the poor gastrointestinal absorption of phenytoin in rats. The doses were chosen, however, to produce plasma drug levels in the human therapeutic range for phenytoin, that is, 10–20 μg/ml. After the last dose was given on day 18 of gestation, plasma phenytoin levels were measured on a separate subgroup of dams for each dose level.

RESULTS

The results of the maternal drug plasma determinations may be seen in FIGURE 1. Note that all three doses of phenytoin produced clinically relevant maternal plasma concentrations.

For purposes of this presentation, emphasis has been placed on the results of tests of behavioral performance. Rats tested early in life for the maturation of locomotor behavior were found to be significantly more active when exposed prenatally to phenytoin compared to controls (TABLE 6). This effect was consis-

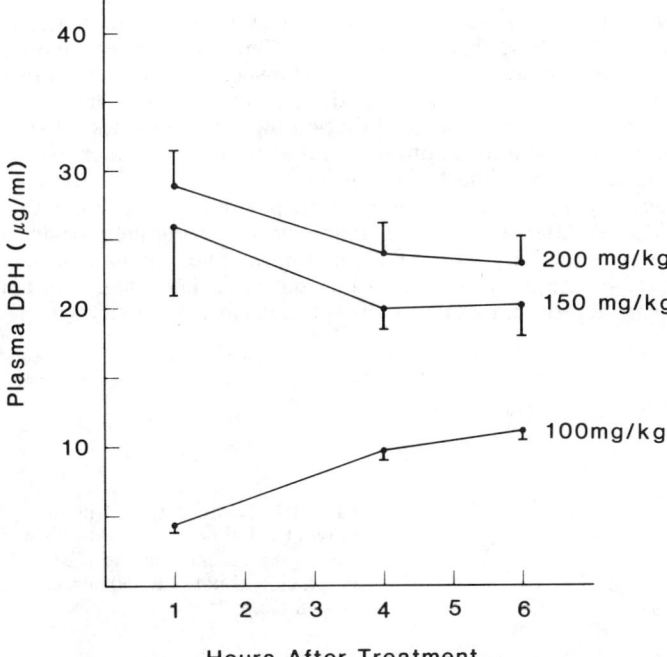

FIGURE 1. Mean (±SE) plasma phenytoin (DPH) levels in rats at various time intervals after their last dose of phenytoin on day 18 of gestation (treatment was on days G7–18).

TABLE 6. Early Locomotion in Rats Exposed Prenatally to Phenytoin or
Trimethadione: Locomotion Was Measured on Days 7, 9, and 11 as Time
Spent Pivoting or Photocell Interruptions

	Exp. 1	Exp. 2			Exp. 3		
Group	7–18	7–10	11–14	15–18	Dose	Pivoting	Locomotion
Control	2.0 (0.6)[a]	1.2 (0.3)	1.3 (0.5)	1.3 (0.5)		2.0 (0.4)[b]	79.4 (5.0)
Phenytoin	5.0 (1.2)[c]	0.9 (0.2)	2.1 (0.6)[c]	2.4 (1.0)[c]	200	4.9 (0.6)[c]	104.5 (5.7)[c]
					150	4.0 (0.8)[c]	97.2 (5.2)[c]
					100	3.2 (0.5)	88.0 (6.7)
Trimethadione	2.0 (0.5)	1.0 (0.3)	1.1 (0.3)	1.7 (0.5)			

[a] Values are the group mean (est. SEM) by litter across all 3 days of testing and both
sexes.
[b] Exposure period is the same as in Exp. 1, that is, days 7–18 of gestation.
[c] $p < 0.05$ by *a posteriori* comparison to controls following a significant treatment main-
effect or treatment × days interaction or both using a split-plot analysis of variance.

tent across all three experiments. In Experiment 2, it was found to occur only
during the middle- and late-exposure periods. In Experiment 3, the effect was
found to be dose-dependent. The behavior measured was termed pivoting. Pivoting occurs in rats as a result of the earlier development of forelimb compared to
hindlimb coordination. Pivoting is a transient behavior that usually peaks between
days 9 and 11 after birth. Thereafter, locomotion is characterized by quadrupedal
locomotion.

Early quadrupedal locomotion shows a characteristic peak in activity shortly
before weaning at around three weeks of age. The same phenytoin offspring that
showed increased pivoting on days 7–11, showed no difference in quadrupedal
locomotion during the period of peak activity on days 15–17. This may not be too
surprising, however, if it is assumed that during this brief burst of activity most
animals exhibit maximal locomotion. If this is true, further increases in activity
may be difficult or impossible during this interval. Such an effect could even mask
any hyperactivity that might be present in the phenytoin offspring. If this assumption is valid, then after weaning, when activity levels normally begin to decline,
pathological hyperactivity should reemerge in the phenytoin animals.

FIGURES 2–4 show that this is in fact the case, and phenytoin animals are
hyperactive in all three experiments at 40–45 days in a figure-eight-shaped activity

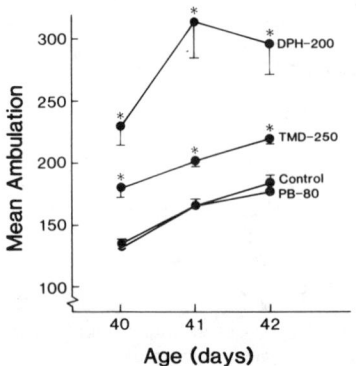

FIGURE 2. Mean (±SE) figure-eight activity
in rats treated prenatally with phenytoin (DPH,
200 mg/kg), trimethadione (TMD, 250 mg/kg),
or phenobarbital (PB, 80 mg/kg), all on days
G7–18 (Exp. 1).

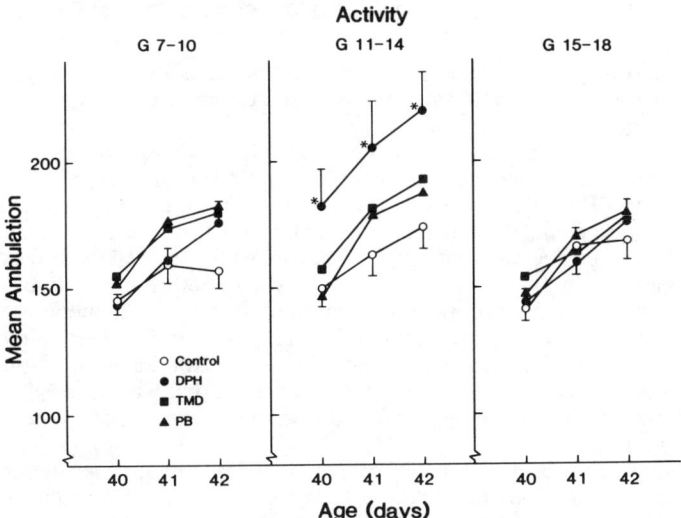

FIGURE 3. Mean (±SE) figure-eight activity in rats treated prenatally with phenytoin (DPH, 200 mg/kg), trimethadione (TMD, 250 mg/kg), or phenobarbital (PB, 80 mg/kg), on the gestation (G) days shown (Exp. 2).

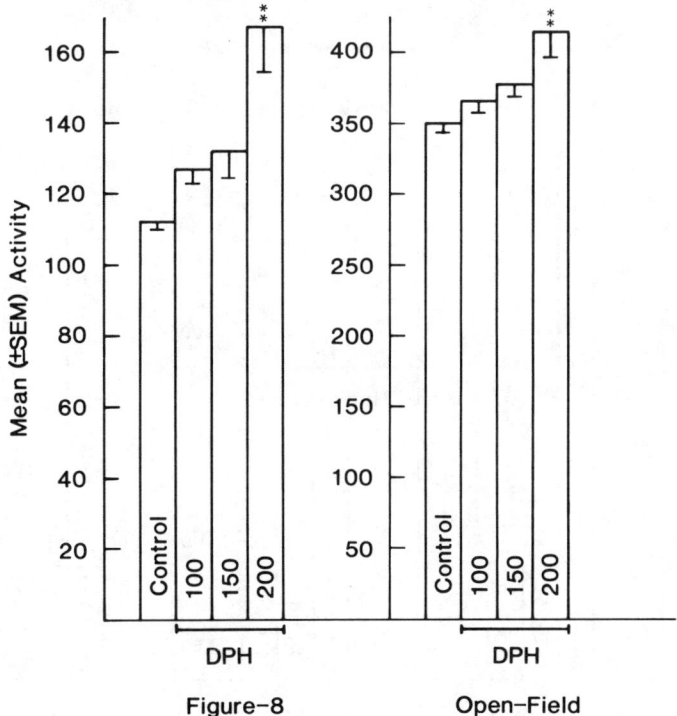

FIGURE 4. Mean (±SE) figure-eight and open-field activity in rats treated prenatally with 100, 150, or 200 mg/kg of phenytoin (DPH) on days G7–18 (Exp. 3).

monitor. Offspring prenatally exposed to trimethadione were also hyperactive in Experiment 1 at this age, but not when exposed to the drug for shorter intervals, as in Experiment 2.

Animals that are made hyperactive pharmacologically by treatment with stimulants are known to exhibit enhanced rates of learning in certain types of tests. One such test is the Biel water maze.[16] Since the phenytoin and trimethadione offspring are hyperactive, they might be expected to exhibit facilitated learning on this test. As can be seen in FIGURES 5–7, however, the opposite was found. In all cases the phenytoin and trimethadione groups made more errors than controls, and of the two treatment groups, the phenytoin offspring were the most severely affected. In Experiment 2, the impaired learning induced by phenytoin occurred only in those exposed during the middle period. In Experiment 3, a dose-dependent increase in maze errors among phenytoin offspring was seen. No effect of trimethadione on maze learning was found with the shorter exposure periods used in Experiment 2.

One important aspect of mastery of any maze-learning task is adequate memory of events on previous maze trials. Animals that cannot remember their previous successes as well as normals cannot repeat appropriate responses, and animals that cannot remember their previous errors cannot eliminate them. In order to determine if phenytoin and trimethadione offspring showed evidence of a memory deficit, we tested them in a passive avoidance test. In this test the animals were taught to refrain from entering a preferred compartment by adminis-

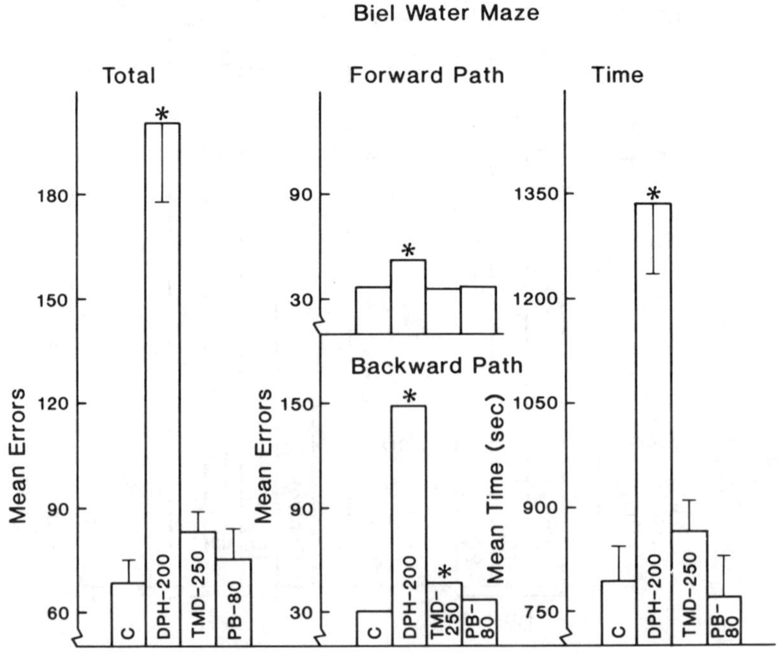

FIGURE 5. Mean (±SE) Biel maze errors and times in rats treated prenatally with phenytoin (DPH, 200 mg/kg), trimethadione (TMD, 250 mg/kg), or phenobarbital (PB, 80 mg/kg) on days G7–18 (Exp. 1).

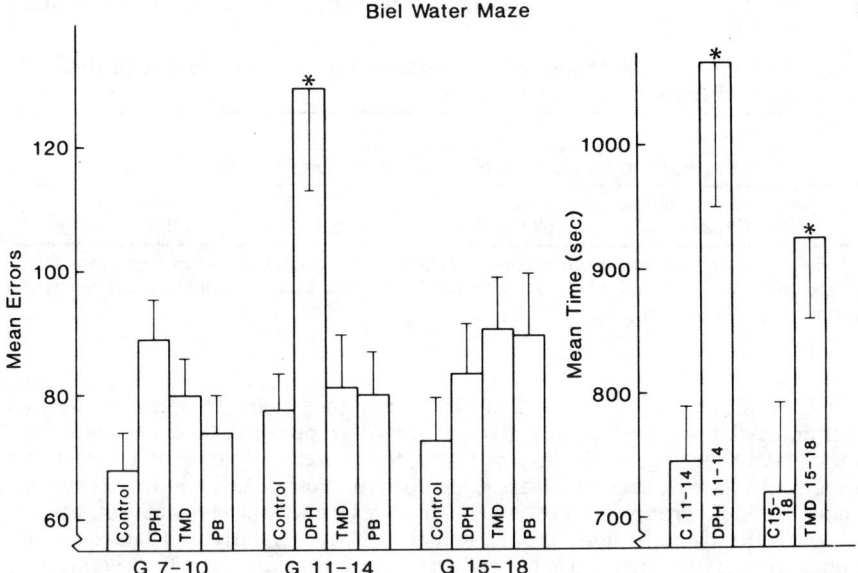

FIGURE 6. Mean (±SE) Biel maze errors and times in rats treated prenatally with phenytoin (DPH, 200 mg/kg), trimethadione (TMD, 250 mg/kg), or phenobarbital (PB, 80 mg/kg) on the gestation (G) days shown (Exp. 2).

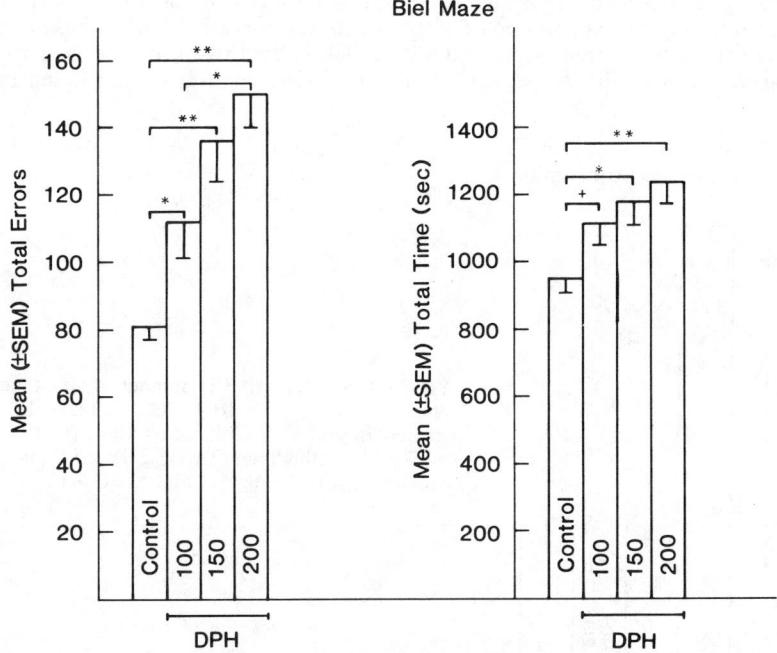

FIGURE 7. Mean (±SE) Biel maze errors and times in rats treated prenatally with 100, 150, or 200 mg/kg of phenytoin (DPH) on days G7–18 (Exp. 3).

TABLE 7. Two-Week Retention of the Passive-Avoidance Response in Rats Exposed to Phenytoin during Gestation

| | Mean ± SE (sec.) | | |
Experiment	Control	Phenytoin	Significance
Experiment 1	165.3 ± 4.4	145.0 ± 10.6	$p < 0.05$
Experiment 2[a]	169.6 ± 6.9	136.3 ± 14.2	$p < 0.05$

[a] Phenytoin and control groups from Exp. 3 treated on days 11–14 of gestation. Those treated on days 7–10 and 15–18 showed no differences between controls and phenytoin groups.

tering a brief electric shock to them each time they failed to show the desired restraint. All groups of rats learned this simple response at the same rate. Two weeks after learning, each rat was tested for its remembrance of the restraint response. As can be seen in TABLE 7, the phenytoin offspring in both Experiments 1 and 2 (middle group) did not remember to restrain themselves as well as controls. All the trimethadione groups and the phenytoin early and late exposure groups from Experiment 2 are not shown because they did not differ significantly from controls in their performance.

Another reason some animals might not do as well in the maze test as others is if they are less flexible and tend to perseverate on previously established response patterns, even when such responses are unproductive. A test that partially addresses this is called spontaneous alternation. In this test the left-right choice behavior of rats is tested in a simple T-maze on a succession of trials. There is no right or wrong response, and no reinforcements are provided. Under these conditions normal rats alternate right and left turns on about two-thirds of their trials. As can be seen in FIGURE 8, our control rats alternated at the expected rate.

Spontaneous Alternation

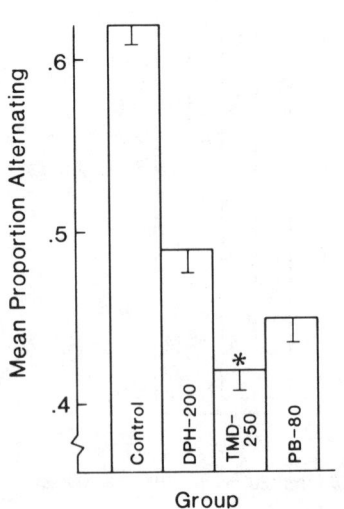

FIGURE 8. Mean (±SE) number of spontaneous alternations on four trials in a T-maze in rats treated prenatally with phenytoin (DPH, 200 mg/kg), trimethadione (TMD, 250 mg/kg), or phenobarbital (PB, 80 mg/kg) on days G7–18.

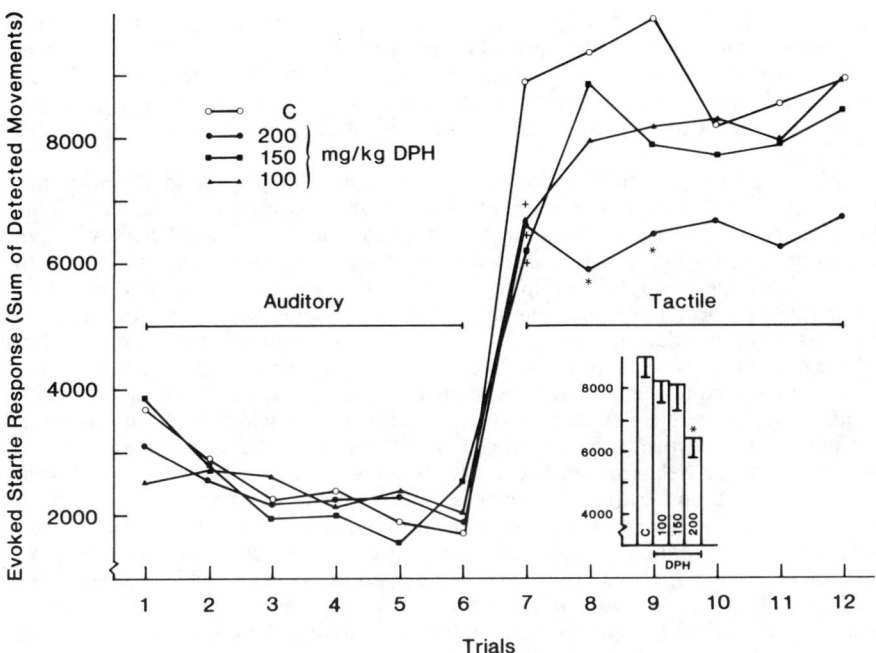

FIGURE 9. Mean of summed evoked startle responses in rats treated prenatally with 100, 150, or 200 mg/kg of phenytoin (DPH) on days G7–18 (Exp. 3).

Offspring exposed prenatally to trimethadione, however, alternated only about 40% of the time, significantly less often than controls. The phenytoin animals showed a similar, but less extreme reduction in alternation frequency.

Animals that are hyperactive because of the influence of stimulants are also usually hyperreactive. Does this apply to animals that are hyperactive by virtue of prenatal exposure to anticonvulsants? We sought to test this in a task that measures startle responsiveness. In Experiment 3, rats were given a series of auditory and tactile startle stimuli and their responses were measured in terms of response time (latency), peak response, and total response. As can be seen in FIGURE 9, auditory and tactile responses differ in strength even among controls at the stimulus settings we used. Prenatal phenytoin exposure had no effect on the low-baseline pattern seen with evoked auditory responses, but significantly reduced total response strength to the high-baseline pattern evoked by the tactile stimulus. Peak response showed the same pattern, but no difference between groups was found for response latency. Thus, the phenytoin offspring were not hyperreactive, as might be expected, but were in fact hyporeactive.

COMMENTS

The results of the three experiments described support the idea that prenatal exposure to phenytoin and trimethadione produces long-term CNS dysfunction in rat offspring at therapeutically relevant plasma levels in the absence of an existing seizure disorder. The data support the view that the developing CNS is particularly vulnerable to these two anticonvulsants and are consistent with clinical evidence for fetal hydantoin and trimethadione syndromes. The data suggest that these syndromes include long-term psychological deficits as part of their pattern of effects.

More specifically, the experimental data show that phenytoin offspring are hyperactive and impaired in their ability to learn a complex spatial task. This learning deficit does not appear to be primarily due to increased perseverative behavior. The cognitive defect cannot be attributed to a motor performance deficit, because the phenytoin offspring are hyperactive, and because in a test of swimming performance measured in a straight channel in which little or no learning was involved (data not shown), they performed no differently than controls. There did appear to be a memory defect in the phenytoin offspring, and this may have contributed to their reduced learning in the maze. By contrast, the trimethadione offspring, while also more active, were not as active as phenytoin offspring. Trimethadione offspring did not show a memory deficit but did exhibit an increase in perseverative behavior on a test of spontaneous alternation; thus, although both phenytoin and trimethadione offspring were more active and made more errors learning a complex maze than controls, the specific features of their cognitive impairments appeared quite different. Further research will undoubtedly reveal other effects of these drugs. The few behavioral studies that have been done with prenatal phenytoin in animals are in general agreement with our findings.[17,18] No behavioral data in animals exist on trimethadione, save ours.[14] Nevertheless, the current findings on trimethadione in two separate experiments were consistent and suggest that this drug is behaviorally teratogenic. More such studies are needed.

In conclusion, we suggest that CNS effects may be one of or perhaps the most important long-term adverse effect resulting from prenatal exposure to phenytoin and trimethadione. We hope that future research will give greater attention to CNS effects, and that the relative CNS toxic potential of phenytoin, trimethadione and the many other anticonvulsants currently in use will be systematically compared in future research. It appears that the neuroembryopathic potential of these agents has been heretofore underappreciated.

REFERENCES

1. KALTER, H. & J. WARKANY. 1983. Congenital malformations: Etiologic factors and their role in prevention. N. Engl. J. Med. 308: 424–431 and 491–497.
2. KELLY, T. E. 1984. Teratogenicity of anticonvulsant drugs. I: Review of the literature. Am. J. Med. Genet. 19: 413–434.
3. HANSON, J. W. & D. W. SMITH. 1975. The fetal hydantoin syndrome. J. Pediatr. 87: 285–290.
4. HANSON, J. W., N. C. MYRIANTHOPOULOS, M. A. SEDGWICK-HARVEY & D. W. SMITH. 1975. Risks to the offspring of women treated with hydantoin anticonvulsants, with emphasis on the fetal hydantoin syndrome. J. Pediatr. 89: 662–668.
5. ZACKAI, E. H., W. J. MELLMAN, B. NEIDERER & J. W. HANSON. 1975. The fetal trimethadione syndrome. J. Pediatr. 87: 280–284.
6. SEIP, M. 1976. Growth retardation, dysmorphic facies, and minor malformations following massive exposure to phenobarbitone in utero. Acta Paediatr. Scand. 65: 617–621.
7. RUDD, N. L. & R. M. FREEDOM. 1979. A possible primidone embryopathy. J. Pediatr. 94: 835–837.

8. SHIH, L. Y., N. DIAMOND & T. KUSHNICK. 1979. Primidone-induced teratology—clinical observations. Teratology 19: 47A.
9. KELLY, T. E., P. EDWARDS, M. REIN, J. Q. MILLER & F. E. DREIFUSS. 1984. Teratogenicity of anticonvulsant drugs. II. A prospective study. Am. J. Med. Genet. 19: 435–443.
10. DiLIBERTI, J. H., P. A. FARNDON, N. R. DENNIS & C. J. R. CURRY. 1984. The fetal valproate syndrome. Am. J. Med. Genet. 19: 473–481.
11. NAKANE, Y. 1979. Congenital malformation among infants of epileptic mothers treated during pregnancy—The report of a collaborative study group in Japan. Folia Psychiatr. Neurol. Jpn. 33: 363–369.
12. SHAPIRO, S., D. SLONE, S. C. HARTZ, L. ROSENBERG, V. SISKIND, R. MONSON, A. A. MITCHELL, O. P. HEINONEN, J. IDANPAAN-HEIKKILA, S. HARO & L. SAXEN. 1976. Anticonvulsants and parental epilepsy in the development of birth defects. Lancet 1: 272–275.
13. FINNELL, R. H. 1981. Phenytoin-induced teratogenesis: A mouse model. Science 211: 483–484.
14. VORHEES, C. V. 1983. Fetal anticonvulsant syndrome is rats: Dose- and period-response relationships of prenatal diphenylhydantoin, trimethadione, and phenobarbital exposure on the structural and functional development of the offspring. J. Pharmacol. Exp. Ther. 227: 274–287.
15. VORHEES, C. V. 1985. Fetal hydantoin syndrome in rats: Effects on postnatal behavior and brain amino-acid content. Neurobehav. Toxicol. Teratol. 7: 471–482.
16. KINNEY, L. & C. V. VORHEES. 1979. A comparison of methylphenidate-induced active avoidance and water maze performance facilitation. Pharmacol. Biochem. Behav. 10: 437–439.
17. ELMAZAR, M. M. A. & F. M. SULLIVAN. 1981. Effect of prenatal phenytoin administration on postnatal development of the rat: A behavioral teratology study. Teratology 24: 115–124.
18. MULLENIX, P., M. S. TASSINARI & D. A. KEITH. 1983. Behavioral outcome after prenatal exposure to phenytoin in rats. Teratology 27: 149–157.

DISCUSSION

QUESTION: Can you do this in a rat and do you pair-feed these animals with your controls?

CHARLES VORHEES: With the exception of the highest dose, there is no change in maternal weight when the animals are exposed to phenytoin or trimethadione, so that obviates the need to do pair-feeding. There doesn't seem to be any nutritional change.

GENE FISCH: You only showed the effect of the highest dosage of phenytoin on maze performance. Can you tell us the effects at a lower dose?

CHARLES VORHEES: All groups had increased maze errors.

WILLIAM BAILEY: You pointed out a similarity between the fetal alcohol syndrome and the teratological effects of anticonvulsants. Would you comment on the role of enzyme induction in the liver, in the effects produced by both compounds? I also was wondering whether exposure to anticonvulsants had any effect on the direction in which animals rotated during activity tests. It is well known that prenatal exposure to some anticonvulsants affects gonadal steroid metabolism; therefore, one might expect, in turn, an influence on cerebral lateralization of function.

CHARLES VORHEES: In fact we have looked at their turning preference; they show a turning preference that is different from that of controls. Each individual animal that had been exposed to phenytoin, not to trimethadione, showed a turning bias that is much stronger than in controls, but you cannot predict who will be predominantly a left turner or a right turner. It is as though each animal is affected so that its turning preference is markedly different from that of controls, but some of them turn to the right and some of them turn to the left.

As far as the question about enzyme induction, we have not specifically looked at that. Our current research is concerned with various neurotransmitters, and we're also using a synaptic membrane probe to look at membrane rigidity to see whether these kinds of parameters are changed in these animals, because gross histologic examination of the brains of these animals shows no reduction in brain weight. There is no reduction in brain protein content, and no obvious gross pathological changes in these animals' brains. The next step is to do more detailed histopathologic and neurochemical study, which is what we are pursuing now, but I cannot really address the possible role of enzyme induction. The question is a very interesting one, but not one we're pursuing at this moment.

QUESTION: On your second slide, it looked like there were differences in the symptom pattern between the primidone and the phenobarbital syndromes, even though primidone is metabolized in the body similarly to phenobarbital.

CHARLES VORHEES: I cannot answer that directly. It's a fascinating point that from a pharmacological point of view you would predict these two to look alike, but they don't look alike. Other than having been struck by that, as you are, I have no current explanation for it.

QUESTION: Was there any effect on the social behavior of the animals?

CHARLES VORHEES: I haven't looked at their social behavior, but I'd like to do that soon.

Studying Alcohol Teratogenesis from the Perspective of the Fetal Alcohol Syndrome: Methodological and Statistical Issues[a]

ANN PYTKOWICZ STREISSGUTH,[a] PAUL D. SAMPSON,[b]
HELEN M. BARR,[a] STERLING K. CLARREN,[c] AND
DONALD C. MARTIN[d]

[a] *Department of Psychiatry and Behavioral Sciences*
School of Medicine
[b] *Department of Statistics*
School of Arts and Sciences
[c] *Department of Pediatrics*
School of Medicine
[d] *Department of Biostatistics*
School of Public Health
University of Washington
Seattle, Washington 98195

INTRODUCTION

The study of alcohol teratogenesis is facilitated by the existence of one end-point of damage called fetal alcohol syndrome (FAS), which demarcates a specific, severe cluster of morphological and behavioral manifestations of prenatal alcohol exposure in humans. Awareness of the behavioral characteristics of FAS can help delineate the type of behavioral endpoints to use in evaluating the behavioral teratology of alcohol. The methodology developed for studying the behavioral teratology of alcohol should be useful in studies of the effects of other teratogens and toxic conditions in human beings.

FETAL ALCOHOL SYNDROME—AN OVERVIEW

FAS is a recognized pattern of major and minor malformation, growth deficiency, and developmental disability caused by heavy alcohol exposure *in utero*.[1,2] Since international awareness of this pattern of malformation in 1973,[3,4] over 2,000 papers have been published on alcohol and pregnancy, according to the National Clearing House on Alcohol Information of NIAAA.

Children diagnosed as having FAS have three types of manifestations as depicted in TABLE1: (1) growth deficiency (which is usually of prenatal onset and continues postnatally); (2) a particular pattern of malformations (including a char-

[a] This work was partially supported by Grant No. AA01455-11 from the National Institute of Alcohol Abuse and Alcoholism.

63

TABLE 1. Diagnostic Features of Fetal Alcohol Syndrome[a]

Growth Deficiency
  Prenatal and postnatal growth retardation for height and weight
Central Nervous System Dysfunction
  Neurologic abnormalities, developmental delay, structural anomalies
  (especially microcephalus)
Craniofacial Anomalies
  Short palpebral fissures, frontonasal alterations (epicanthal
  folds, flat nasal bridge, short and turned nasal tip, hypoplastic
  philtrum), thin upper vermilion, flat midface

[a] Data from Clarren and Smith.[2] Table from Clarren.[5] Reprinted with permission.

acteristic facies as pictured in FIG. 1–3 and diagrammed in FIG. 4; and (3) some evidence of central nervous system abnormality often manifested as microcephaly, mental retardation, motor problems, tremulousness, hyperactivity, and so forth. Skeletal anomalies, heart defects, and other abnormalities are also observed with increased frequency. FAS occurs in some offspring of chronically alcoholic mothers, drinking during pregnancy. Alcohol is the causative agent, but individual factors of susceptibility influence the manifestation of effects.

After a decade of research, it is clear that FAS and alcohol teratogenicity are reciprocal terms. A woman who drinks alcohol heavily in pregnancy carries some risk of producing a child with FAS. Conversely, identifying a child with all the features of FAS strongly suggests that the child was affected by alcohol *in utero*. It is also possible for alcohol teratogenicity to result in a partial or "milder" FAS[1] phenotype, an alternate unusual phenotype, or a normal appearance with isolated growth deficiency or brain dysfunction. These conditions may be referred to as "possible" fetal alcohol effects although they are *not* reciprocal with alcohol teratogenicity. While *in utero* alcohol exposure may produce these effects, other environmental agents or genetic problems could produce similar manifestations (see FIG. 5). When examining the individual patient, the examiner cannot be sure that alcohol produced a "possible" fetal alcohol effect, even when a maternal history is positive for alcohol exposure. "Possible" fetal alcohol effects can only be causally related to alcohol teratogenesis in controlled studies in which alcohol exposure and adverse pregnancy outcomes can be statistically linked.

Severity of physical manifestations (growth deficiency and malformations) is associated with severity of neurobehavioral deficits. TABLE 2 demonstrates the decreasing mean IQ scores associated with increasing severity of symptoms.[6] A recent 10-year follow-up of the first children given this diagnosis revealed that those with the most severe early manifestations were all severely handicapped on

TABLE 2. IQ in 20 Children with Fetal Alcohol Syndrome: A Clinical Sample[a]

Severity of Diagnosis	(No.)	Mean IQ	Range of IQ
Mild and very mild	(4)	82	60–105
Moderate	(6)	68	59–81
Moderately severe	(5)	58	15–89
Severe	(5)	55	41–69
Summary	(20)	65	15–105

[a] Data from Streissguth, Herman, and Smith.[6]

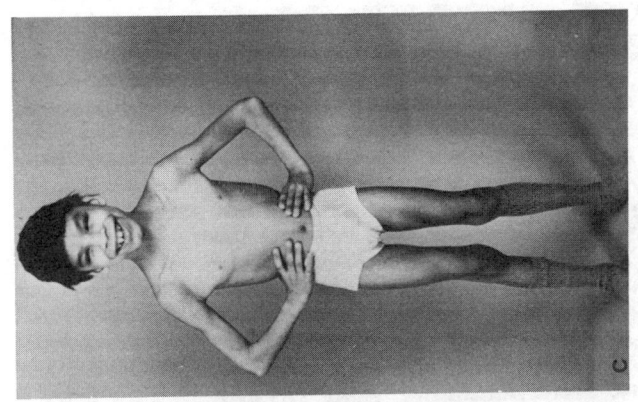

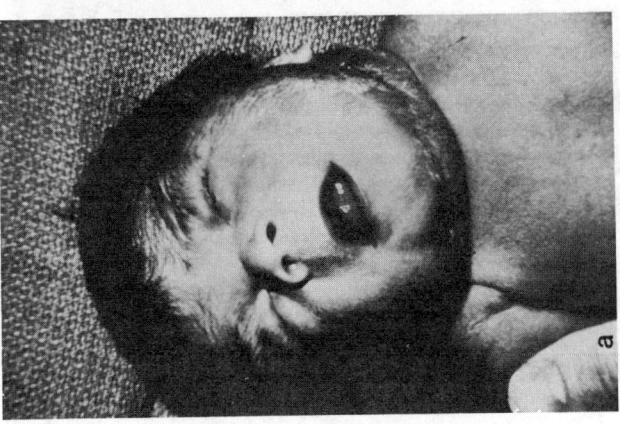

FIGURE 1. Patient with fetal alcohol syndrome photographed at (a) birth, (b) five years, and (c) eight years. Note the short palpebral fissures, epicanthal folds, short upturned nose, long and hypoplastic philtrum, thin upper lip vermilion, flat midface, hirsutism, and characteristic emaciated appearance of the prepubescent FAS child. This child has had a testable IQ in the 40 to 45 range from 2 years through 10 years of age, despite having been raised throughout life in one excellent foster home with appropriate schooling and intervention. (Photograph a from Jones and Smith;[4] photograph b from Streissguth *et al.*;[7] photograph c from Streissguth *et al.*[17] Reproduced by permission.)

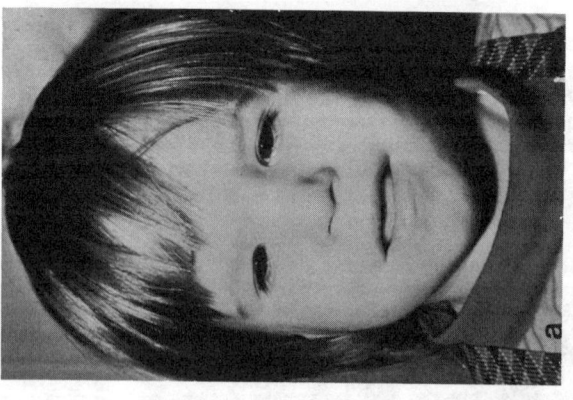

FIGURE 2. Boy with FAS photographed at ages (a) 2 years, 6 months, and (b and c) at age 12 years, 2 months. Note the short palpebral fissures, epicanthal folds, flat midface, hypoplastic philtrum and thin upper vermilion border. Note also the short, lean prepubertal stature characteristic of young adolescent boys with FAS. This child showed a steady increase in IQ scores from less than 50 when first tested to 80 when tested ten years later. The biggest improvement in test scores coincided with a decrease in hyperactivity at the time the parents were divorced, the mother left the home, and the boy began attending school. Home life from then on was stable but not excessively stimulating. (Photograph a from Jones et al.;[3] photographs b and c from Streissguth et al.[7] Reproduced by permission.)

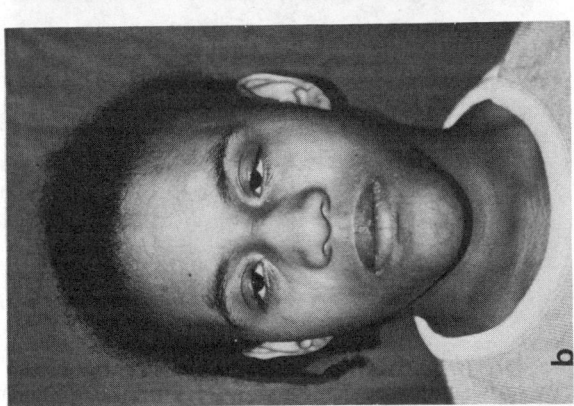

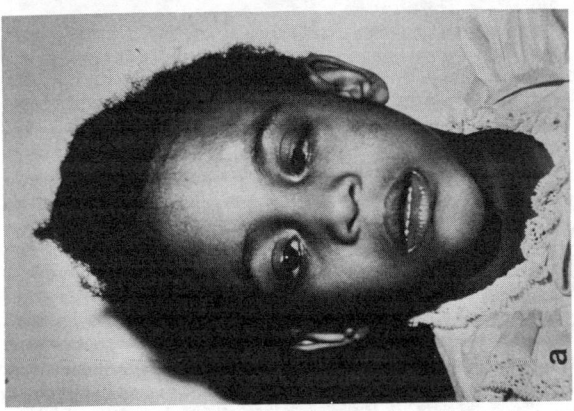

FIGURE 3. Girl with FAS photographed at ages (a) 3 years, 9 months and (b and c) at 14 years, 2 months. Note the persistence across ages of the short palpebral fissures, hypoplastic philtrum, strabismus, and ptosis. Note also the increased growth of the nose and mandible, and the short, stocky stature often associated with puberty in girls with FAS. This patient has had consistent IQ scores in the range of 40–57 over the past ten years, despite marked improvement in living conditions when her mother died when she was six. (Photograph a from Jones *et al.*;[3] photographs b and c from Streissguth *et al.*[7] Reproduced by permission.)

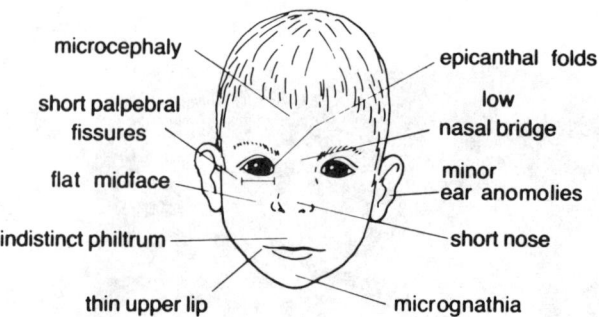

FIGURE 4. This figure describes the typical facial features in fetal alcohol syndrome. The features on the left are those that are the most characteristic and delineating; those on the right are less frequently observed and are not as differentiating. (Figure from Little and Streissguth. 1982. Alcohol. Pregnancy and the Fetal Alcohol Syndrome. A slide/teaching library unit available from Milner-Fenwick, Inc. Timonium, MD 21093.)

follow-up while those with milder manifestations had IQ scores in the borderline range.[7] A birth prevalence of 1/750 to 1/1,000 live births has been estimated.[8] FAS is thought to be the leading teratogenic cause of mental retardation and the third among all known causes of retardation.[2] We view FAS as the teratogenic endpoint against which to assess the more subtle behavioral effects associated with lower levels of exposure.

TERATOLOGY AND BEHAVIORAL TERATOLOGY OF ALCOHOL

Direct evidence for the teratogenicity of alcohol comes from laboratory studies of many species of animals, where alcohol, in the absence of other drugs and in the presence of adequate nutrition, has been found to have many types of deleterious effects on offspring.[8,9]

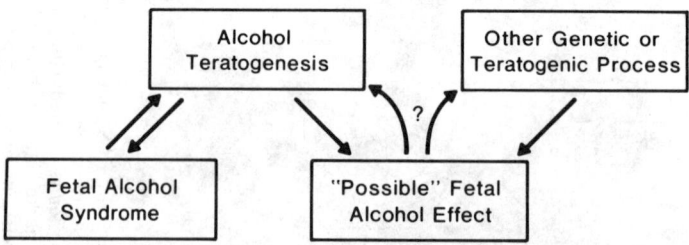

FIGURE 5. Alcohol teratogenesis produces a spectrum of adverse outcomes. Among these outcomes is FAS, which is a distinct pattern of malformation unique to alcohol teratogenesis. Other adverse outcomes of alcohol teratogenesis may be called "possible" FAE, but these outcomes do not necessarily imply gestational alcohol exposure. (Figure from S. Clarren, unpublished.)

Teratogenic agents are those causing adverse offspring effects as a result of gestational exposure. Originally teratology was the study of malformations, but it is now recognized that there are other types of teratogenic endpoints. In the classic *Handbook of Teratology*,[10] Wilson defines four principle manifestations of teratogenic agents: death, malformation, growth deficiency, and functional deviations. *In utero* alcohol exposure can cause all four of these manifestations, depending on the dose, timing, and circumstances of exposure. In general, the heavier the exposure, the more severe the effects on the offspring. On the other hand, careful timing of even one or two heavy binge doses during gastrulation, can produce a pattern of frontonasal dysmorphogenesis in mice that is similar to that found in the fetal alcohol syndrome.[11] As not all exposed offspring are affected, individual genetic differences in maternal alcohol metabolism are reflected in differential outcomes.[12] Factors associated with susceptibility have been less systematically investigated in humans.

A growing field of research on behavioral teratology is emerging that is focusing on the functional or central nervous system (CNS) effects of prenatal exposures to teratogenic substances. Vorhees[13] has set forth principles for behavioral teratology and proposed a dose-response/dose-effects relationship between the teratogenic agent and the four main manifestations, as shown in FIGURE 6. Here we see not only that increased dose leads to increased probability of adverse effects within a type of outcome, but that the effects can be ordered in terms of the level of exposure necessary to produce them. The highest dose produces death of the offspring; functional effects are produced at lower doses.

Animal studies have revealed a wide variety of functional effects associated with prenatal alcohol exposure, including decreased learning ability, decreased response inhibition, increased activity, increased reactivity, increased seizure susceptibility, and suckling defects in the newborn.[14] Recent primate research confirms earlier rodent studies, showing alcohol-related behavioral effects on offspring in the absence of gross morphologic or growth effects.[15]

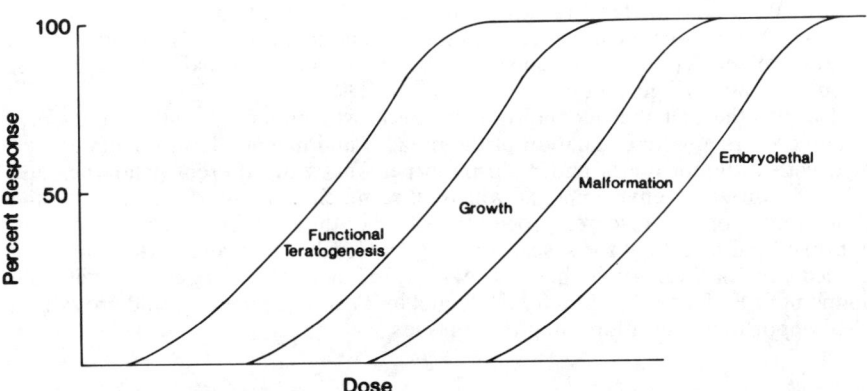

FIGURE 6. Idealized dose-response curves for the major manifestations of teratogenesis. The slope and spacing of the curves are dependent on the agent under investigation. The position of the curves is only valid in their ordinal relationship to one another and if and only if the agent in question is capable of producing all four types of embryotoxicity shown. (Figure from Vorhees.[13] Reproduced by permission.)

Animal studies of alcohol teratogenesis indicate: (1) Differences in dose and timing of exposure seem related to differences in outcome and severity; (2) a wide variety of behavioral effects are manifest that are similar in man and animals; (3) behavioral effects can appear in the absence of morphologic and growth effects and are produced at lower doses; and (4) teratogenic effects represent the interaction of *in utero* exposure and individual genetic factors.

SPECIAL PROBLEMS IN ASSESSING THE BEHAVIORAL TERATOLOGY OF ALCOHOL IN HUMANS

Compared to other teratogenic manifestations such as death and malformations, several aspects of behavioral manifestations make them particularly difficult to study. These include:

(1) Behaviors are usually measured on a continuum.

(2) There are no specific differentiating behavioral characteristics observable at birth when the relationship to teratogens is most easily established.

(3) The behavioral phenotype changes with increasing maturation of the individual.

(4) Behavioral deficits may not be readily observable in affected individuals, and often need specialized test conditions to elicit.

(5) Only the most severely affected individuals will be clinically observable as deviant. The detection of subtle behavioral effects, which characterize the majority of persons, will require groups of subjects tested under standardized conditions and analyzed with appropriate statistical procedures.

(6) Behavior is strongly affected by postnatal conditions and genetic determinants, and these must be carefully incorporated into behavioral teratology studies.

(7) As with other teratogenic manifestations, not all exposed individuals are affected. Until the protective and exacerbating factors are understood, we will be unable to establish a close relationship between exposure and effects.

(8) With alcohol teratogenesis studies in humans, there is the additional problem that exposure can never be precisely quantified, not only because of the vagaries of self-report, but because of actual variations in drinking patterns over the nine months of gestation.

Despite the fact that alcohol is now the most frequently studied behavioral teratogen, systematic evaluation of the breadth and magnitude of effects associated with different doses and patterns in persons with different genotypes and raised in different environments will require much more study as well as the development of new research procedures. Two of the most important of these are improved methodology for assessing the behavioral effects and better statistical procedures for evaluating them. The study of alcohol teratogenesis, with the endpoint of FAS and the ubiquity of alcohol in Western society, should provide an ideal opportunity for addressing these issues.

THE SEATTLE LONGITUDINAL STUDY ON ALCOHOL AND PREGNANCY: RATIONALE, DESIGN, AND FINDINGS

For the past 11 years we have been conducting a longitudinal prospective study on alcohol teratogenesis. We planned this study with assessment of exposures during pregnancy; with a large sample size to permit appropriate statistical adjustment for covariates and the interaction between covariates and exposure; with both outcomes and covariates assessed at several developmental ages; and with behavioral endpoints assessed on both precise laboratory tests as well as clinical scales.

Detailed descriptions of the study have appeared elsewhere;[16] a diagram of the longitudinal prospective design appears in FIGURE 7. In brief, a consecutive sample of 1,529 pregnant women were asked about their smoking, drinking, and drug use. Alcohol was quantified in several ways as shown in TABLE 3. We selected a follow-up cohort of 500 children born to most of the heaviest drinkers and smokers in the sample and a range of moderate drinkers, infrequent drinkers, and abstainers. Follow-up examinations of offspring were conducted at five developmental stages as shown in FIGURE 7. Sample maintainance has been excellent with 95% of the subjects seen at four years of age recently evaluated at seven years. TABLE 4 lists some of the key covariates obtained at different stages of the child's development. Summaries of major findings and papers deriving from the study have been published previously,[16,17] and a full list of the 65 papers is available from the senior author. A summary of some of the primary behavioral outcomes appears in TABLE 5.

Selection of appropriate outcome variables has been of particular interest to us. Our rationale has involved the derivation of hypotheses about the behavioral teratology of alcohol based on clinical observations of children with FAS. At the

TABLE 3. Alcohol Measures Obtained by Prenatal Self-Report of Mothers: Seattle Longitudinal Study on Alcohol and Pregnancy

Measures Obtained for Each of Two Prenatal Periods (Before Pregnancy Recognition and During the Fifth Month of Pregnancy)	
AA	Average ounces of absolute alcohol per day (Jessor *et al.*[28])
VV	Volume variability (Cahalan): 11-point, two-dimensional scale of average amount × maximum amount.
MOCC	Monthly occasions: Number of times per month that the fetus is exposed (ignores amount).
DOCC	Average drinks per occasion: (When the fetus is exposed, how much s/he is usually exposed.)
MAX	Maximum number of drinks the fetus was ever exposed to (0, 1.5, 3.5, 6, 9.5, 13).
QFV	Quantity-frequency-variability (Cahalan): Five-point scale resulting from categorizing cells of a three-dimensional table. (1 = heavy, 5 = abstainer).
AUP	Alcohol use pattern: Four-point scale of how frequently fetus is exposed to five or more drinks.
Other Alcohol Measures Obtained Prenatally	
ORDEXC	Ordered experimental code: Five-point scale (0–4) combining timing, pattern, and amount of exposure based on *a priori* estimate of risk to the fetus.
NINTOX	Number of intoxications.
NPE	Number of personal effects. 25 reasons for drinking were screened. Fourteen are considered to be indicators of a problem. NPE is a count from 0–14.
NPROB	Number of problems due to drinking. Count from 0–4 of indicators of one or more: Arrest, divorce, hospitalization or job loss due to drinking.

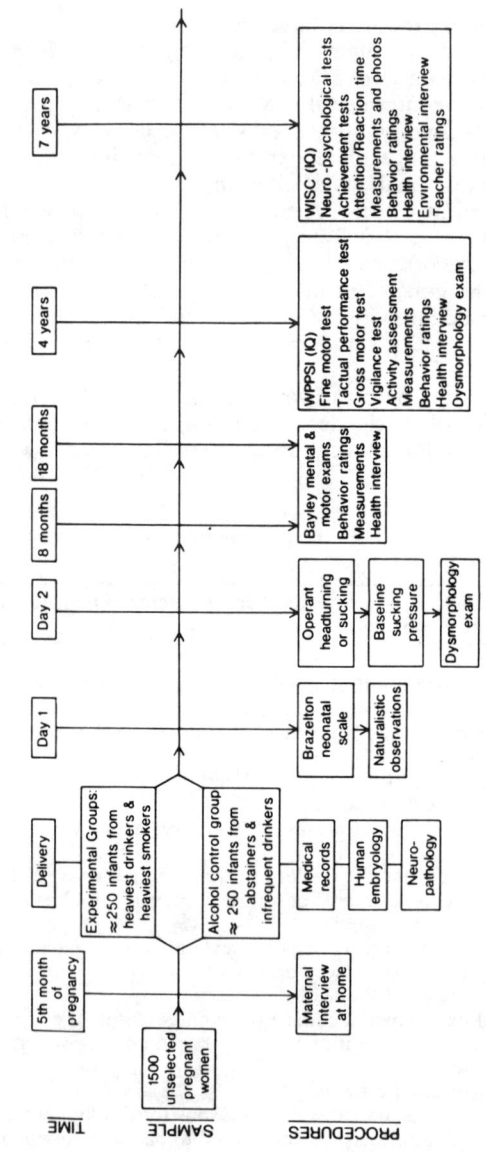

FIGURE 7. The experimental design for the Pregnancy and Health Study, a longitudinal prospective study on the effects on the offspring of maternal drinking in pregnancy. Sample sizes for substudies: Brazelton, 469; naturalistic observation, 124; operant head-turning, 225; operant sucking, 80; base-line sucking study, 151; neonatal dysmorphology, 163; eight-month follow-up, 468; 18-month follow-up, 496; four-year follow-up, 465; seven-year follow-up, 487.

extreme end of the alcohol continuum, we see a direct relationship between morphology, growth, function, and exposure. This observation serves as the core of our hypotheses regarding alcohol teratogenesis, particularly at levels of exposure below maternal alcoholism. By attempting to measure alcohol on a continuum, we can then assess the exposure conditions associated with morphologic effects, effects on growth, and effects on behavior. Furthermore, through the longitudinal format we can also assess the duration of different types of effects and their interacting and modifying conditions.

The specific behavioral deficits we observed in infants and children with FAS, and which we hypothesized would be observable in groups of infants and children experiencing lower levels of exposure, are presented in FIGURE 7. These behavioral endpoints are ordered according to the developmental stages of the child at

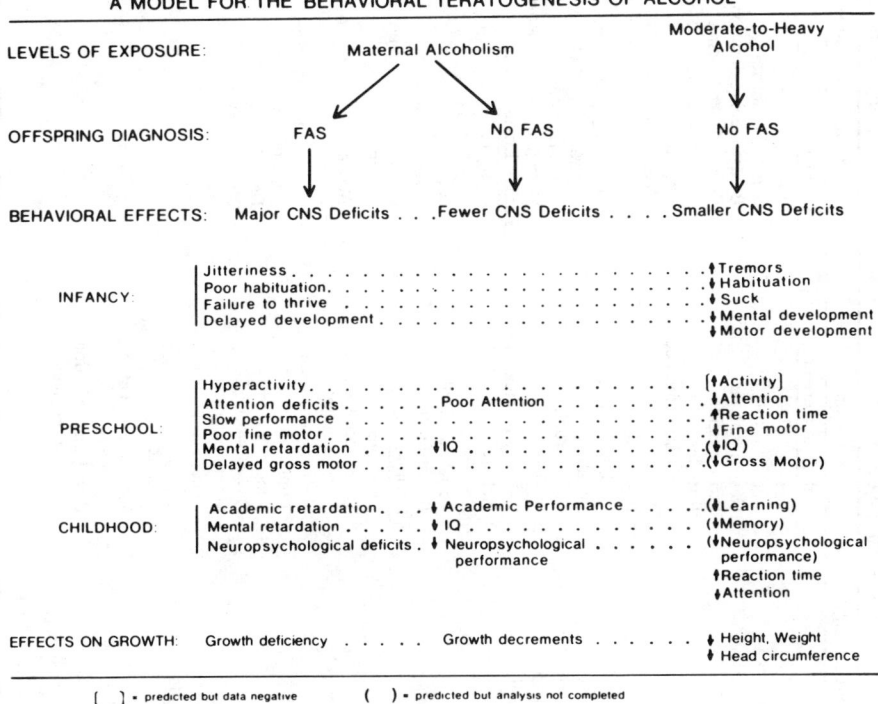

FIGURE 8. A model for the behavioral teratogenesis of alcohol. *Left,* some of the primary behaviors suggesting CNS problems in infants and children with FAS. *Center,* some of the problems in children without the full FAS, who are born to alcoholic mothers, compared to matched controls (for example, Jones *et al.*[29]). *Right,* the behavioral and performance decrements predicted to be associated with prenatal alcohol exposure across a moderate to heavy range of exposures and examined in the Pregnancy and Health Study. The findings for activity did not support the prediction (perhaps because of the procedures used.) For variables in parentheses, analyses of the results have not yet been completed. For the variables without brackets or parentheses, the analyses in the Pregnancy and Health Study support the prediction. (Revision of original figure published in Streissguth, *et al.*[17])

TABLE 4a. Primary Covariates According to Age at Which Assessed: Seattle Longitudinal Study on Alcohol and Pregnancy

Fifth Month of Pregnancy	Neonatal	8 Mo. and 18 Mo.	Four Years
Maternal Characteristics	Infant Characteristics	Infant Characteristics	Child Characteristics
Age	Sex	Age at testing	Age at testing
Race	Age at testing	Sensory problems	Sensory problems
Education	Gestational age	Breast feeding duration	Illness, fever, trauma
Marital status	Birth weight	Illness, fever, trauma	Hospitalizations
Parity	Birth length	Hospitalizations	Medications
Gravidity	Head circumference	Vitamins	Maternal Characteristics
SES	Apgar	Medications	Illness, hospitalization
Occupation	Presentation	Maternal Characteristics	Separation from child
Working status	Birth trauma	Illness, hospitalization	Working status
Hx alcohol problems	Jaundice	Separation from child	Caretaking arrangements
Hx pregnancy complications	Other problems	Working status	Family Environment
Hx deceased children	Environment	Mother/infant interaction	Major life changes in household (No.)
Prepregnancy weight	Delivery medications	Family Environment	School
Height	Isolette	Major life changes in household (No.)	Preschool attendance
Other Exposures	Days/hospital	H.O.M.E. scale of environment	Daycare attendance
Cigarette/nicotine	To whom discharged	Caretaking arrangements	Situational
Caffeine	Breast feeding	Situational	Accompanying person
Aspirin	Situational	Accompanying person	Examiner
Tylenol	Examiner	Examiner	Room of testing
Valium	Time since feeding	Room of testing	Temperature
Other teratogenic drugs	State at outset	Time of day	Time of day
Marijuana	Hospital of testing	Day of week	Day of week
Other illegal drugs	Temperature	Month of year	Month of year
All other drugs	Time of day		
Pregnancy	Day of week		
Nutritional intake	Month of year		
Eating habits	Prenatal Hx		
Vitamin supplements	No. of prenatal visits		
Mother vegetarian	Threatened miscarriage		
Illness/infection	Mother's weight at delivery		
Family Environment	Illness and infection during pregnancy		
Source of income	Atypical conditions		
No. of adults in household			
No. of children < 5 yrs			
No. of children > 5 yrs			

TABLE 4b.[a]

		Seven Years	
Home Environment	Biological Parents	Family Environment	Situational
Biol. mother at home	Mother: race	Family environment scales	Accompanying person
Biol. father at home	education	(Moos)	Examiner
Any father at home	height, weight	cohesion	Room of testing
No. of siblings/children < 5 yr	Father: race	expressiveness	Temperature
No. of siblings/children > 5 yr	education	conflict	Time of day
No. of adults in household	height, weight	independence	Day of week
Ratio of children to adults	Family Hx:	achievement orientation	Month of year
Marital status	alcoholism	intellectual-cultural	
Major life changes (No. and type)	learning disability	orientation	Child Characteristics
death, illness	psychiatric illness	active-recreational	Age at testing
divorces, marriages	chronic illness	moral-religious emphasis	Sleep last night (hours)
births		organization	Handedness
moves	Caretaking Parents	control	Medications/exam day
job changes, etc.	Education		Caffeine on exam day
Welfare status (source/income)	Working status	Family adaptability and cohesion	Illness on exam day
Religion and religious activity	Years surrogate parenting	Scales (FACES II)	Food intake/day prior
Computer in home		Family cohesion	Sensory problems
Video games in home	School	+ 8 subscales	Hx illness, fever, trauma
Smokers in household	Grade in school	Family Adaptability	Hx hospitalizations
Language in the home	Type of school	+ 6 subscales	Medical problems
	Type of class		Hours TV/day
	Demogr. char. of school		Hours video games/month
	Median test scores of		Regular medical/dental care
	school		Immunizations

[a] These 150 variables are only a representative group of those obtained for this study. Other covariates are derived scores calculated from the above (such as weight change during pregnancy). Some of the covariates are informally classified at the beginning of analysis of an outcome as either high priority, low priority, or variables to be examined by deleting cases from final analysis.

TABLE 5. Major Behavioral Findings of the Seattle Longitudinal Study on Alcohol and Pregnancy to Date

Outcome	Test Procedure	Age Assessed	R	F Test for Alcohol[a] F (df)	p
1. Habituation	Brazelton	Day 1	0.32	7.07(1,323)	0.008
2. Opened eyes	Naturalistic observations	Day 1		4.15(1,115)	0.044
Body tremors				4.21(1,115)	0.042
Head to left				4.00(1,115)	0.048
High level body activity				5.43(1,115)	0.022
Hand to face				5.16(1,115)	0.025
3. Sucking pressure	Pressure transducer	Day 2	0.25	4.62(1,147)	0.03
Latency to suck		Day 2	0.23	2.90(1,147)	0.09
4. Mental Dev. Index	Bayley scales	8 Months	0.33	3.22(2,452)	0.04
Psychomotor Dev. Index		8 Months	0.28	5.15(1,453)	0.024
5. Attention					
Errors of omission	Vigilance task	4 Years	0.23	6.32(1,351)	0.012
Errors of commission	with	4 Years	0.26	5.35(1,351)	0.021
Ratio correct	microcomputer	4 Years	0.33	7.98(1,351)	0.005
Trials oriented		4 Years	0.23	.01(1,276)	0.936
Reaction time		4 Years	0.57	11.71(1,066)	0.001
6. Time in movement	Motion detector	4 Years	0.15	.00(1,362)	0.984
7. Balance	Gross motor	4 Years	0.26	4.62(1,446)	0.032
Distance		4 Years	0.24	4.27(1,433)	0.039

8. Word reading	Stroop	7 Years	0.80	4.59(1,216)	0.033
Color naming		7 Years	0.51	4.08(1,216)	0.045
9. Attention					
Errors of omission-X Task	Vigilance task with microcomputer	7 Years	0.27	5.09(1,444)	0.025
Errors of omission-AX Task		7 Years	0.32	4.35(1,443)	0.026
Errors of commission-X Task		7 Years	0.40	4.06(1,444)	0.044
Errors of commission-AX Task		7 Years	0.45	9.68(1,442)	0.002
Reaction time		7 Years	0.38	7.22(1,443)	0.007

[a] The F statistic presented here is for the effect of alcohol after adjusting for all the other variables in the multiple regression model. Regression models were developed individually for each set of outcomes: Nicotine is adjusted for in each analysis. All models except the sucking variables and naturalistic observations also adjust for caffeine, mother's diet during pregnancy, and mother's education. Other covariates were included as appropriate for each set of outcomes.

The alcohol measures and the parameterizations are as follows:

Habituation (Day 1): Average ounces of absolute alcohol per day: linear in midpregnancy exposure.

Naturalistic observations (Day 1): 5-point ordinal alcohol scale defined by *a priori* hypothesized risk to the fetus. It combines amount, timing, and pattern of drinking.

Stroop (7 Years): Binary indicator of highest risk exposure from the 5-point scale described above.

Reaction time (7 Years): Average number of drinks per drinking occasion (early pregnancy).

All other results report various parameterizations of exposure described by average ounces of absolute alcohol per day before pregnancy recognition:

(1) Mental development index (8 mo.) and errors of omission (4 years): quadratic
(2) Errors of omission (X and AX Task) and distance (4 years): linear threshold at 1.5 ounces
(3) Errors of commission (X Task: 7 years): step function at 2.0 ounces
(4) All else: linear in log ounces

which they are observed. For example we saw that infants with FAS whose mothers were alcoholics were jittery, hyperexcitable, and habituated poorly, often had a weak suck and failure to thrive, and had delayed development. We predicted that infants with moderate to heavy alcohol exposure would manifest similar behaviors but of a lesser magnitude. As FIGURE 8 and TABLE 5 indicate, all of these hypotheses were confirmed. Similarly, hypotheses about the preschool and primary years were developed based on observation of children with FAS and tested with large samples of appropriately aged children from the Seattle Longitudinal Study. As FIGURE 8 and TABLE 5 indicate, many of the hypotheses have been confirmed and others are still under investigation. Although the manifestations change with different developmental stages, they may reflect latent underlying deficits whose structure can be studied with appropriate longitudinal statistical technique. Behaviors that seem to be most affected by varying levels of gestational alcohol exposure across several developmental ages are those reflecting attention, speed of central processing, and motor development.

The next sections of this paper discuss methodologic and statistical issues relevant to the study of behavioral teratology in general and alcohol teratogenesis in particular. We begin by describing the specific problems involved in assessing the three main classes of variables: the outcome variables, the exposure variables, and the confounding variables or covariates. FIGURE 8 depicts our view of how these variables interact in our model of behavioral teratogenesis.

ASSESSMENT OF VARIABLES

Outcome Variables

Three factors are important to consider with respect to assessment of outcome variables in human behavioral teratology: (1) precision of assessment, (2) timing of assessment, and (3) the parameters of the measured variables.

While severe neurobehavioral endpoints (such as clear mental retardation) can be observed clinically, very precise measurement seems to be necessary to assess subtle effects resulting from lower dose exposures. The significance of such subtle effects must be examined with large samples and sophisticated statistical procedures. Unlike traditional "risk" outcomes such as death, cancer, and mental retardation, the medical significance of such subtle deviations in groups of children can always be questioned. We recently compared two different types of sucking measures obtained on the same sample of newborn infants exposed to varying levels of alcohol *in utero*.[18] While the two measures were significantly related (t=3.04, df 99, $p=0.003$), only sucking pressure assessed with mm of mercury from a pressure transducer reflected the effects of the prenatal alcohol exposure, not the standard clinical assessment allowing the infant to suck on the examiner's thumb. We obtained similar results when measuring attention with laboratory tests of vigilance compared to maternal rating scales of attention. Computerized data collection, which we employ when possible, can provide precision of assessment with relatively large samples of subjects.

Timing of assessments is extremely important in behavioral teratology. If timing is not carefully considered, one can reach erroneous conclusions that effects are "washing out" over time, when in fact, the test may not be age appropriate, or the behavior may not be salient at that age. For example, in assessing gross motor behavior, skill at walking will only be relevant up to the age at which most children walk. The best time to assess a developmentally linked behavior is probably when it is emerging. The best time to measure deviant clinically significant endpoints is after most children have acquired the behavior.

Testing procedures must also be evaluated for "ceiling" or "basal" effects; that is, tasks must be sufficiently difficult so that not all but the worst performers will score perfectly. Similarly they must not be too hard. Without careful pretesting of the outcome at the test age, one could preclude the delineation of effects at either end of the distribution.

In evaluating the effects of exposure on outcomes, we heed the advice of Cochran[41] that investigators conducting exploratory studies not concentrate only on tests of significance, but that they also think in terms of estimating the size of the effect and its practical importance. This latter process should make the investigator more aware of potential sources of bias, lead to a more informative statistical analysis, and permit more interpretive results (see Cochran,[41] page 48).

Exposure Variables

Unfortunately, to date, the only feasible way to assess alcohol exposure is through self report. In an earlier paper we discussed special problems with assessing alcohol exposure during pregnancy.[19] Although we show good test-retest reliability of our exposure measures, they nevertheless represent attempts to average disparate consumption patterns over a several-month period. They may also represent a subject's reticence to admit to potentially damaging behavior.

The animal literature has suggested that different patterns and doses of alcohol are related to different outcomes. Human behavioral teratology studies (particularly those of alcohol teratogenesis) are characterized by the absence of a true "exposure" variable. Dose, timing, and exposure conditions can only be estimated. Therefore, such studies would do well to assess as many parameters of exposure as possible.

In addition to dose, which may be gathered as a continuous variable, but later grouped for analysis (cf. Cochran,[41] sect. 3.4; and Bellinger *et al.*[23]) we also attempt to assess timing and pattern of exposure (defined both by maximum quantity and by effect on the consumer, that is, intoxication). Questions about problems with alcohol were also included to differentiate the chronic abusers. Timing of exposure is more difficult to assess in that periods of exposure are not mutually exclusive. We measured drinking over two periods: before pregnancy recognition and during early pregnancy. (See TABLE 3 for a listing of the primary exposure variables in the present study.) In an earlier paper we have shown that the women classified as "heavy drinkers" vary considerably depending on which of these alcohol measures is used;[21] this finding prepares us for results that vary according to the exposure variable employed in the analysis. The effects associated with each pattern of alcohol exposure would not be expected to be the same.

Covariates

Thoughtful assessment of a multitude of covariates is essential in human behavioral teratology research, not only for appropriate statistical control of potential confounders, but also for clarification of the conditions that modify teratogenic effects. Assessment of appropriate covariates will ideally occur at each developmental stage at which outcomes are assessed.

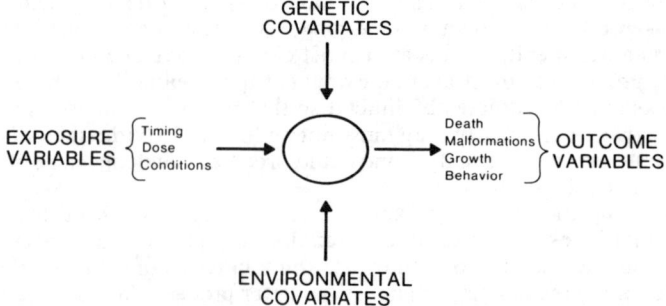

FIGURE 9. Diagram depicting the four main types of variables under consideration in behavioral teratology studies. Exposure variables combine with genetic and environmental covariates to produce certain outcomes of which behavior is one.

TABLE 4a and 4b lists some of the covariates considered in our study, grouped according to time of assessment and type. During the prenatal period, maternal ingestants, diet, and health variables can act either directly or in interaction with teratogens to produce perturbations in embryo/fetal development that can later be reflected in functional deficits in the child. Covariates around the time of delivery, particularly obstetric medication, will be important in evaluating neonatal variables. Mother-infant interactions and family environment variables will be important in assessing later behavioral outcomes. School attendance and educational setting will certainly impact academic performance. Appropriate covariates for each analysis will depend on the outcome being assessed.

As FIGURE 9 depicts, we think of two primary types of covariates impacting the organism: environmental and genetic. Unfortunately, at this stage of our research, we can observe genetic variables only indirectly at best, even though we know from our clinical work with patients with FAS and from the animal literature, that genetic susceptibility factors have a profound effect on teratogenic manifestations. Some surrogate genetic variables that are highly related to outcomes, such as IQ scores, are parental educational levels. Other surrogate variables that we are presently exploring include familial history of alcoholism, learning disability, and other diseases. When the salient genetic variables for alcohol teratogenesis have been identified clinically, we will be in a better position to address them in human behavioral teratology studies. For the time being, we use surrogate or indirect measures.

STATISTICAL ANALYSIS AND MODEL BUILDING

In this section we sketch our regression model building process with particular regard for behavioral teratology studies. In the following sections we suggest a new factor-analytic approach for regression models in this field.

Behavioral teratology studies are characterized by very large numbers of covariates, as suggested in previous sections. Our studies begin with about 150 variables in the categories indicated in TABLE 4. In fact, a complete list of candidate variables is never satisfactorily fixed at the beginning of a study, but rather the process of analysis often leads to searches for new potential covariates not originally considered or written down. Few of the many recent textbooks on regression analysis describe studies faced with such large numbers of variables. One book that does propose methods (similar to the scheme we follow) for such large data sets is by Mosteller and Tukey (Chap. 15).[22] Another similar strategy is outlined by Bellinger *et al.*[23] for a study of the effects of lead exposure on infant development.

The stages of analysis we employ include the following. One should note that these stages are often reconsidered iteratively.

Descriptive Analysis of Data

This involves a thorough descriptive, exploratory analysis of three classes of variables: dependent (outcome) variables, alcohol exposure variables, and covariates or potentially confounding variables. The latter are initially considered as either high or low priority. Stem and leaf plots or histograms, bivariate scatter plots, and cross tabulations are generally useful at this stage. These analyses lead to preliminary decisions about transformations of variables and appropriate methods of analysis.

Development of a "Base Model"

Beginning with the high-priority covariates (derived from literature reviews and past experience), a "base model" is constructed. This model building involves exploratory regression methods that consider a number of models. All subsets regression and informal stepwise techniques are considered here.

Exploration of Exposure Variables

Next, alcohol exposure variables are considered for addition to the base model (specific issues are discussed below). Following inclusion of one or more alcohol variables, the list of covariates is reconsidered.

Diagnosing the Model

Fitted models are assessed and criticized using current diagnostic methods in regression analysis (partial residual plots, measures of leverage, influence, collinearity, etc.).

SPECIAL MODELING ISSUES FOR ALCOHOL TERATOGENESIS

The next two sections focus on issues that are not always considered in a basic regression modeling strategy for a given set of variables, but that are especially important in the study of alcohol teratogenesis.

Treatment of Behavioral Outcomes

Analysis of Multiple Correlated Outcomes and Data Reduction

We are often faced with multiple recorded responses, and we must first decide (partly on the basis of the bivariate exploratory analyses) whether to analyze single response measures, or to combine measures (e.g. dominant and nondominant hand responses, groups of behavior ratings or neuropsychological tests) into new, perhaps more informative, response measures. Alternatively, in some cases we propose using partial least squares methods (discussed below) to compute appropriate dependent-factor scores; that is, we would aggregate individual test scales into a linear combination most strongly associated with variation in alcohol exposure.

Categorization of Outcome Variables

While most effects are best modeled with continuous outcomes, it is sometimes useful to categorize continuous outcome variables in order to answer certain clinically important questions concerning the exposure conditions associated with extreme outcomes (for example, focusing on IQ scores less than 80). In such cases the cutpoints for categorization could be determined from clinical experience, from the literature, or by arbitrarily selecting the lower range of the distribution of a given outcome. Once a cutpoint reflecting deviancy is defined, the probability of this deviant outcome can be estimated (from the model for the continuous outcome) for various exposure conditions.

Parameterization of Alcohol Effects

Analysis of Nonlinear Exposure Effects

Interest in the level of exposure at which significant alcohol effects appear translates into special concern for the form of the relationship between alcohol exposure and an outcome. We generally begin by log-transforming exposure variables with highly skewed distributions to limit the influence of extreme high-exposure cases. We also assess a variety of transformations of alcohol exposure variables, including thresholds and step functions. We are now incorporating new nonparametric statistical methods to estimate arbitrary nonlinear transformations. One such tool is the "ACE algorithm."[24] The estimated transformations can suggest fitting a particular threshold model.

Step functions can be used to assess differential thresholds for alcohol effects on different outcomes, or for the same outcome at different ages; however, the errors of measurement in the self-reported alcohol variables, together with the great natural variability in most behavioral outcomes, make precise interpretation of step-function models difficult. Categorization of exposure variables is often considered, not to draw precise inferences about threshold effects, but rather to draw inferences from a model that is robust with respect to the effects of extreme cases. (See also discussions in Cochran,[20] Sect. 3.4, and Mosteller and Tukey.[23])

Interaction Effects

In some cases alcohol effects are manifest only through interactions. For example, alcohol exposure may have a greater teratogenic effect for one race rather than another, or the effect of alcohol exposure may be heightened in the presence also of nicotine exposure. As a rule we consider interactions of each potentially significant alcohol variable with other exposures, with sex, and with other child-specific and environmental covariates.

Analysis of Multiple Exposure Variables

The spectrum of alcohol exposure variables measuring quantity, timing, and pattern (TABLE 3) presents a number of alternative modeling strategies. First, we may simply look for the most significant individual variables from the long list of candidates, sometimes hoping that the inclusion of one variable rather than another in a model will be a reflection of the mechanism of alcohol effects. We sometimes investigate the relevance of multiple dimensions of alcohol exposure by considering interactions between alcohol variables; for example, there may be different effects of alcohol exposure during pregnancy depending on the drinking pattern of the mother before recognition of pregnancy.

With multiple alcohol measures, we face a large number of correlated variables, and only a small number of these will find their way into a model due to the intercorrelations. This is also the case with other groups of covariates. In such cases it may help to combine variables into components, replacing bundles of significantly intercorrelated variables by linear combinations chosen by judgment, or by a method of principal components or factor analysis. We have, for example, analyzed multiple home environment scales in this way in our model building for IQ scores at age seven. Later we discuss the PLS approach to combining variables into factors in regression models.

Interpretation of Competing Models

In model building we are often left with several different models that are "statistically significant," but that might have different interpretations if considered individually. This point is particularly important with regard to the various parameterizations of the exposure (alcohol) variables. In such cases, although we can designate a preferred model (as judged by the standard error of the fitted model), we must exercise caution in emphasizing the final model when it has been chosen from among many competitors. When comparing a linear and a quadratic model, we can carry out a simple t- or F-test for the significance of adding the quadratic term. When one model is not a special case of another (the models are not "nested"), we may test the significance of the difference in fits, measured by mean square error, by other methods.[25] Such tests help us to decide whether or not conclusions can be drawn about the timing or level of exposure at which significant alcohol effects are observed.

NEW METHODOLOGY: FACTOR MODELS AND PARTIAL LEAST SQUARES

Ordinary multiple regression models (with special attention to the issues raised above) suffice for modeling a single outcome variable with well-defined measures of teratogenic exposure; however, in complex studies of alcohol teratogenesis, where many concepts can be measured only indirectly, multiple regression methodology is limiting. It fails to take advantage of all the information available in highly multivariate data collected with certain factors in mind. In these studies we measure many groups of variables not because we are interested in each of the variables singly, but because we think of them as joint manifestations of an underlying concept or factor.

This structure applies to all three classes of variables in behavioral teratology studies: exposure variables, outcome variables, and covariates. With alcohol exposure, for example, we may believe that timing and level of exposure during gestation simply cannot be precisely measured and averaged considering the inter- and intra-individual differences in consumption and reporting of consumption among a large group of women. In such cases we prefer to determine whether our multiple alcohol variables represent an underlying exposure factor that is more clearly related to the outcome of interest than is any one measure alone.

Similarly with respect to a block of covariates like the home environment scales, we are not interested in the effects of the predefined scales (such as intellectual-cultural orientation, and conflict) per se, but whether or not they reflect an underlying factor, or "latent variable," that explains variation in, say, IQ. In contrast with the use of judgment or principal components for combining variables (as suggested above), we want to combine variables into weighted averages as estimates of underlying factors that are optimal for predicting an outcome of interest.

Finally, in the case of outcome variables we are, for example, not interested specifically in "errors of omission" and "reaction time" on a given task, but whether or not these (and other) responses jointly reflect an "attention" factor that is related to alcohol exposure.

Partial least squares (PLS) is a statistical method, developed recently by Herman Wold (Joreskog and Wold,[26] part II), that provides a systematic search for properly weighted averages of variables to be taken as new measurements of latent variables underlying the choice of multiple measures and the aggregation of these measures into blocks. Among its strengths are:

(1)   It provides a new approach, based on a factor model, to the multicollinearity problem (highly correlated predictors) in multiple regression. We need not settle for including in a model just one of a list of correlated predictors, and we need not choose relatively arbitrary linear combinations (such as principal components) without regard for the outcome variable being predicted.

(2)   PLS averages out measurement error in a block of variables (the alcohol exposure variables, for example) while preserving the "signal" in these variables (the correlation between the latent variables), thus providing increased predictive power for the outcome when the latent variable model is true.

(3)   In contrast to LISREL (see Joreskog and Wold,[26] part I), the most common alternative to PLS for latent variable modeling, PLS handles indefinitely many variables (and latent variables), it provides explicit estimates of the values of the latent variables (useful for diagnosing models), and it requires no assumptions of joint normality for the distributions of the measured variables. For further discussion see Bookstein.[27]

We are currently developing, for our studies of alcohol teratogenesis, an application of PLS that we expect will prove an important extension of regression methodology for the analysis of large data sets of hundreds of variables, and thus be particularly useful for the field of behavioral teratology.

SUMMARY

Alcohol is a teratogenic drug and the effects appear to be grossly dose related. The severest effects are observable clinically as the fetal alcohol syndrome and are associated with heavy prenatal alcohol exposure and a history of chronic maternal abuse of alcohol. Hypotheses for subtler behavioral effects associated with lower levels of exposure can be generated from observation of the behavioral effects in FAS.

Behavioral effects associated with various levels of prenatal alcohol exposure in humans include poor sucking and poor habituation in the newborn, poorer mental and motor development in infancy, and attentional and reaction time effects at four and seven years of age.

Human behavioral teratology studies are necessarily complex due to the large number of covariates that affect behavior, modify the effects of teratogens, and influence interpretation. Challenging problems exist in assessing exposure, outcomes, and covariates. Common to assessment of all these classes of variables are the multiplicity of measurement and the indirect nature of the measurement. Answering specific questions about timing and dose effects demands careful statistical modeling procedures and large, complex data bases. Large data bases and indirect measurement problems suggest factor analytic extensions of current regression methodology, which we have proposed in this paper.

ACKNOWLEDGMENTS

We thank the many colleagues who have contributed to this research over the past years and extend our gratitude to Maryann Luse and Rebeca Olson, of our office staff, for help in the preparation of this manuscript.

REFERENCES

1. SMITH, D. W. 1982. Recognizable Patterns of Human Malformation. Third Edition. W. B. Saunders Company. Philadelphia, PA.
2. CLARREN, S. K. & D. W. SMITH. 1978. The fetal alcohol syndrome. N. Engl. J. Med. **298:** 1063–1067.
3. JONES, K. L., D. W. SMITH, C. N. ULLELAND & A. P. STREISSGUTH. 1973. Pattern of malformation in offspring of chronic alcoholic mothers. Lancet **2:** 1267–1271.
4. JONES, K. L. & D. W. SMITH. 1973. Recognition of the fetal alcohol syndrome in early infancy. Lancet **2:** 999–1001.
5. CLARREN, S. K. 1982. The diagnosis and treatment of fetal alcohol syndrome. Compr. Ther. **8**(10): 41–46.
6. STREISSGUTH, A. P., C. S. HERMAN & D. W. SMITH. 1978. Intelligence, behavior, and dysmorphogenesis in the fetal alcohol syndrome: A report on 20 patients. J. Pediatr. **92**(3): 363–367.
7. STREISSGUTH, A. P., S. K. CLARREN & K. L. JONES. 1985. Natural history of the fetal alcohol syndrome: A ten-year follow-up of eleven patients. Lancet **2:** 85–92.
8. STREISSGUTH, A. P., S. LANDESMAN-DWYER, J. C. MARTIN & D. W. SMITH. 1980. Teratogenic effects of alcohol in humans and laboratory animals. Science **209:** 353–361.

9. ABEL, E. L. 1982. Fetal Alcohol Syndrome, Vol. III: Animal Studies. CRC Press Inc. Boca Raton, FL.
10. WILSON, J. 1977. Current status of teratology. In Handbook of Teratology: General Principles and Etiology. J. G. Wilson & F. C. Fraser, Eds. 1: 47–74. Plenum Press. New York.
11. SULIK, K. K. 1984. Critical periods for alcohol teratogenesis in mice, with special reference to the gastrulation stage of embryogenesis. In Mechanisms of Alcohol Damage in Utero. CIBA Foundation Symposium No. 105: 124–141. Pitman Publishing Co. London.
12. CHERNOFF, G. F. 1980. The fetal alcohol syndrome in mice: Maternal variables. Teratology 22: 71–75.
13. VORHEES, C. V. Principles of behavioral teratology. In Handbook of Behavioral Teratology. E. P. Riley & C. V. Vorhees, Eds. Plenum Press. New York. In press.
14. MEYER, L. S. & E. P. RILEY. Behavioral teratology of alcohol. In A Handbook of Behavioral Teratology. E. P. Riley & C. V. Vorhees, Eds. Plenum Press. New York. In press.
15. CLARREN, S. K. & D. M. BOWDEN. 1984. Measures of alcohol damage in utero in the pigtailed macaque (Macaca nemestrina). In Mechanisms of Alcohol Damage in Utero. CIBA Foundation Symposium #105:157–172. Pitman Press. London.
16. STREISSGUTH, A. P., D. C. MARTIN, J. C. MARTIN & H. M. BARR. 1981. The Seattle longitudinal prospective study on alcohol and pregnancy. Neurobehav. Toxicol. Teratol. 3:223–233.
17. STREISSGUTH, A. P., H. M. BARR & D. C. MARTIN. 1984. Alcohol exposure in utero and functional deficits in children during the first four years of life. In Mechanisms of Alcohol Damage in Utero. CIBA Foundation Symposium No. 105: 176–196. Pitman Publishing Co. London.
18. STOCK, D. L., A. P. STREISSGUTH & D. C. MARTIN. 1985. Neonatal sucking as an outcome variable: Comparison of quantitative and clinical assessments. J. Early Hum. Dev. 10: 251–256.
19. STREISSGUTH, A. P. & R. E. LITTLE. 1985. Alcohol-related morbidity and mortality in offspring of drinking women: Methodological issues and a review of pertinent studies. In Alcohol Patterns and Problems. M. A. Schuckit, Ed. Rutgers University Press. New Brunswick, NJ. pp. 113–155.
20. COCHRAN, W. G. 1983. Planning and Analysis of Observational Studies. John Wiley and Sons. New York.
21. STREISSGUTH, A. P., D. C. MARTIN & V. E. BUFFINGTON. 1977. Identifying heavy drinkers: A comparison of eight alcohol scores obtained on the same sample. In Currents in Alcoholism, Vol. 2. F. A. Seixas, Ed. Grune & Stratton, Inc. New York.
22. MOSTELLER, F. & J. W. TUKEY. 1977. Data Analysis and Regression. Addison-Wesley Publishing Co. Reading, MA.
23. BELLINGER, D., A. LEVITON, C. WATERNAUX & E. ALLRED. 1985. Methodological issues in modeling the relationship between low-level lead exposure and infant development: Examples from the Boston Lead Study. Environ. Res. 38: 119–129.
24. BREIMAN, L. & J. H. FRIEDMAN. 1985. Estimation of optimal transformations for multiple regression and correlation. J. Am. Stat. Assoc. 80: 580–619.
25. EFRON, B. 1984. Comparing non-nested linear models. J. Am. Stat. Assoc. 79: 791–803.
26. JORESKOG, K. G. & H. WOLD. 1982. Systems Under Indirect Observation: Causality-Structure-Prediction, Part I & II. North Holland Publishing Co. Amsterdam.
27. BOOKSTEIN, F. L. 1986. The elements of latent variable models: A cautionary lecture. In Advances in Developmental Psychology, Vol. 4. M. Lamb et al., Eds. :203–230. Lawrence Erlbaum. Hillsdale, NJ.
28. JESSOR, R., T. D. GRAVES, R. C. HANSON & S. L. JESSOR. 1968. Society, Personality and Deviant Behavior: A Study of a Tri-Ethnic Community. Holt, Rinehart & Winston. New York.
29. JONES, K. L., D. W. SMITH, A. P. STREISSGUTH & N. C. MYRIANTHOPOULOS. 1974. Outcome in offspring of chronic alcoholic women. Lancet 1: 1076–1078.

Significant Determinants of Susceptibility to Alcohol Teratogenicity[a]

ROBERT J. SOKOL,[b] JOEL AGER, AND SUSAN MARTIER

Department of Obstetrics and Gynecology and
C.S. Mott Center for Human Growth and Development
Hutzel Hospital/Wayne State University
Detroit, Michigan 48201

SARA DEBANNE

Department of Biometry
Case Western Reserve University
School of Medicine
Cleveland, Ohio 44106

CLAIRE ERNHART

Department of Psychiatry
and Perinatal Clinical Research Center
Cleveland Metropolitan General Hospital/
Case Western Reserve University
Cleveland, Ohio 44109

JAN KUZMA

Department of Biostatistics and Epidemiology
Loma Linda University
Loma Linda, California 92354

SHELDON I. MILLER

Sheppard and Enoch Pratt Hospital
Baltimore, Maryland 21204

INTRODUCTION AND BACKGROUND

Fetal Alcohol Syndrome and Alcohol-Related Birth Defects

The potential for damage to the offspring from prenatal alcohol exposure is now well established. A link between alcohol and birth defects has been strongly suspected since the 1700s, with early scientific evidence gradually being generated during the 19th and early 20th centuries.[1,2] Modern recognition of alcohol's potential as a teratogen, however, awaited the appearance of two key articles in 1973 by Jones, Smith, and their colleagues[3,4] describing growth retardation,

[a] Supported in part by USPHS Grants AA 03282, AA 00688 S1 and 2, HDAA 14883, RR 00210, and AA 06334.
[b] Address for correspondence: Robert J. Sokol, M.D., Professor, Chairman and Chief, Department of Obstetrics and Gynecology, Hutzel Hospital/Wayne State University, 4707 St. Antoine, Detroit, MI 48201.

craniofacial, and cardiac defects, as well as developmental delay, in 11 children born to alcoholic women. The impact of these two publications was, perhaps, greáter than might have been expected since they appeared during a period of increased public and professional awareness and concern with maternal/fetal/infant health. Indeed, in retrospect, the most important contribution of these two articles may have been the authors' coining of the term "fetal alcohol syndrome (FAS)." This dramatically refocused interest on an important perinatal risk.

Approximately 500 examples of FAS have been reported from the northern hemisphere.[5] Cases have also been reported from the southern hemisphere. As might be anticipated, no standardized criteria were used in making the diagnoses in these cases. In 1980, however, the Fetal Alcohol Study Group of the Research Society on Alcoholism proposed such criteria.[6] This standard requires that an abnormality in each of three general categories for diagnosis of FAS to be made. These criteria are shown in TABLE 1. None of the listed features is pathognomonic of prenatal alcohol exposure. In addition, there are other nonspecific abnormalities that may be seen in conjunction with FAS. These include ocular retinal tortuosity; cardiac abnormalities, particularly septal defects; renal anomalies, such as hydronephrosis; genital anomalies, such as hypospadias and undescended testes; hemangiomas and dermatoglyphic abnormalities; as well as other anomalies, such as hernias.[7] In the presence of evidence of heavy or frequent drinking during pregnancy, such abnormalities have been referred to as "alcohol-related birth defects" or "fetal alcohol effects" (FAE).

FAS occurs with a prevalence in the range of one to three per 1,000 births, with alcohol-related birth defects having been reported considerably more frequently.[3,4,8] It has been calculated that approximately 5% of all congenital anomalies might be attributable to prenatal alcohol exposure.[2] Additional adverse pregnancy consequences that have been attributed to prenatal alcohol exposure include increased risk for spontaneous abortion,[9–12] possibly increased risk for stillbirth[13] or preterm delivery,[14] and a range of neurobehavioral abnormalities, including abnormal electroencephalographic activity,[15] lower developmental quotients on the Bayley scales,[16,17] mental retardation,[18] cerebral palsy,[8] and attention-deficit syndrome with learning difficulties.[19] Abnormal neuroanatomic development, possibly underlying such neurobehavioral abnormalities, has also been observed.[20]

Susceptibility to Effects of Prenatal Alcohol Exposure

As noted above, FAS has been reported from around the world with a frequency of one to three per 1,000 births. Very heavy and/or frequent drinking during pregnancy has, however, been reported to occur much more frequently. It has been estimated that approximately 60% of American women drink alcoholic beverages, with approximately 3% classified as "problem drinkers." The proportion of women of reproductive age drinking an average of at least two drinks per day, that is, one ounce of absolute alcohol per day, approximates 5.5%.[21] During pregnancy, the proportion of women drinking at least at this level decreases to about 2%. Different populations of gravidas, however, have been reported to behave differently. In a study from California,[10] 0.5%, and in a study from Buffalo,[22] 16% of pregnant women reported that they drank 14 or more drinks per week; thus, it may be concluded that there is at least a 10-fold difference in the rate of FAS and that of considerable fetal alcohol exposure. It would follow that only a limited proportion of exposed fetuses exhibit FAS. Indeed, in an epidemiologic investigation of over 12,000 consecutive gravidas performed by our group in

TABLE 1. Minimal Criteria for the Diagnosis of Fetal Alcohol Syndrome As Proposed by the Fetal Alcohol Study Group of the Research Society on Alcoholism[a]

1. Prenatal and/or postnatal growth retardation (weight, length and/or head circumference for gestational age)
2. CNS involvement (signs of neurological abnormality, developmental delay, or intellectual impairment < 10th percentile)
3. Characteristic facial dysmorphology (at least two of three)
A. Microcephaly (head circumference < third percentile)
B. Microphthalmia and/or short palpebral fissures
C. Poorly developed philtrum, thin upper lip, and flattening of the maxillary area

[a] Adapted from Rosett.[6]

Cleveland, Ohio, 1.7% were identified as abusive drinkers. Among the offspring of these 204 abusive drinkers, only 5 (2.5%) were identified as exhibiting FAS.[9]

When the full range of alcohol-related birth defects are considered, a discrepancy between the rates of heavy exposure and adverse outcomes persists; thus, of the 204 offspring from the Cleveland study cited above, a maximum of 50% had any abnormality possibly attributable to prenatal alcohol exposure. Similarly, in a study of infant neurobehavioral development in which the Bayley scales were administered at age eight months, among the 5% of pregnancies complicated by the heaviest drinking, only 5 to 10% of the infants had abnormally low (< 85) scores on the mental and psychomotor scales, respectively.[16]

The cited results appear to indicate that only a proportion of fetuses exposed to the risk of heavy *in utero* alcohol exposure seem to be adversely affected. This suggests that it is reasonable to hypothesize that some fetuses may be more susceptible than others to the adverse consequences of alcohol exposure. There may be maternal characteristics that have a protective effect. Alternatively, pregnancy risk factors and/or other substance use could act synergistically with alcohol in producing fetal damage. It must be noted that there might be alternative explanations for failure to observe adverse pregnancy consequences in all exposed offspring, such as the lack in some cases of exposure of the embryo/fetus during critical periods of development.

When one reviews the available literature, results that speak directly to the issue of susceptibility must be considered very sparse; few studies have been reported. Our purpose in this paper is to describe two studies, performed in different populations, that address the issue of susceptibility to the effects of prenatal alcohol exposure. The first study focuses on the risk for intrauterine growth retardation, that is, lowered birth weight for gestational age; the second focuses on susceptibility to the full fetal alcohol syndrome. In presenting these two studies, the limited pertinent data available in the human and animal alcohol and pregnancy literature will be noted.

SUSCEPTIBILITY OF THE PRENATALLY EXPOSED FETUS TO INTRAUTERINE GROWTH RETARDATION

Literature Review

Intrauterine growth retardation is the most consistently reported effect attributed to prenatal alcohol exposure.[23] Three reports of human twins support the hypothesis that genetic differences in fetal susceptibility and/or genetically determined maternal metabolic capacity may explain why some offspring of women who drink heavily through pregnancy are severely affected while others are not. In 1974, Palmer reported a set of identical twins born to a chronic alcoholic mother.[24] The morphologic features and growth of these twins were remarkably similar. In contrast, Christoffel and Salafsky reported a pair of fraternal twins born to a very heavy-drinking mother.[25] Both of the male offspring had alcohol-related birth defects, but their growth parameters were very different. The weight, length, and head circumference of one fraternal twin were at the 30th, 15th, and 60th percentiles, respectively, while these parameters for the other twin were all at or below the 10th percentile. A similar discrepancy in a set of fraternal twins was also reported by Santolaya et al.[26]

Added to these twin studies in humans are results indicating that different strains of mice showed different susceptibilities to retarded intrauterine growth at the same maternal doses of alcohol.[27,28]

Susceptibility Study Using the Loma Linda NIAAA Fetal Alcohol Database

An early epidemiologic prospective observational study was supported by the National Institute on Alcohol Abuse and Alcoholism (NIAAA) at Loma Linda University. Using data from this very large study, 44 potential determinants of birth weight were examined in 5,093 pregnancies. Of these 44 determinants, 10, including the frequency of beer use, were found to make significant and independent contributions to decreased birth weight for gestational age.[29] A decrement in birth weight attributable to alcohol of approximately 100 grams was observed only in the 3% of pregnancies (No. = 176) complicated by very frequent beer drinking, that is, greater than 20 days per month.

Methods

A follow-up study was performed to look for any susceptibility factors in the 176 pregnancies that had been identified to be at risk.[30] The strategy adopted was to use discriminant analysis to contrast pregnancies that yielded affected infants ($< 2,700$ grams, No. = 27) with those that yielded infants of greater weight (No. = 149). The cut point of 2,700 grams was chosen, inasmuch as infants of greater weight are greater than the 10th percentile, even at full term, according to a standardized definition for small, appropriate, and large-for-gestational-age infants.

Results

Results of the discriminant analysis indicated that mothers of affected infants were more likely to be black (26% vs 7%), to have weighed approximately 11 pounds less before pregnancy and to have gained an average of five fewer pounds during pregnancy ($p < 0.0001$). An analysis of variance (ANOVA) of birth weight adjusted for gestational age, revealed no two-way interactions among these risk factors. Further, an additional ANOVA of birth weight adjusted for gestational age was performed, using data from all 2,233 pregnancies in the study population sample in which the mother drank any beer. Significant main effects of maternal weight, weight gain, cigarette smoking, ethnicity, and frequency of beer drinking were detected ($p < 0.001$), but again, there were no significant two-way interactions.

Comment

These results suggested that if birth weight, lowered below a given threshold, is used as a marker for intrauterine growth retardation, an effect of abusive drinking will be more likely to be observed in offspring of pregnancies with concomitant risks for intrauterine growth retardation. Such risks include black race, low maternal weight, and cigarette smoking. One interpretation, then, would be that pregnancies with these additional risk factors should be considered more susceptible to alcohol-related intrauterine growth retardation; however, such a conclusion would have been strengthened had interactions been found to be significant. They were not. Thus, while it could be argued that the offspring of underweight women who failed to gain weight during pregnancy and who smoke are more susceptible to the growth-retarding effects of prenatal alcohol exposure, synergistic adverse effects were not demonstrated. Frequent drinking was associated with decreased infant birth weight, regardless of the presence or absence of the other factors.

Whether interactions of abusive drinking might occur with other risks that were not measured in this study, for example, genetic predisposition and whether susceptibility to other alcohol-related birth defects occurs, remained open questions at the conclusion of this study. A detailed analysis of a larger sample is under way. We were able to conclude from the above preliminary analysis, however, that since no factors were identified that appeared to lessen the significant effect of alcohol on decreased intrauterine growth, from a clinical perspective, abusive maternal drinking must be considered a risk for any pregnancy.

DETERMINANTS OF FETAL ALCOHOL SYNDROME

Literature Review

Rosett and Weiner have recently reviewed evidence for genetic susceptibility to alcohol-related dysmorphologic effects, citing the human twin studies noted above, as well as evidence from animal experiments.[31] Riley and Lochry[32] have reported that even with substantially lower blood alcohol levels, CBA mice showed greater effects than C3H mice. Similarly, Sulik and her colleagues[33] have reported that C57 mice have a genetic predisposition to alcohol-induced facial anomalies.

Very recently, Martinez *et al.*[34] have performed a study in which maternal ethanol administration was found to cause biochemical alterations in the fetal tissues of both normally ovulating and superovulated mice; however, morphologic alterations occurred predominantly in fetuses of superovulated animals. The prevalent morphologic alteration identified at day 19 of gestation was a hemorrhagic lesion dorsal to the lumbar or dorsal vertebrae. It is suggested by these investigators that superovulation appears to increase the susceptibility of fetal mouse tissues to the teratogenic effects of ethanol.

Abel has approached the issue of susceptibility to fetal alcohol syndrome from a different perspective.[35] He reviewed more than 300 cases of FAS in the literature. These were mainly clinical reports. From these reports, he extracted factors, in addition to maternal alcoholism and fetal alcohol exposure, that could be considered possible risks for FAS. The following factors were so identified: increased maternal age and/or parity, poor nutrition, lower maternal socioeconomic status, and increased cigarette smoking and other drug use. Other possibilities included "genetic factors," female sex of the offspring, and paternal factors. While it is impossible to sort out the possible independent and interactive effects of these factors in producing fetal alcohol syndrome on the basis of case reports, this listing provides a source of variables that should be considered covariates in an analysis of risk for and susceptibility to FAS. Unfortunately, no such studies in humans featuring blinded prospective ascertainment and the necessary extensive statistical control using these covariates had been performed. Our purpose in this study was to further explore the issue of susceptibility, with the outcome being fetal alcohol syndrome in a series of clear-cut cases prospectively ascertained, blinded for all prenatal information, with extensive statistical control.

Susceptibility Study Using the Cleveland NIAAA Fetal Alcohol Database

Methods

Data were collected in a prospective observational matched-pair design in the Cleveland Fetal Alcohol Study supported by the NIAAA. Methods of patient ascertainment and data collection have been reported in detail elsewhere.[36,37] It is enough to say here that a total of 8,331 consecutive antenatal gravidas were screened during a 33-month period at their first prenatal visit. In addition to identifying information, each patient was administered the Michigan Alcoholism Screening Test (MAST)[38] and a two-week recall of alcohol intake on a daily basis. Nutritional data and information concerning other substance use, as well as medical/obstetric risks, were also assessed. Over the 33-month period, the total MAST positivity rate was approximately 11%. Six hundred MAST-positive patients were successfully recruited for study. Six hundred MAST-negative patients were matched for seven factors to the 600 MAST positives. All patients were then reinterviewed at each successive prenatal visit; there was an average of approximately five such visits. Successive two-week drinking histories were obtained. After delivery each neonate was examined, blinded for all prenatal information, as close to 72 hours of age as possible. The Ballard gestational age,[39] weight, length, head circumference, and palpebral fissure measurements of the infant[40] were recorded. The infant was carefully examined for possible alcohol-related birth defects. The total list of these defects has been previously published[41] and was based on abnormalities, previously reported in the literature to be associated with alcohol use during pregnancy,[7] including, for example, microcephaly, facial dysmorphology, genitourinary abnormalities such as hypospadias, and so forth.

All antenatal and postnatal data were stored in a computerized relational database for later analysis.

From the 8,331 consecutively screened pregnancies, 25 cases of full fetal alcohol syndrome, meeting precisely the minimal criteria as suggested by the Fetal Alcohol Study Group of the Research Society on Alcoholism (see TABLE 1),[6] were identified; thus, the rate of FAS in this large, unselected sample of antenatal gravidas was approximately three per 1,000. For the current analysis, to maximize statistical efficiency, a synthetic case-control study design was chosen.[42] Fifty non-FAS control infants were selected. These were the preceding and succeeding births matched on a one-to-one basis with the 25 cases of FAS. The FAS and non-FAS groups were contrasted for infant characteristics, as a method of describing the membership of each of the study groups. To examine the issue of susceptibility to the development of FAS, the two groups were contrasted for prenatal risks, including reported maternal alcohol use throughout pregnancy. All data management and analyses were carried out on a VAX 11/750 computer with a UNIX operating system and an INGRES relational database management system, as well as the BMDP statistical software package. Major analyses were performed using stepwise discriminant analysis and logistic regression.

In this study, effect size was described by the ratio, R^2/R^2 max.[43] A procedure for determining r max for any pair of marginal distributions has been described by Carroll[44] and takes into account the fact that the highest possible correlation (r) between two variables and hence, the variance (R^2) is less than unity when marginal distributions are dissimilar. Thus, the ratio R^2/R^2 max expresses explained variance as a proportion (percent) of total explainable variance.

Results

Significant differences between the case and control infants are shown in TABLE 2. Cases were more likely to be female and were considerably smaller in all measurements than controls. In addition, cases had significant excesses of anomalies of the eyes, nose, mandible, lips, total face, neurologic, cutaneous, muscular, total nonneurologic, nonfacial, and total anomalies.

Comparing the groups of nonalcohol prenatal risks, there were no differences in socioeconomic status, educational level, cigarette smoking, narcotic use, prepregnancy weight, nine nutritional indices, gestational age at screening, obstetric risk score, previous abortions, or father's weight. The FAS group, however, was characterized by increased maternal age, an increased frequency of black race, higher gravidity, and higher parity.

For the alcohol variables examined, no differences were identified in the proportion of days reported during pregnancy or in the percent alcohol obtained from wine or liquor. The FAS group was characterized by a significantly higher frequency of MAST positivity (> 5), higher MAST scores, a higher proportion of drinking days, absolute alcohol per drinking day, and absolute alcohol per day, as well as a higher volume index and a pattern index[45] and a higher proportion of alcohol from beer.

TABLE 2. Significant Differences between Case and Control Infants[a]

Characteristic	FAS (No. = 25)	Non-FAS (No. = 50)
Female sex	72%	36%
Weight (g)	2167 ± 430	3108 ± 606
Percentile	4 ± 2	48 ± 34
Length (cm)	45.2 ± 3.2	49.1 ± 3.3
Head circumference (cm)	31.4 ± 2.0	33.9 ± 2.1
Palpebral fissures (mm)	14.6 ± 1.4	15.9 ± 2.1

[a] $p < 0.05$ for all comparisons.

A preliminary, unforced, stepwise discriminant analysis, contrasting FAS and non-FAS groups and including all prenatal risk factors, indicated that race and MAST positivity significantly differentiated the two groups ($F(2,72) = 17.87$ R^2/R^2 max $= 45.6\%$, $p \ll 0.0001$)

Following the concept of hierarchical analysis, each of these two explanatory variables was then carefully examined to explore what they "really" represented. Sixty-seven percent of the race variance (black/non-black, R^2 max $= 63.2\%$) was explained by maternal age, narcotic use, proportion drinking days, and other variables including alcohol variables. Sixty-nine percent of the MAST ($+/-$, R^2 max $= 61.6\%$ was explained by the pattern index, gestational age at screen, and other variables including alcohol variables. These analyses made it clear that "race" and "MAST" were absorbing considerable variance, possibly attributable to alcohol and other factors; thus, a parsimonious model explaining as high a percent of explainable variance as possible was sought using discriminant analyses. These results are shown in TABLE 3. Jointly, four factors explained 63.6% of the explainable variance in FAS/non-FAS outcomes.

To obtain estimates of the probability of development of FAS in the absence of any of these risks and in their presence singly or jointly, a logistic regression analysis was performed. The results are shown in TABLE 4. It is estimated from the current sample that, in the absence of any of the four risks, the probability of the offspring being affected with FAS is less than 2%. In the presence of all four of the factors, the chance of an FAS-affected offspring rises to 85.2%, an odds ratio of 314.

Extraordinarily detailed and comprehensive analyses were performed to rule out the possibility that various biases might have affected these results. Considering prenatal risks, it was found that blacks reported more persistent drinking during pregnancy. This, however, did not account for the findings, inasmuch as this factor is directly adjusted for in the analyses. Also, blacks with FAS children may have been more likely to deny any drinking during pregnancy. This occurred in three cases, but would tend to bias the results in a direction opposite from the current findings.

The possibility of bias in ascertainment of the outcome was also examined. From the perspective of infant size, black infants are indeed smaller than white infants. The current data set was reanalyzed with the elimination of five black infants who were of borderline weight for gestational age, using criteria from a nonexposed sample of whites. Similarly, the analysis was performed with an additional two white infants who would have been included had weight for gestational age criteria based on nonexposed black infants been used. The results of these analyses were the same as those detailed above.

The possibility of bias in examination for central nervous system features was evaluated by adjusting head circumferences for weight. Again, the results did not

TABLE 3. Four-Variable Model of Prenatal Risks for FAS[a]

Variable	F to Enter	F to Remove	Cumulative R^2/R^2 max
Drinking days (percent)	16.23	3.62	30.5%
Mast +/−	3.13	5.13	36.2%
Parity	6.96	2.00	48.0%
Race	10.52	10.52	63.6%

[a] $F(4,70) = 10.69$; $p \ll 0.000001$.

TABLE 4. Risk for FAS from Four Factors in Logistic Regression Analysis

Risk	Probability of FAS	Odds Ratio (X)
None	0.018	
High percentage of drinking days	0.052	3.0
Positive mast	0.074	4.4
High parity	0.054	3.1
Black race	0.123	7.6
All four	0.852	314.1

change. Further, no differences between black and nonblack infants for CNS features other than head circumference were identified in the study sample or in a nonexposed group of infants (No. = 263).

It was considered possible that the faces of black and white infants might appear different to the examiners, so this possible source of ascertainment bias was examined. No significant differences in the defining characteristics for FAS (eyes, nose, and lips) were identified between black and nonblack infants either among the FAS cases or among the control cases. A few small differences between black and nonblack infants for facial characteristics were identified in a nonexposed group (No. = 263); however, these differences were balanced between the groups.

The results of analyses seeking bias as a source of the differences observed between pregnancies resulting in FAS and non-FAS offspring indicated that the results could not be explained by bias. Race was a significant determinant of FAS outcome regardless of all adjustments.

Comment

The results of this study suggest that persistent exposure of the fetus through pregnancy to alcohol is a major determinant of FAS. Proportion drinking days, along with chronic alcohol problems, accounted for over one-third of the explainable and over one-half of the explained variance. The probability of an offspring exhibiting FAS was found to increase nearly 50-fold (85.2% vs 1.8%) in the presence of four risk factors—high proportion of drinking days, MAST positivity, high parity, and black race. Expressed as an odds ratio, the risk was increased over 300 times, though it must be noted that odds ratios may be overestimated in case-control studies.

COMMENTS AND CONCLUSIONS

Based on the observation that a limited proportion of infants exposed during pregnancy to alcohol exhibit FAS or other possible alcohol-related birth defects, including intrauterine growth retardation, it was reasonable to hypothesize that it might be possible to detect susceptibility factors that may either increase the probability of abnormality or protect the fetus. Limited support for this concept could be derived from twin studies in humans and from effects observed in differing strains in animal experiments already in the literature. The two studies presented in this paper were performed in entirely different population samples using different outcomes and analytic strategies. It is reassuring that measures of alcohol use were found to contribute significantly to the occurrence of both intrauterine growth retardation and full FAS. In the first study, the contributions of black race, low maternal weight, and weight gain as well as smoking were not unexpected. These are well-recognized risks for intrauterine growth retardation.[47] In the second study, the explanation for the contributions of MAST positivity and high parity are speculative. These variables may be measuring maternal physiologic change related to prolonged exposure of the maternal organism to abusive levels of alcohol intake. These physiologic changes could, in turn, contribute to adverse pregnancy outcome.

The most striking finding in both studies presented here, however, is that black race may be a susceptibility factor with regard to prenatal alcohol exposure. In the first study, black race and frequent maternal beer drinking were found to additively increase the chances for lowered birth weight for gestational age. Based on results of the second study, it appears that fetuses of black gravidas may dose per dose, that is, adjusted for frequency of maternal alcohol intake, be at considerably increased risk for fetal alcohol syndrome. This risk appears to be about seven-fold higher than for white infants. This finding is not inconsistent with the limited human and animal studies available in the literature. Moreover, in a very recent report from Iosub and her colleagues,[46] familial fetal alcohol syndrome was compared in blacks and Hispanics. These investigators had previously observed a higher incidence of FAS in blacks versus Hispanics and in this study of repeated cases in the same family, they concluded that for unexplained reasons single and multiple cases of FAS are more frequent in black than Hispanic families, a finding certainly consistent with that reported here.

The findings reported here, particularly if confirmed by others, have potential public health and clinical practice implications. Advice to abstain from alcohol from the time of conception throughout the entire perinatal period has been disseminated through public and professional education efforts.[48] Broad media coverage has been obtained for public health advisories regarding alcohol and pregnancy. In a survey performed in 1981, 90% of the respondents knew that drinking during pregnancy might be harmful, but three-quarters of the respondents who did not recommend abstinence believed that an average of more than three drinks was safe for daily use.[49] This suggests that public education programs may not be as successful as desirable in modifying attitudes toward drinking during pregnancy. Supporting this contention are the findings of Streissguth and her associates who surveyed drinking patterns over a six-year period.[50] The results of that study indicated that although the proportion of women drinking during pregnancy had decreased, the proportion of women drinking at least two drinks per day was relatively constant. It is reasonable to suggest that the heaviest drinkers incur the greatest risk for alcohol-related birth defects, so these results may be taken to suggest that mass media–based public education efforts may not modify either attitudes or behavior sufficiently and, therefore, are not the answer to the problem of abusive drinking during pregnancy. An approach stressing focused prevention in the clinic or physician's office[51,52] might be more appropriate. The findings presented here would support the implementation of focused prevention efforts among disadvantaged black women, particularly those with previous children and/or low weight and weight gain. Even without confirmation of the current findings, such an approach would follow the medical dictum *primum non nocari*– first do no harm. Such a focus, with appropriate funding of such efforts, would not be expected to have adverse consequences.

The finding that there are very probably factors that modify the susceptibility of the fetus to alcohol teratogenicity has research, as well as clinical, implications. As previously noted, the current literature focusing on this issue is very limited. Further studies in animals, examining strain differences and susceptibility, and most importantly, mechanisms that may explain such differences, are certainly needed. Studies of the impact of alcohol on human pregnancy outcome, particularly those related to neurobehavioral development, would also do well to focus on factors that may additively or interactively modify the impact of alcohol on the offspring. Susceptibility warrants further intensive study.

SUMMARY

Typically, the rate of abusive drinking during pregnancy considerably exceeds the rates of fetal alcohol syndrome (FAS) and alcohol-related birth defects, suggesting that other factors may modify the impact of alcohol on the developing organism. Data in the literature supporting this susceptibility hypothesis are sparse. In this paper, two studies in different samples, using different analytic strategies to examine susceptibility to different adverse outcomes are presented. Among 176 pregnancies in which lowered birth weight for gestational age was detected as an effect attributable to frequent beer drinking, 27 infants weighted $< 2,700$ grams and 149 weighed more. Using discriminant analysis to contrast these groups, lowered birth weight for gestational age was associated with black race and lower maternal weight and weight gain. The effects of these factors were additive with that of persistent alcohol exposure; no interactions were detected, but pregnancies with risks in addition to alcohol were more likely to yield growth-retarded infants. In a second study, pregnancies resulting in 25 FAS cases were contrasted with 50 controls. A four-factor model accounted for nearly two-thirds of the explainable variance in the occurrence of FAS. Adjusted for frequency of maternal drinking, chronic alcohol problems and parity, there was a sevenfold increase in risk for FAS among black infants. The findings from both studies are consistent with the susceptibility hypothesis and have potentially important implications for public health and clinical approaches to prevention, as well as for future research.

ACKNOWLEDGMENT

We wish to acknowledge the helpful suggestions of Dr. Michael J. Lewis, Ph.D., of Howard University, Washington, DC, with respect to the analyses of potential bias.

REFERENCES

1. WARNER, R. H. & H. L. ROSETT. 1975. The effects of drinking on offspring: An historical survey of the American and British literature. J. Stud. Alcohol **36:** 1395–1420.
2. SOKOL, R. J. 1981. Alcohol and abnormal outcomes of pregnancy. Can. Med. Assoc. J. **125:** 143–148.
3. JONES, K. L., D. W. SMITH, C. N. ULLELAND & A. P. STREISSGUTH. 1973. Pattern of malformation in offspring of chronic alcoholic mothers. Lancet **1:** 1267–1271.

4. JONES, K. L. & D. W. SMITH. 1973. Recognition of the fetal alcohol syndrome in early infancy. Lancet **2:** 999–1001.
5. RENWICK, J. H. & R. L. ASKER. 1983. Ethanol-sensitive times for the human conception. Early Hum. Devel. **88:** 99–111.
6. ROSETT, H. L. 1980. A clinical perspective of the fetal alcohol syndrome. Alcohol: Clin. Exp. Res. **4:** 119–122.
7. CLARREN, S. K. & D. W. SMITH. 1978. The fetal alcohol syndrome: A review of the world literature. N. Engl. J. Med. **298:** 1063–1067.
8. OLEGARD, R., K. G. SABEL, M. ARONSSON, B. SANDIN, P. R. JOHANSSON, C. CARLSSON, M. KYLLERMAN, K. IRUSON & A. HRBEK. 1979. Effects on the child of alcohol abuse during pregnancy. Retrospective and prospective studies. Acta Paediatr. Scand. **S275:** 112–121.
9. SOKOL, R. J., S. I. MILLER, G. REED. 1980. Alcohol abuse during pregnancy: An epidemiologic study. Alcohol: Clin. Exp. Res. **4:** 135–145.
10. HARLAP, S. & P. H. SHIONO. 1980. Alcohol, smoking, and incidence of spontaneous abortions in the first and second trimester. Lancet **3:** 173–176.
11. KLINE, J., P. SHROUT, Z. STEIN, M. SUSSER & D. WARBURTON. 1980. Drinking during pregnancy and spontaneous abortion. Lancet **2:** 176–180.
12. SOKOL, R. J. 1980. Alcohol and spontaneous abortion. Lancet **2:**(8203): 1079.
13. KAMINSKI, M., C. RUMEAU-ROUQUETTE & D. SCHWARTZ. 1978. Alcohol consumption in pregnant women and the outcome of pregnancy. Alcohol: Clin. Exp. Res. **2:** 155–163.
14. BERKOWITZ, G. S. 1981. An epidemiologic study of preterm delivery. Am. J. Epidemiol. **110:** 355.
15. HAVLICEK, V., R. CHILDIAEVA & V. CHERNICK. 1977. EEG frequency spectrum characteristics of sleep rates in infants of alcoholic mothers. Neuropadiatrie **8:** 360–373.
16. STREISSGUTH, A. P., H. M. BARR, D. C. MARTIN & C. S. HERMAN. 1980. Effects of maternal alcohol, nicotine and caffeine use during pregnancy on infant mental and motor development at eight months. Alcohol: Clin. Exp. Res. **4:** 152–164.
17. GOLDEN, N. L., R. J. SOKOL, B. R. KUHNERT & S. BOTTOMS. 1982. Maternal alcohol use and infant development. Pediatrics **70:** 931–934.
18. STREISSGUTH, A. P., C. S. HERMAN & D. W. SMITH. 1978. Intelligence, behavior, and dysmorphogenesis in the fetal alcohol syndrome: A report on 20 clinical cases. J. Pediatr. **92:** 363–367.
19. SHAYWITZ, S. E., D. J. COHEN & B. A. SHAYWITZ. 1980. Behavior and learning deficits in children of normal intelligence born to alcoholic mothers. J. Pediatr. **96:** 978–982.
20. WEST, J. R., C. A. HODGES & A. C. BLACK. 1981. Prenatal exposure to ethanol alters the organization of hippocampal mossy fibers in rats. Science **211:** 957–959.
21. ABEL, E. L., ED. 1983. Marihuana, tobacco, and alcohol effects on reproduction. CRC Press, Boca Raton, FL.
22. RUSSELL, M. & L. R. BIGLER. 1979. Screening for alcohol-related problems in an outpatient obstetric-gynecologic clinic. Am. J. Obstet. Gynecol. **34:** 4–12.
23. SOKOL, R. J. 1983. The effects of alcohol on pregnancy outcome. Chapter V of the Fifth Special Report To The U.S. Congress On Alcohol And Health From The Secretary of Health And Human Services. December 1983. pp 69–82. Also published in Alcohol Health Res. World **9:** 27–31, 67–68, 1984.
24. PALMER, H. P., E. M. OUELLETTE, L. WARNER & S. R. LEICHTMAN. 1974. Congenital malformations in offspring of a chronic alcoholic mother. Pediatrics **53:** 490–494.
25. CHRISTOFFEL, K. K. & I. SALAFSKY. 1975. Fetal alcohol syndrome in dizygotic twins. J. Pediatr. **87:** 963–967.
26. SANTOLAYA, J. M., G. MARTINEZ, E. GOROSTIZA, J. AIZPIRI & M. HERNANDES. 1978. Alcoholismo fetal (fetal alcohol syndrome). Drogalchol **3:** 183–192.
27. CHERNOFF, G. F. 1977. The fetal alcohol syndrome in mice: An animal model. Teratology **15:** 223–230.
28. YANAI, J. & B. E. GINSBURG. 1977. A developmental study of ethanol effect on behavior and physical dependence in mice. Alcohol: Clin. Exp. Res. **1:** 325–333.

29. KUZMA, J. & R. J. SOKOL. 1982. Maternal drinking behavior and decreased intrauterine growth. Alcohol: Clin. Exp. Res. **6:** 396–402.

30. KUZMA, J. & R. J. SOKOL. 1983. Fetal alcohol exposure and lowered birth weight: Do maternal characteristics protect the fetus? (Abstract). Presented at the Annual Meeting of the Society of Perinatal Obstetricians, San Antonio, TX. January 1983. Proceedings, p. 278. Also presented as: Fetal alcohol exposure and lowered birth weight. How protective are maternal characteristics? (Abstract). Presented at the 111th Annual Meeting of the American Public Health Association, Dallas, TX, November 1983.

31. ROSETT, H. L. & L. WEINER. 1984. Alcohol and the Fetus: A Clinical Perspective. Oxford University Press, New York and Oxford.

32. LOCHRY, E. A., C. L. RANDALL, A. A. GOLDSMITH & P. B. SUTKER. 1982. Effects of acute alcohol exposure during selected days of gestation in C3H mice. Neurobehav. Toxicol. Teratol. **4:** 15–19.

33. SULIK, K. K., M. C. JOHNSTON & M. A. WEBB. 1981. Fetal alcohol syndrome: Embroyogenesis in a mouse model. Science **214:** 936–938.

34. MARTINEZ, F., J. HAPPA & F. ARIAS. 1985. Biochemical and morphologic effects of ethanol on fetuses from normally ovulating and superovulated mice. Am. J. Obstet. Gynecol. **151:** 428–33.

35. ABEL, E. L. 1984. Fetal Alcohol Syndrome and Fetal Alcohol Effects. Plenum Press, New York and London.

36. SOKOL, R. J., S. I. MILLER, S. DEBANNE, N. GOLDEN, G. COLLINS, J. KAPLAN & S. MARTIER. 1981. The Cleveland NIAAA prospective alcohol-in-pregnancy study: The first year. Neurobehav. Toxicol. Teratol. **3:** 203–209.

37. SOKOL, R. J., S. MARTIER & C. ERNHART. 1985. Identification of alcohol abuse in the prenatal clinic. Alcohol Res. Monogr, N. Chang & H. Chao, Eds.: 209–227. U.S. Dept. of Health & Human Services Publication No. 85-1258. Washington, D.C.

38. SELZER, M. L. 1971. The Michigan Alcoholism Screening Test: The quest for a new diagnostic instrument. Am. J. Psych. **127:** 89–94.

39. BALLARD, J. L., K. NOVAK & M. DRIVER. 1979. A simplified score for assessment of fetal maturation of newborn infants. J. Pediatr. **95:** 769.

40. DEBANNE, S., R. J. SOKOL, S. MARTIER, N. GOLDEN & S. I. MILLER. 1984. Are shortened palpebral fissures really an alcohol-related birth defect? Alcohol: Clin. Exp. Res. **8:** 87.

41. ERNHART, C., A. W. WOLF & P. L. LINN. 1985. Alcohol-related birth defects: Syndromal anomalies, intrauterine growth retardation, and neonatal behavioral assessment. Alcohol: Clin. Exp. Res. **9:** 447–453.

42. SOKOL, R. J., J. AGER, S. MARTIER, S. DEBANNE, C. ERNHART, N. GOLDEN & S. I. MILLER. 1984. Determinants of fetal alcohol syndrome. Presented at the annual meeting of the Fetal Alcohol Study Group of the Research Society on Alcoholism. Sante Fe, N.M. June 1984.

43. SOKOL, R. J., S. DEBANNE, J. AGER, S. MARTIER, C. ERNHART & L. CHIK. 1985. Data collection/analytic strategy dissociation: Maximizing the efficiency of a study of an alcohol-related birth effect. Am. J. Perinat. **2:**245–249.

44. CARROLL, J. B. 1961. The nature of the data, or how to choose a correlation coefficient. Psychometrika **26:** 347–372.

45. BOWMAN, R. S., L. I. STEIN & J. R. NEWTON. 1975. Measurement and interpretation of drinking behavior. Q. J. Stud. Alcohol **36:** 1154–1172.

46. SOKOL, R. J., P. JONES & L. CHIK. 1983. Statistical enhancement of intrauterine growth retardation clinical risk assessment. Presented at the Society for Gynecologic Investigation. Washington, DC. March 1983. Scientific Program and Abstracts. p. 254.

47. IOSUB, S., M. FUCHS, N. BINGOL, H. RICH, R. K. STONE, D. S. GROMISCH & E. WASSERMAN. 1985. Familial fetal alcohol syndrome: Incidence in blacks and hispanics. Alcohol: Clin. Exp. Res. **9:** 185.

48. UNITED STATES DEPARTMENT OF HEALTH AND HUMAN SERVICES. 1981. Surgeon General's advisory on alcohol and pregnancy. FDA Drug Bull. **11:** 9–10.

49. LITTLE, R. E., H. L. GRATHWOHL, A. P. STREISSGUTH & C. MCINTYRE. 1981. Public

awareness and knowledge about the risks of drinking during pregnancy in Multnomah County, Oregon. Am. J. Pub. Health **71:** 312–314.
50. STREISSGUTH, A. P., B. D. DARBY, H. M. BARR, J. R. SMITH & D. C. MARTIN. 1982. Comparison of drinking and smoking patterns during pregnancy over a six-year period. Alcohol: Clin. Exp. Res. **6:** 154.
51. ROSETT, H. L. & L. WEINER. 1981. Identifying and treating pregnant patients at risk from alcohol. Can. Med. Assoc. J. **125:** 149–158.
52. ROSETT, H. L., L. WEINER & K. C. EDELIN. 1981. Strategies for prevention of fetal alcohol effects. Obstet. Gynecol. **57:** 1–7.

DISCUSSION

MARIA MICHEJDA: I understand that you used certain arbitrary criteria for the studies of the intrauterine growth retardation of newborns and for the small-for-gestation babies that were below the 10th percentile. This study was carried out on a large population of pregnant mothers; however, I have difficulty in accepting the racial factors. Did you take into account the teenager's reproductive conditions and malnutrition or hypertension and alcoholism? Did you put factors like hypertension or high altitude, and age into the group or did you sort them out?

ROBERT SOKOL: With respect to altitude, Cleveland is 500 feet above sea level and there is only a slight difference in altitude in Loma Linda. While we did consider altitude, we did not code it into the analyses because all study subjects in both samples lived near sea level.

The other factors were taken into account statistically. In the Loma Linda sample, this was done entirely by adjustment in the multiple regression setting. In the Cleveland sample, we used two methods. The first was that the cases and controls in the sample were matched for seven factors, including, for example, gestational age at ascertainment, smoking, race, and drug use. Despite this matching strategy we were still able, as we have previously reported, to detect the impact of smoking on some outcomes, such as grown retardation. Therefore, although matching decreased the amount of attributable variance, we still put it into the multivariate model to adjust for any possible effect related to confounders as fully as possible. We only controlled for medical risk using an antenatal risk scale that we have worked on extensively during the last decade. We used this approach because if these factors had been considered singly, we would have had more factors than patients, leading to a major multiple comparison problem; therefore, all medical risk was expressed in a single number.

SUMNER YAFFE: There are some preliminary data from the intramural research program of NICHD that suggest a genetic susceptibility to the effects of alcohol. This would explain the marked interindividual response in the occurrence of the fetal alcohol syndrome. The findings have to do with the effects of alcohol in thiamine metabolism and the relationship to carbohydrate utilization.

What is the reversibility or potential for reversibility of the effects of alcohol on the fetus? What happens when a mother who is drinking and pregnant stops drinking?

ROBERT SOKOL: The potential reversibility of the prenatal alcohol effect is an important issue. A number of specific alcohol variables might be examined in this regard. You could try to break down drinking histories by trimester, for example, but this is difficult to do. The alcohol measures used in the studies I have presented here are averaged across pregnancy.

We will, however, be reporting next week at the Research Society on Alcoholism on some new work in a series of 500 patients in Detroit done on a March of Dimes grant. Overall, drinking behavior of pregnant women in Detroit was very similar to what we observed in Cleveland. Examined longitudinally, we found in the Detroit sample that alcohol intake typically decreases during pregnancy, confirming the early observation of Ruth Little in 1977. Further, this decrease begins early in the first trimester, but is highly variable on a case-by-case basis. The pattern reported by Nancy Day, who is now doing a study in Pittsburgh, is the same. Now, if you think about what is going on in the brain and with brain development, it really should be possible to protect the central nervous system if you can get women to stop drinking early in pregnancy, because major brain development and growth occurs in the third trimester. Henry Rosett and his colleagues in Boston have provided evidence that women who stop drinking before the third trimester have healthier, less damaged babies.

GENE FISCH: In doing the discriminate analysis did you look at the classification functions to see what the probability of assignment would be?

ROBERT SOKOL: This question relates to a concern about the stability. We examined both the "jackknifed" and "nonjackknifed" classifications from the discriminant analyses. The classifications obtained were very similar, suggesting that the rule by which we were classifying is reasonably stable. I do not have the printouts with the actual correct classification rates with me, but, as I remember, they were in the 80% range.

GENE FISCH: For both groups?

ROBERT SOKOL: The correct classification rate was a little bit better for the FAS group than the control group.

GENE FISCH: Did you consider using the tests from California with the Cleveland group to see how well it is assigned?

ROBERT SOKOL: We're talking about two different things here. The Loma Linda outcome studied was growth retardation, but there were no FAS children identified in the Loma Linda study, so that you could not come up with a comparable rule of thumb for this outcome for the Cleveland data. We did a split half, however, in Cleveland for growth retardation and applied the rule we got from the first half to the second half; it discriminated similarly. You will never discriminate well for growth retardation because alcohol makes up a very small part of the explainable variance (2–3%) in fetal growth. There are so many things that affect intrauterine growth so the rules for discriminating among them are lousy.

JOHN KIELY: I agree with your final policy recommendations, and I also agree with some of your recommendations for future research; however, I have a problem with the way you made judgments about interactions. As an epidemiologist, I don't see how you can make statements about susceptibility in an epidemiological design, unless you have tested for and found significant two-way interactions.

ROBERT SOKOL: That is a good observation, and we have thought about that very carefully. When I first presented this in a preliminary form at R.S.A., particularly the Loma Linda data, the conclusion we reached was that alcohol is an equal-opportunity risk. Nonetheless, we typically think that growth retardation is a threshold phenomenon, and it is reasonable to say that if you have a patient who has a number of other characteristics that accompany growth retardation you are indeed more likely to see growth retardation if the mother drank.

Now it is a matter of what you mean by susceptibility. As an "on-the-job" trained biostatistician myself, I would put it this way. If you have interaction, it is reasonable to conclude that you have a susceptibility factor. Interaction is a sufficient, but not necessary, condition to infer susceptibility. In the Loma Linda data on intrauterine growth, we have examined the sample and it may be possible to detect interactions using appropriate multivariate techniques. As I noted before, with alcohol and intrauterine growth we are dealing with small effect sizes, which complicate the detection of interactions. The Cleveland FAS logistic regression results are more clear-cut, which also suggests that it is reasonable. When you get a sevenfold difference with black race, adjusted for the other three factors in a multiplicative model, it is difficult not to call it susceptibility. The biostatistician in this study, Joel Ager, felt strongly, particularly on the FAS study, that there was no other reasonable conclusion.

RAUL RUDELLI: Did you look at placenta?

ROBERT SOKOL: We did not look at placentas in this series, as we had in a previous series. Overall we have studied more than 20,000 consecutive cases for alcohol effects. In the previous series of 12,000 cases, we examined every placenta for a subset of 2500 deliveries, all blindly and prospectively. We looked at about 200 placental factors and even taking into account Tukey's multiple comparision, we should have found ten differences in 400 factors. There was not one statistically significant difference. This was a very small sample—we had only about 50 placentas from pregnancies where there was clear heavy maternal drinking. We also looked at estrogen metabolism; a study published by Sheehan could detect no effect on placental estrogen metabolism. However, Shenker and some other investigators have found differences in placental amino-acid transport. There should be gross and microscopic findings, but we were not able to detect any. We don't have them in either of these two series.

RAUL RUDELLI: What about placental weight?

ROBERT SOKOL: Placental weight should be reduced; in the series we did, we did not have the statistical power to detect any decrease, however.

Publication Trends for Alcohol, Tobacco, and Narcotics in MEDLARS[a]

ERNEST ABEL

Research Institute on Alcoholism
Buffalo, New York

There are many environmental agents that increase the risk to which an embryo, fetus, or newborn baby is exposed. For instance, the effects of smoking have been of interest to perinatologists since about 1950. The effects of narcotics on newborns have also been of interest for some time. Dr. Haddad mentioned Dr. Rodier's work with mercury. In addition to mercury, many researchers like Dr. Rodier have been investigating lead and its actions on the developing brain. In addition to looking at antiepileptics and alcohol, Dr. Vorhees has done work with caffeine, aspirin, and food additives. Dr. Streissguth has worked with smoking and caffeine. Dr. Sokol has been studying the effects of tobacco, PCBs, and many other compounds and is the proverbial "walking encyclopedia" on pregnancy complications.

I recently completed a study in which I examined publication trends for three very common compounds—alcohol, tobacco, and narcotics. I counted the number of publications for these three compounds starting in 1973, which was the year the first publication on fetal alcohol syndrome appeared. Alcohol, as you might expect, outdistanced not only these drugs in terms of number of citations, but any drugs you would want to look at. The other two are among the next most commonly studied. The rate of growth was pretty much similar for alcohol and tobacco, but there were considerably more articles about alcohol than either tobacco or narcotics.

I next examined the number of articles relating to pregnancy and these three drugs. The rate of growth for alcohol and pregnancy publications was much greater than the rate of growth of publication for pregnancy and either smoking or narcotics. I was rather surprised to see this for narcotics because it was my impression that the work on narcotics was almost as active as that for alcohol.

Now I stress that the data base for this particular study was MEDLARS. That means that nearly all of the articles I looked at were in English. Articles in other languages are generally omitted from MEDLARS so the sample is somewhat biased, but it is not biased in terms of the entry of articles into that data base—no one selects particular articles on the basis of their subject matter.

I next looked at the percentage of growth for the various articles for these three agents. The narcotics literature has grown very little in percentage terms since 1973. The literature for alcohol and for tobacco, on the other hand, has been increasing, almost parallel to the total number of articles for alcohol or tobacco.

I don't think I'm off base to suggest that our scientific literature reflects the concerns of our society. The area of alcohol abuse is receiving more and more attention not only from behavioral teratologists and teratology itself, of course,

[a] This paper was supported in part by Grants #03496 from the National Institute on Drug Abuse and #AA05631 from the National Institute on Alcoholism and Alcohol Abuse.

but also from the DWI (driving while intoxicated) researchers and from researchers studying homicide and other forms of violent death. In each case, they have
found that there is a very high percentage of involvement of alcohol in homicides,
suicides, and traffic accidents.

Alcohol has been around for a very, very long time, but it is only recently that
it has been getting the attention it presently is receiving. I don't know of any
compound that is coming on board that is even going to come close to receiving
the attention that alcohol or tobacco or narcotics have received in the past, and I
think that the attention alcohol has received is going to remain strong for many,
many years to come.

Human Teratogens: How Can We Sort Them Out?[a]

THOMAS H. SHEPARD

Central Laboratory for Human Embryology
University of Washington
Seattle, Washington 98195

There are approximately five million chemicals to which our population has significant exposure. Of these, approximately 1,600 have been tested in animals for teratogenicity. About one-half of these agents have been shown to produce some form of teratogenicity. Teratogenicity can be defined as any significant postnatal change in function or form in an offspring after prenatal treatment. Of the 30 human teratogens thus far identified, twenty are chemicals or drugs (TABLE 1).

In addition to the increasing numbers of chemicals, the total amount of synthetic chemical production has increased at a tremendous rate since the 1940s. In FIGURE 1 this increase is illustrated in billions of pounds produced yearly. This estimate was made by the National Toxicology Program—a Federal group that oversees and coordinates much of the toxicology testing performed in the government.[1]

The reason there are so many more animal than human teratogens is explained by the way we are able to test animals. Experimental animals have close to ten fetuses per litter, so with ten animals we may accumulate 100 fetuses to examine in a rather short time and at a relatively low cost. A similar human "experiment" depends on serendipity and might be expected to cost a million dollars and consume three years' time. The dosage administered to animals can be much greater than most human exposures, and it is common to give enough to produce some signs of maternal toxicity. In fact, the problem of separating the effect of maternal toxicity from the effect of the teratogen is a real one. Good animal experiments are usually planned to show a concentration dose effect—human teratogens seldom have enough data to calculate dosage effects. Controls can be rigorously planned using sham treatment, pair feeding, and blind examinations. Finally, the animal experiment may be designed to expose the embryo/fetuses to the agent at their most sensitive time in development. In summary, the animal tests have the advantage of allowing for higher numbers, higher doses, and the use of scientific methods.

Most teratologists accept the principle that any agent can be shown to be teratogenic in an animal, providing enough is given at the right time. For instance, both sodium chloride and sucrose have been shown to produce animal teratogenicity. This concept is used so often by teratologists that some have found it useful to call it Karnofsky's principle. Some years ago the membership of the Teratology Society was polled on the question "should animal teratogens be excluded from the diet of humans?" (similar to the Delaney clause as it applies to carcinogens). Fortunately, the membership expressed their disapproval of any such exclusion and this stance was made public.[2]

[a] This work was supported in part by Grant No. HD00386 from the National Institutes of Health.

Synthetic Organic Chemical Production

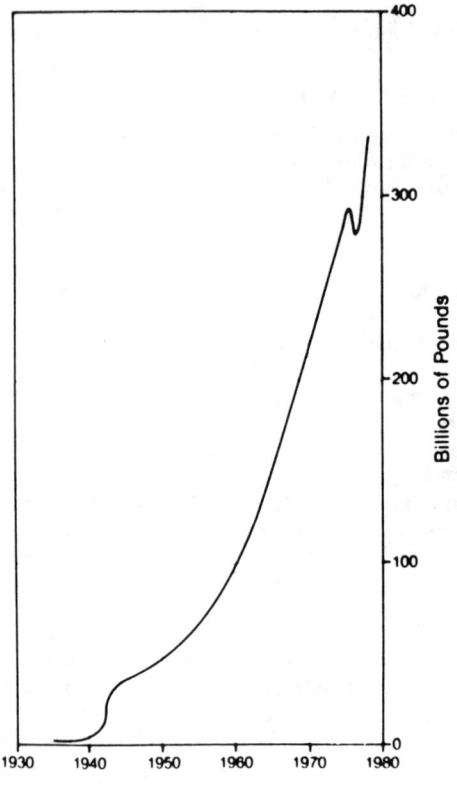

FIGURE 1. Billions of pounds of synthetic chemicals produced yearly. (Illustration courtesy of the National Toxicology Program).

NTP 81-6

Have the results of teratogenicity in animals helped to identify human teratogens? Until three or four years ago, the candid answer was no. Now at least three examples can be given. Valproic acid, a medication introduced for use in epilepsy, was noted by two pharmacologically oriented teratologists to have a high ratio of developmental to maternal toxicity.[3,4] This increased ratio as compared to other drugs used in treating epilepsy helped to draw the attention by monitoring groups who have noted an increase in neural tube defects among the offspring of users.[5] A second example is hyperthermia about which Marshall Edwards and his associates for some time have developed animal modes.[6] Shortly after David Smith visited Edward's laboratory, he initiated studies of hyperthermia in the human. A number of studies have supported evidence that hyperthermia is related to congenital malformations of the central nervous system and eye.[7-11] One study[12] did not show an association and sauna bathing does not appear to be associated with an increase in the rate of congenital defects.[13] The third example of animal teratogen work contributing to the identification of a human teratogen is that of Isotretenin, which is a form of retinoic acid. It was known from small-

TABLE 1. The Teratogen Problem

Chemicals	5,000,000
Tested in animals	1,600
Teratogenic in animals	800
Teratogenic in humans	20

animal experiments[14] and studies in primates[15,16] that retinoic acid had a specific predeliction for the craniofacial structures, and in particular, the external ear. Accordingly, after the medication was introduced for oral therapy of cystic acne, Franz Rosa at the Food and Drug Administration was alerted by adverse drug reactions consisting of craniofacial defects.[17] There have now been over 150 inadvertent pregnancies. Of the approximate 40 infants who went to term, one-half have significant craniofacial defects and a high proportion had additional defects of the heart and hind brain.[18]

The discovery that coumarin derivatives produce congenital warfarin embryopathy was made without any animal experimental data since this agent does not produce significant teratogenicity in experimental animals. What drew our attention was the association of a very rare human defect (nasal hypoplasia) with an equally rare form of treatment (anticoagulants). Several early reports of this association[19,20] went largely unnoticed until Judith Hall presented another similar case to a group of dysmorphologists who among them had a significant number of additional associations in the human. This example emphasizes the important role that an alert clinician or dysmorphologist can play in establishing causes of congenital malformations.

PRESCREEN TERATOGENICITY TESTING

A broad outline of the testing methods that seem to be evolving is shown in TABLE 2. In recent years a good deal of effort has gone into attempting to develop and validate a prescreening set of tests that should allow us to select the most likely candidates for further tests in animals and then perhaps in humans. Is it possible to identify a simple, inexpensive short-term test or tests for teratogenicity similar to the Ames test for mutagens? There are a number of teratologists who believe that this goal is not possible since the mechanisms for teratogenicity are much more diverse than those for mutagenesis. In any case a number of summary

TABLE 2. Levels of Teratology Testing

Level	Description	No. of Tests per Year
I. Prescreen (mass testing)	Short-term, simple inexpensive, mostly *in vitro*	10,000
II. Animal testing	Small-laboratory-animal tests	200
III. Relevancy testing	In-depth tests on mechanisms behavioral and primate, epidemiologic study in humans, risk and hazard assessment	20–50

TABLE 3. Features of Prescreen

Essential
  Few false positives
  Very few false negatives
  Inexpensive, readily available biologic unit
  Easily quantitated end point
  Reproducibility from laboratory to laboratory
Desirable
  Involvement of as many teratogenic mechanisms as possible
  Capability for testing water-soluble, water-insoluble, and gaseous agents
  Ability to determine concentration-effect curve
  Ability to incorporate biotransforming enzyme systems (such as hepatic monooxy-
    genase systems from different species—especially man)

articles on the subject have appeared.[21-24] Forty-seven candidate compounds for the validation of *in vitro* teratogenesis tests have been selected and proposed.[25]

Some essential and desirable features of a prescreen are listed in TABLE 3. The false positives and false negatives must be based ultimately on the results as seen in humans. The other features listed need little amplification except for perhaps the last item. In the common case where the chemical must be bioactivated in order to produce teratogenicity, there is hope that by the use of hepatic-derived systems from animals and from humans we will be able to understand and then predict the interspecies differences in sensitivity to toxicity.[26,27] Often a public health decision must be made in the absence of strong evidence for teratogenicity. A suspicion of teratogenicity is transformed into a drug warning, which then causes anguish to patients and doctors and sometimes even unnecessary therapeutic abortions. Unfortunately, there can be a long time between the observation of the defect and the testing and proof of the correct hypothesis regarding cause. Even in the case of the oral anticoagulants, which gave a very specific malformation syndrome, it took approximately six years from the initial case observation in 1968 until a sufficient number of cases were collected by 1974 to give a convincing positive association.

CRITERIA FOR HUMAN TERATOGENESIS

Criteria for deciding whether or not an agent is teratogenic should be developed. It may be possible to modify Koch's postulates and derive some useful criteria for human teratogenicity[28] (TABLE 4).

The fulfillment of the first two conditions is sufficient to define a teratogenic agent. The third may be considered desirable but not essential. Few animal teratogens satisfy all three of these criteria. Surprisingly, teratogenic agents in the human generally do fill all three criteria—for instance, rubella virus, radiation, and androgens that masculinize the female fetus. Thalidomide, although accepted universally as a teratogen, does not fit the third criterion because the compound in its unaltered state has not been demonstrated to affect the conceptus directly. Although this third criterion may seem unnecessary to many, a more complete knowledge of these important molecular mechanisms may generate the means for preventing malformations.

Further comments on the associations of Koch's postulates with human teratogenicity have been made by Brent.[29] He recommends that a dose:response

TABLE 4. Criteria for Human Teratogenesis

Koch's Postulates	Application to Teratology
1. A specific microorganism must be present in each case.	1. The agent must be present during the critical period of development.
2. A pure culture of the organism should produce a similar disease in the experimental animal.	2. The agent should produce congenital defects in an experimental animal. The defect rate should be statistically higher in the treated group than in the control animals receiving same vehicle or sham procedure.
3. Organisms from the experimental animal must be recovered and grown in pure culture.	3. Proof should be obtained that the agent in an unaltered state acts on the embryo/fetus either directly or indirectly through the placenta. In this area, biochemistry and organ culture are most often used instead of bacteriology.

relationship should exist between the agent and the incidence of malformation, that a nongenetic type of malformation should be increased in the exposed individuals, and that the malformation should be unique. These points are helpful but specificity and concentration response may not always be present or available from epidemiologic data. Stein *et al.*[30] have discussed the epidemiologic criteria for establishing human teratogenicity. Our understanding of confounding variables and statistical methods, especially the concept of numbers, needed to exclude a causative factor (power), have become much more sophisticated in recent years.

HUMAN TERATOGENS

In TABLE 5 are listed some teratogenic agents in humans along with some that are possible teratogens and another group that are unlikely teratogens.

Lithium has been shown to be associated with an approximate tenfold increase in the rate of congenital heart disease in the offspring of exposed mothers.[31-33] The fact that the animal teratogen dose level was reasonably close to that used in humans helped to focus attention on this important public health problem. Since none of the animal models are associated with heart disease, this is an example where findings in the human indicate a need for more animal studies. If an animal model that produces congenital heart defects can be established, the basic mechanism may be understood. This could allow us to interrupt or prevent human congenital heart defects.

In TABLE 4 there are listed five possible human teratogens and the reason for their inclusion will be described briefly in this paragraph. A serious problem is the possible association of heavy cigarette smoking with reduced intellectual function in the offspring. There are three fairly extensive studies, which after removal of confounding variables, support an association between heavy cigarette smoking and intellectual function.[34-36] One study could find no intellectual impairment in the offspring of heavy smokers.[37] Diazepam (Valium) has been associated with a low incidence of facial clefts.[38-40] Another study with more specific focus and a better control for confounding variables failed to show an association between

TABLE 5. Teratogenicity of Various Agents in Humans

Teratogenic Agents in Humans	Possible Teratogens	Not Likely Teratogenic
Radiation	?cigarette smoking	aspirin
therapeutic	?Valium	marihuana
radioiodine	?Zn deficiency	LSD
atomic weapons	?high vitamin A	anesthetics
Infections	?varicella virus	antinauseants (Bendectin)
rubella virus		oral contraceptives
cytomegalovirus		Rubella vaccine
herpes virus hominis? I and II		videodisplay terminals
toxoplasmosis		ultrasound
Venezuelan equine encephalitis virus		
Maternal metabolic imbalance		
endemic cretinism		
diabetes		
phenylketonuria		
virilizing tumors		
alcoholism		
hyperthermia		
Drugs and environmental chemicals		
androgenic hormones		
aminopterin and methylaminopterin		
cyclophosphamide		
busulfan		
thalidomide		
mercury, organic		
chlorobiphenyls		
diethylstilbestrol		
diphenylhydantoin and trimethadione		
coumarin anticoagulants		
?penicillamine		
valproic acid		
tetracyclines		
13-cis-retinoic acid (Accutane)		
lithium		

facial clefts and diazepam ingestion during pregnancy.[41] Since zinc deficiency can be produced easily in animals and a very short period during gestation produces defects,[42] it was of some interest that maternal zinc levels were found to be significantly lower in the anencephalics born to zinc-deficient Turkish mothers.[43] Since a form of vitamin A, retinoic acid, is teratogenic in humans (see above), and there are some case reports of defects associated with high vitamin A intake,[44–46] it would seem appropriate at this time to make a careful study along with appropriate controls of mothers who ingest as part of the food-fad craze more than 25,000 units of vitamin A per day. Postnatal behavioral[47,48] studies might be of special significance here. Varicella is on the list of possible teratogens in spite of a prospective study by Segal and First[49] of 135 mothers with varicella during pregnancy where no elevated defect rate was found. Since then there have been a number of case reports showing some specificity in terms of focal skin and muscle defects. These have been reviewed by Bai and John.[50]

In TABLE 5 the group of agents that are not likely to be human teratogens will not be discussed because lack of space, but these can be looked up in references of texts described below.

SOURCES OF INFORMATION ON TERATOGENIC AGENTS

An excellent general introduction to and discussion of teratology is available in *Wilson's Environment and Birth Defects*.[51] The annotated *Catalog of Teratogenic Agents*[52] includes over 1,500 agents that have been studied for teratogenesis in animals and man. Briggs *et al.*[53] summarize human teratogenic agents in their book, *Drugs During Pregnancy and Lactation. The Handbook of Teratology* edited by J. G. Wilson and F. C. Fraser[54] is a comprehensive set of four volumes summarizing the general and specific topics as well as techniques important in teratology. Heinonen *et al.*[55] reported on the outcome of more than 50,000 pregnancies in which drug exposures were known and listed their risk rates. Since these pregnancies occurred between 1959 and 1965, many of the newer drugs were not included.

If there is time for a careful literature search, the Environmental Teratology Information Center, Research Triangle Park, North Carolina or Oak Ridge, Tennessee maintains a fairly current listing of the references to reproductive effects of different agents. In the case of chemicals and drugs in addition to the name, they are listed by unique chemical numbers (CAS, Chemical Abstract Service Numbers).

LINKAGE OR ASSOCIATION BETWEEN ENVIRONMENTAL AGENTS AND HUMAN DISEASE OR DISABILITIES

In the past 50 years, disease states with short incubation have been fairly easily linked with causative factors. There is now a challenge to associate early exposures to long-term health events. For instance, women who smoke heavily appear to enter menopause at an earlier age than nonsmokers. Is this also true for women who have their only exposure to smoking during prenatal life? Are there any prenatal determinants to the onset and course of atherosclerosis and hypertension in old age? What percent of childhood malignancies are causally related to prenatal exposures? Besides diethylstilbestrol and possibly ionizing radiation, there is some evidence appearing that suggests that diphenylhydantoin may be a transplacental carcinogen.

TRIANGULATION FROM EXISTING DATA BASES

Triangulation is a navigational technique that allows a traveler to plot his location and subsequent course by determination of his position in respect to certain known points such as stars or coastal markers. Similarly, scientists are finding that the cause and prevention of certain disease states can be determined by linkage of three fixed, but expanding, data bases (FIG. 2). These data bases can be labeled as universal identifiers since they are world-wide. They consist of (1) individuals, (2) agents, and (3) disease syndromes. At birth, an individual should receive a unique health identification number to be used throughout life for all

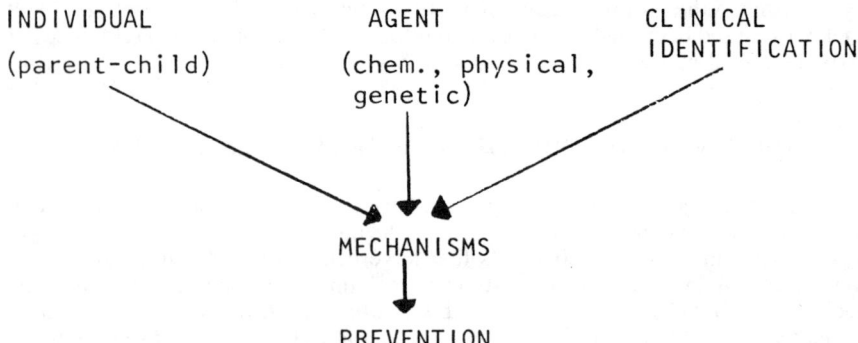

FIGURE 2. Diagram of a scheme whereby three data bases (individual, agents, and clinical syndromes) might be used to determine the mechanisms that cause congenital defects and their prevention.

health-related records. Such a system, must, of course, incorporate proper safe-guards for protection of privacy. An example of how this system could be useful would be the identification and recording of the health numbers of offspring of women exposed to workplace agents of unrealized long-term toxicity during pregnancy. There is a reasonable chance that some existing workplace toxicants will be identified later as agents that can cross the placenta and initiate changes that, over a long term, will produce cancer or other pathology in the offspring. Such a mutigenerational identification could foster public health measures to protect the unborn young of exposed pregnant women.

The second universal identifier system, causative agents, is partly in place since unique numbers are being assigned to chemicals (Chemical Abstracts Service Registry Numbers, CAS). There is a delay in the assignment of CAS numbers to many physical and infectious agents.

The area of congenital syndrome identification (syndromology) is providing a great deal of useful and important data. An example of the utility of specific syndrome identification is the fetal alcohol syndrome, which appears to account for a major part of our population in which a specific cause of mental retardation can be assigned. McKusick's *Catalog of Mendelian Inheritance in Man* (1983) represents a good updated annotated system for labeling human mutations and their associated syndromes.

Some disease states have been controlled without knowledge of their cause because of the discovery of effective treatment. An example of this is the surgical correction of congenital pyloric stenosis. Congenital rubella is partly controlled by immunizing young women, but we do not know the exact mechanism by which the fetus is in some cases protected by natural defenses. Since we lack the intimate knowledge of mechanisms of pathogenesis, these two congenital syndromes are still not completely preventable.

REFERENCES

1. SHEPARD, T. H. 1981. Letter to the editor, National Toxicology Program. Teratology **23:** 133.

2. STAPLES, R. L. 1974. Teratogens and the Delaney clause. Teratology **10:** 1.
3. NAU, H., D. RATING, S. KOCH, I. HAUSER & H. HELGE. 1981. Valproic acid and its metabolites: Placental transfer, neonatal pharmacokinetics, transfer via mothers milk, and clinical status in neonates of epileptic mothers. Pharmacol. Exp. Ther. **219:** 768–777.
4. FABRO, S., G. SCHULL & N. A. BROWN. 1982. The relative teratogenic index and teratogenic potency: Proposed components of developmental toxicity risk assessment. Teratogen. Carcinogen. Mutagen. **2:** 661–676.
5. ROBERT, E. & P. GUIBAND. 1982. Maternal valproic acid and congenital neural tube defects. Lancet **2:** 937.
6. EDWARDS, M. J. & R. A. WANNER. 1977. Extremes of temperature. *In* Handbook of Teratology, **1:** 421–444. J. G. Wilson & F. C. Fraser, Eds. Plenum Press. Chicago.
7. MILLER, P., D. W. SMITH & T. H. SHEPARD. 1978. Hyperthermia as one possible etiology of anencaphaly. Lancet **1:** 519–521.
8. LAYDE, P. M., L. D. EDMONDS & J. D. ERICKSON. 1980. Maternal fever and neural tube defects. 1980. Teratology **21:** 105–108.
9. FRASER, F. C. & J. SKELTON. 1978. Possible teratogenicity of maternal fever. Lancet **2:** 634.
10. SHIOTA, K. 1982. Neural tube defects and maternal hyperthermia in early pregnancy: Epidemiology in a human embryo population. Am. J. Med. Genet. **12:** 281–288.
11. KLEINBRECHT, J., H. MICHAELIS, J. MICHAELIS & S. KOLLER. 1979. Fever in pregnancy and congenital defects. Lancet **1:** 403.
12. CLARREN, S. K., D. W. SMITH, M. A. S. HARVEY, R. H. WARD & N. C. MYRIANTHO-POLOUS. 1979. Hyperthermia—a prospective evaluation of a possible teratogenic agent in man. J. Pediatr. **95:** 81–83.
13. SAXEN, L., P. C. HOLMBERG, M. NURMINEN & E. KOOSMA. 1982. Sauna and congenital defects. Teratology **25:** 309–313.
14. KOCHHAR, D. M. 1967. Teratogenic activity of retinoic acid. Acta Pathol. Microbiol. Scand. **70:** 398–404.
15. FANTEL, A. G., T. H. SHEPARD, L. L. NEWELL-MORRIS & B. C. MOFFETT. 1977. Teratogenic effects of retinoic acid in pigtail monkeys (macaca nemestrina). Teratology **15:** 65–72.
16. WILSON, J. G. 1971. Use of primates in teratological research and testing. *In* Malformations Congenitales Des Mammiferes. H. Tuchman-Duplessis, Ed. Masson. Paris. pp. 377–280.
17. ROSA, F. W. 1983. Teratogenicity of isotretinoin. Lancet **2:** 513.
18. LAMMER, E. J., D. T. CHEN, R. M. HOAR, N. D. AGNISH, P. J. BENKE, J. T. BRAUN, C. J. CURRY, P. M. FERNHOFF, A. W. GRIX, JR., I. T. LOTT, J. M. RICHARD & C. C. SUN. 1985. Retinoic acid embryopathy: A new human teratogen and a mechanistic hypothesis. New Engl. J. Med. **313:** 837–841.
19. KERBER, I. J., O. S. WARR & C. RICHARDSON. 1968. Pregnancy in a patient with prosthetic mitral valve. J. Am. Med. Assoc. **203:** 223–225.
20. HOLMES, B., H. W. MOSER, S. HALLDORSSON, C. MACK, S. S. PANT & B. MATZILEVICH. 1972. Mental retardation: An atlas of disease with associated physical abnormalities. Macmillan Co. New York. pp. 136–137.
21. WILSON, J. G. 1978. Review of *in vitro* systems with potential for use in teratogenicity screening. J. Environ. Pathol. Toxicol. **2:** 149–167.
22. KIMMEL, G. L., K. SMITH, D. M. KOCHHAR & R. M. PRATT. 1982. Overview of *in vitro* teratogenicity testing: Aspects of validation and application to screening. Teratogen. Carcinogen. Mutagen. **2:** 221–229.
23. SHEPARD, T. H., A. G. FANTEL, P. E. MIRKES, J. C. GREENAWAY, E. FAUSTMAN-WATTS & M. CAMPBELL. 1983. Teratology testing: I. Development and status of short-term prescreens. II. Biotransformation of teratogens as studied in whole-embryo culture. *In* Developments in Pharmacology. S. McLeod, A. Oakey & S. Spielberg, Eds. Alan R. Liss. New York, pp. 147–164.
24. BEAUDOIN, A. R. 1985. Validity of *in vitro* methods *In* New Approaches in Toxicity Testing and then Application to Risk Assessment. A. P. Li, Ed. Raven Press. New York. pp. 203–211.

25. SMITH, M. K., G. L. KIMMEL, D. M. KOCHHAR, T. H. SHEPARD, S. P. SPIELBERG & J. G. WILSON. 1983. A selection of candidate compounds for *in vitro* teratogenesis test validation. Teratogen. Carcinogen. Mutagen. **3:** 461–480.
26. FANTEL, A. G., J. C. GREENAWAY, M. R. JUCHAU & T. H. SHEPARD. 1979. Teratogenic bioactivation of cyclophosphamide *in vitro*. Life Sci. **25:** 67–72.
27. SHEPARD, T. H. 1983. Teratology: Information useful in advising health personnel and at-risk individuals. *In* Developments in Pharmacology. S. McLeod, A. Oakey & S. Spielberg, Eds. Alan R. Liss, Inc. New York. pp. 343–349.
28. SHEPARD, T. H. 1986. Catalog of Teratogenic Agents. Editions 1–5. Johns Hopkins Press. Baltimore, Maryland.
29. BRENT, R. L. 1978. Editor's note. Teratology **17:** 183.
30. STEIN, Z., J. KLINE & M. KHARRAZI. 1984. What is a teratogen? Epidemiological criteria. *In* Issues and Reviews in Teratology. **2:** 23–66. H. Kalter, Ed. Plenum Press. Chicago.
31. NORA, J. J., A. H. NORA & W. H. TOEWS. 1974. Lithium, Ebstein's anomaly, and other congenital heart defects. Lancet **2:** 594–595.
32. WEINSTEIN, M. R. 1979. Lithium Teratogenesis in Lithium, Controversies and Unresolved Issues. T. B. Cooper, S. Gershon, N. S. Kline & M. Schou, Eds. Excerpta Medica. Amsterdam. pp. 432–446.
33. KALLEN, B. & A. TANDBERG. 1983. Lithium and pregnancy: A cohort on manic-depressive women. Acta Psychiatr. Scand. **68:** 129–134.
34. DAVIE, R., N. BUTLER & H. GOLDSTEIN. 1972. From Birth to Seven: A report of the National Child Development Study. Longman and the National Children's Bureau. pp. 175–177. London.
35. DUNN, H. G., A. KARAA, S. INGRAM & C. M. HUNTER. 1977. Maternal cigarette smoking during pregnancy and the child's subsequent development II. Neurological and intellectual maturation to the age of six and one-half years. Can. J. Pub. Health **68:** 43–49.
36. NAEYE, R. L. 1978. Relationship of cigarette smoking to congenital anomalies and perinatal death. Am. J. Pathol. **90:** 289–293.
37. HARDY, J. B. & E. D. MELLITIS. 1962. Does maternal smoking have a long-term effect on the child? Lancet **2:** 1332–1336.
38. ARASKOG, D. 1975. Association between maternal intake of diazepam and oral clefts. Lancet **2:** 29.
39. SAXEN, I. 1975. Associations between oral clefts and drugs taken during pregnancy. Int. J. Epidemiol. **4:** 37–44.
40. SAFRA, M. J. & G. P. OAKLEY. 1975. An association of cleft lip with or without cleft palate and prenatal exposure to valium. Lancet **2:** 478–479.
41. ROSENBERG, L., A. A. MITCHELL, J. L. PARSELLS, H. PASHAYAN, C. LONIK & S. SHAPIRO. 1983. Lack of correlation of oral clefts to diazepam use during pregnancy. New Engl. J. Med **309:** 1282–1285.
42. HURLEY, L. S. & R. E. SHRADER. 1975. Abnormal development of preimplantation rat eggs after three days of maternal dietary zinc deficiency. Nature New Biol. **254:** 427–429.
43. CADVAR, A. O., A. ARCASOY, T. BAYCU & O. HIMMETOGLU. 1980. Zinc deficiency and anencephaly in Turkey. Teratology **22:** 14.
44. BERNHARDT, I. B., & D. J. DORSEY. 1974. Hypervitaminosis A and congenital renal anomalies in a human infant. Obstet. Gynecol. **43:** 750–755.
45. MOUNOUD, R. L., D. KLEIN & F. WEBER. 1975. A propos d'un case de syndrome de Goldenhar intoxication aique a la vitamin A chez la mere pendent la grossesse. J. Genet. Hum. **23:** 135–154.
46. VON LENNEP, E., N. E. KHAZEN, G. DEPIERREUX, J. J. AUREY, F. RODESCH & N. V. REGEMORTER. 1985. A case of partial sirenomelia and possible vitamin A teratogenesis. Prenatal Diag. **5:** 34–40.
47. HUTCHINGS, D. E. & J. GASTON. 1973. The effects of vitamin A excess administered during midfetal period on learning and development in rate offspring. Develop. Psychol. **7:** 225–233.

48. VORHEES, C. V., R. L. BRUNNER, R. E. BUTCHER, 1979. Psychotropic drugs as behavioral teratogens. Science **205:** 1220–1225.

49. SIEGEL, M. & H. T. FUERST. 1966. Low birth weight and maternal virus diseases. A prospective study of rubella, measles, mumps, chicken pox, and hepatitis. J. Am. Med. Assoc. **197:** 680–684.

50. BAI, P. V. A. & J. JOHN. 1979. Congenital skin ulcers following varicella in late pregnancy. J. Pediatr. **94:** 65–66.

51. WILSON, J. G. 1973. Environment and Birth Defects. Academic Press. New York.

52. SHEPARD, T. H. 1986. Catalog of Teratogenic Agents. 5th Edition. The Johns Hopkins University Press. Baltimore.

53. BRIGGS, G. G., T. W. BODENDORFER, R. K. FREEMAN & S. J. YAFFE. 1983. Drugs in Pregnancy and Lactation. A Reference Guide to Fetal and Neonatal Risk. Williams and Wilkins. Baltimore.

54. WILSON, J. G. & F. C. FRASER. 1977. Handbook of Teratology, Volume I–IV. Plenum Press. New York.

55. HEINONEN, O. P., D. SLONE & S. SHAPIRO. 1977. Birth Defects and Drugs in Pregnancy. PSG Inc. Littleton, MA.

Environmental Agents and Birth Outcomes

JEANNE M. STELLMAN

School of Public Health
Columbia University
600 West 168th Street
New York, New York 10032

Although the incidence of birth defects, spontaneous abortions, and stillbirths among the human population is quite high, with as many as 7% of all newborns born with birth defects in the United States annually, the cause of the majority of these defects remains unknown.

An important yet often neglected area of research, clinical investigation, and, indeed, clinical practice in the etiology of birth defects is the role of the environment, both at work and in the general environment, on fertility and birth outcome. Little training time, for example, is devoted to teaching the clinician about environmental agents or instructing him or her in taking an environmental exposure history. In the laboratory, etiologic research is generally carried out with experimental substances or conditions that are of intrinsic interest from a mechanistic point of view, rather than because they are substances that may be of industrial or ecologic significance.

It should therefore be of little surprise that only a small amount of data is available to the scientific, medical, or public health communities about what effects, if any, the almost 50,000 chemicals in use in daily commerce have.

In this paper, a brief overview will be presented of the ways in which toxic agents in the workplace or the general environment can interact with the human reproductive system and lead to adverse outcomes. This review is intended to complement the other presentations in this symposium that have focused on other kinds of agents and mechanisms that are related to birth defects.

POTENTIAL MECHANISMS OF REPRODUCTIVE FAILURE

Adverse reproductive outcome can result from exposure of either the male or the female to toxic substances or to toxic physical environmental factors. Sever and Hessol[1] have recently reviewed the toxic effects of occupational and environmental chemicals on the testes. Testicular effects appear to be a main mechanism by which male-mediated adverse outcomes can occur and are likely to result from a direct chemical effect on the testes themselves or possibly from an effect on the steroidogenic functions of the gonads.

A number of toxic agents capable of damaging the anatomical structure and function of the testes have been identified. Among the most well-known of these is the chemical dibromochloropropane, DBCP, which produced testicular atrophy, decreased fertility, and reduced sperm count in a group of workers engaged in its manufacture.[2] DBCP is a potent alkylating agent that appears to have a direct

116

effect on the germinal cells of the testes, a mechanism recently proposed by Mattison[3] as one of the major ways in which the male reproductive system can be affected.

It is interesting that DBCP was observed to affect the extent of production of sperm rather than the shape and motility. This is in contrast to another industrial and environmental chemical, inorganic lead, which has been found to increase the extent of teratospermia as well as the extent of sperm production.[4] Teratospermia appears to be another mechanism by which male-mediated adverse reproductive effects can occur; however, the relationship between abnormally shaped sperm and adverse outcome is not yet established. Wyrobek et al.[5] have demonstrated that there is a substantial correlation between mutagenic substances and substances that can induce teratospermia.

Hormonally based effects in males are not well understood, and it is often not clear whether abnormal hormonal levels are a cause of or a reflection of damage to the reproductive system. Sever and Hessol[1] discuss the perplexity of the evidence and suggest that hormonal analysis is an important component of the "armamentarium" of studies of male reproductive toxicity but that they must be used in conjunction with other evidence and data on testicular function.

In females, similar to the case of males, there can be a direct effect on the germ cells or there can be an alteration in hormonal function. Ionizing radiation is an example of a physical environmental hazard that can alter the female germ cells, although it appears in actuality that the male germ cells may be as much as ten times as vulnerable to ionizing radiation as the female.[6]

It also appears possible that the female menstrual cycle may be affected by estrogen-inducing chemicals, such as DDT, which can possibly lead to decreased fertility or to spontaneous abortions.[7]

Although the experimental and epidemiological evidence that is growing appears to demonstrate that both male and female gonadal function and germ cells can be affected by the environment in similar ways, it is clear that the female is at still greater risk because it is she who carries the developing conceptus; thus, the environmental considerations for females must consider the period of gestation.

The sensitivity of the embryo and the fetus to toxic environmental agents will, of course, vary with the stage of gestation. During the first trimester, there will be a particular vulnerability for structural defects since all organogenesis will occur during this period. The second and third trimesters are the period of functional development and it is during these stages that behavioral and functional effects are likely to occur.

Mutagenic agents and ionizing radiations can induce genetic damage, while other chemicals, such as thalidomide, may not affect the genetic integrity of the conceptus. A woman may be exposed to both kinds of agents in the workplace.

It is important to note that the relationship between toxic maternal exposures and toxic exposures to the conceptus is neither direct nor obvious. The placenta, while of course not a barrier, is also not completely permeable to exogenous agents. For many substances it does appear to exert a protective effect, lowering the extent of exposure of the conceptus to toxic agents in the maternal bloodstream. Ferm et al.[8] for example, observed that placental permeability of the hamster to cadmium varies with gestational age. At the early stages of pregnancy, the level of cadmium in the conceptus is considerably lower than the maternal level. The degree of permeability increases with gestational age but the level of cadmium in the fetus was never found to be that of the mother. This decreased permeability, however, cannot be considered adequate protection against toxic environmental agents.

TABLE 1. Examples of Reproductive Toxins and Industries Where They Can Be Found[a]

Exposure or Chemical	Reported Effect	Example of Industry
Documented Human Effects		
Anesthetic gases	spontaneous abortion low birth weight infertility (?)	medical dental veterinary
Busulfan	major malformations	medical pharmaceutical
Carbon disulfide	decreased libido impotence abnormal sperm morphology menstrual disorders spontaneous abortions prematurity	viscose rayon fumigant
DDT	prematurity	pesticide (banned USA)
DBCP (dibromochloropropane)	infertility due to azoospermia, oligospermia	pesticide
DES (diethylstilbesterol)	vaginal & cervical adenocarcinoma in female offspring; reproductive tract abnormalities in both sexes	pharmaceutical
EDB (ethylene dibromide)	decreased fertility in wives of workers	chemical fumigant
HCB (hexachlorobenzene)	stillbirth congenital malformations metabolic abnormalities	chemical
Ionizing radiation	lowered fertility increased childhood cancer fetal growth retardation disrupted gametogenesis chromosome aberrations	medical and dental nuclear industry
Kepone	decreased libido decreased sperm count and motility	chemical, pesticide
Laboratory reagents (benzene, xylene, ethers)	prolonged menstrual bleeding spontaneous abortions chromosome aberrations in offspring	laboratory workers
Lead and smelter emissions	sperm abnormalities menstrual disorders prematurity kidney tumor in children behavioral abnormalities chromosome aberrations	smelting battery leaded paint
Methotrexate	congenital malformations	medical pharmaceutical
Methylmercury	severe neurological defects	chemical wastes
PCBs (polychlorinated biphenyls)	menstrual disorders	capacitors in telephone/electrical equipment
Vinyl chloride	chromosome aberrations malformations (?)	PVC manufacturing and processing

118

(TABLE 1. *Continued*)

Exposure or Chemical	Reported Effect	Example of Industry
Human Evidence Developing		
Cadmium	low birth weight malformations in animals	chemical battery
Ethylene oxide	chromosomal abnormalities spermatic abnormalities birth defects in animals	health care food sterilization chemical
Glycol ethers	infertility	widely used solvents
Formaldehyde	menstrual disorders spontaneous abortions	widely used solvent, sterilant, preservative
Mercury (inorganic)	abnormal ovarian function	chemical industry thermometers/meters
Phthalates	anovulation spontaneous abortions	plastics
Selenium	spontaneous abortions congenital malformations	chemical semiconductors
Styrene	menstrual disorders chromosome aberrations	plastic workers

a Taken from Kipen and Stellman[12] and adapted from Barlow and Sullivan[9] and Council on Environmental Quality.[10]

SOURCES OF ENVIRONMENTAL EXPOSURE

There are a large number of workplaces and environmental situations in which there can be potential exposure to agents toxic to the reproductive system. These have been well summarized in several references. (See, for example, Sullivan and Barlow[9] and Council of Environmental Quality.[10] Some of these chemicals are given in TABLE 1, together with the industry in which they are likely to be found.

It can be seen that several large-scale employment sectors are well represented on this list. Health care, for example, is the largest "industry" in the United States today and several substances and conditions associated with adverse outcomes are known. These include cancer chemotherapeutic agents, infectious agents like hepatitis B, ionizing radiation, and ethylene oxide, a commonly used sterilizing agent for heat- and steam-sensitive objects.

COMMENTS

It is clear that there is a great need to understand more about the etiologic bases of the large number of adverse reproductive outcomes experienced by humans. It is also clear that there are many potential agents that can in whole, or in part, be responsible for some of these effects.

A major research need and priority is to learn more about the relationship between exposure to toxic substances and conditions and reproductive outcomes. An important advance would be the extension of training and research into such areas of practical importance. There is insufficient training in basic research or in medical sciences on the nature of toxic exposures or the likely ways in which they can be encountered. It is entirely conceivable that basic researchers and clinicians could continue mechanistic and clinical research while using substances of public health interest.

It is also of importance that environmental exposures be taken into consideration when studies of other agents, such as smoking and drinking, are carried out. It may well be that occupational and environmental toxins are contributing to the effects under observation. Without an adequate exposure and occupational history, which is infrequently taken, it may be that important effects are being overlooked. This is particularly important in settings such as fertility clinics and genetic counseling facilities.

The past several years have seen a growing recognition that male-mediated birth defects and other adverse outcomes are possible. It is necessary that research and clinical programs include both the male and female and do not simply focus on the female and her exposures.

A final point that is of interest is to note the difficulty with which reproductive effects can be observed in the human population, although they may in fact be occurring. The problem is one of the statistical frequency with which an effect can be expected to occur and the likelihood that one has of observing such an effect. This problem has been discussed in more depth elsewhere.[11] Neural tube defects, for example, occur with a normal frequency of between 0.01–1.0/100 births; thus careful scrutiny and record keeping would be essential before an effect could be observed should an environmental agent be suspected. Indeed, unless there were a sufficiently large number of either births (for a case-control study) or fertile parents (for a prospective study), all exposed to sufficiently high levels of the agent in question, there may be insufficient statistical power available to detect an effect.

Limitations of statistical power make the careful design and execution of experimental studies extremely important in the elucidation of the relationship between the environment and adverse reproductive outcome. They also make clear the need for improved record keeping and record linkages, such as the CDC's Birth Defects Monitoring Program in Atlanta.

REFERENCES

1. SEVER, LOWELL & NANCY A. HESSOL. 1985. Toxic effects of occupational and environmental chemicals on the testes. Endocr. Toxicol. pp. 211–248.
2. WHORTEN, D., R. M. KRAUSS, S. MARSHALL & T. H. MILBY. 1979. Infertility in male pesticide workers. Lancet 2: 1259–1261.
3. MATTISON, D. R. 1983. The mechanisms of action of reproductive toxins. Am. J. Indust. Med. 4: 65–79.
4. LANCRANJAN, I., H. I. POPESCU, O. GAVANESCU, I. KLEPSCH & M. SERBANESCU. 1975. Reproductive ability of workmen occupationally exposed to lead. Arch Environ. Health 30: 396–401.
5. WYROBEK, A. J., L. A. GORDON, J. G. BURKHART, M. W. FRANCIS, R. W. KAPP, JR., G. LETZ, H. V. MALLING, J. C. TOPHAM & D. M. WHORTON. 1983. An evaluation of human sperm as indicators of chemically induced alterations of spermatogenic function. Mutat. Res. 115: 74–148.
6. UNSCEAR. 1984. The Effects of Ionizing Radiation on Human Populations. United Nations. New York.
7. STELLMAN, JEANNE M. 1979. The Effects of Toxic Agents on Reproduction. Occup. Health Safety pp. 36–43.
8. FERM, V. H., D. P. HANLON & J. URBAN. 1969. The permeability of the hamster placenta to radioactive cadmium. J. Embryol. Exp. Morphol. 22: 107.

9. BARLOW, SUSAN & FRANK SULLIVAN. 1982. Reproductive Hazards of Industrial Chemicals. Academic Press. New York.
10. COUNCIL ON ENVIRONMENTAL QUALITY. 1981. Chemical Hazards to Human Reproduction. Washington, D.C.
11. STELLMAN, JEANNE M. 1985. Statistical and Practical Problems of Cohort Study Design: Occupational Hazards in the Health Care Industry. Natl. Cancer Inst. Monogr. **67:** 95–100.
12. KIPEN, HOWARD & JEANNE M. STELLMAN. 1985. Core Curriculum: Reproductive Hazards in the Workplace. Atlanta: American Association of Occupational Health Nurses. Atlanta, GA.

DISCUSSION

IRVIN EMANUEL: The recent CDC (Centers for Disease Control) Vietnam study was not about Agent Orange specifically, but it attempted to make different estimates of exposure. This is not the only study about Agent Orange—there is the Air Force "ranch-hand" study, and the results were essentially negative. You don't agree with that, Dr. Stellman. A small study of agricultural applicators in New Zealand also gave negative results. Other studies as well from Australia and this country have also yielded negative findings. When do you stop in the face of essentially negative results?

JEANNE STELLMAN: How many people were in the VA study? How many do they think were exposed? It was a very small number. The study was not an Agent Orange study; it was not designed to be.

IRVIN EMANUEL: I forget the numbers, but there were more than 100 who were probably exposed. But in the absence of someone coming up with a good biologic mechanism, it is time to lay some of these things to rest. Agent Orange has been accused of causing almost every imaginable type of health problem, and the only solid evidence is for chloracne.

JEANNE STELLMAN: There is disagreement here. This study of the Vietnam experience reported on very small numbers of people who, given the exposure opportunity index that was used to calculate it, actually were exposed to Agent Orange; consequently, the authors do not want it to be considered a study of Agent Orange.

I don't know what you mean by biologic mechanism. Are you saying that it is impossible to induce male-mediated birth defects?

IRVIN EMANUEL: Not at all, but it is difficult to imagine a nongenetic effect years after the exposure.

JEANNE STELLMAN: That is why I am suggesting that the Agent Orange question ought to be settled by going to Vietnam where the exposures were clearly much higher.

WILLIAM BAILEY: Let us turn to the issue of video display terminals. You believe that we should look seriously at VDTs. From all of the evidence I have seen, however, there do not seem to be any significant amounts of electromagnetic radiation from VDTs or any evidence that would associate either DC or AC electromagnetic fields with any health hazard.

JEANNE STELLMAN: I agree; it is extremely low frequency (ELF) that is in question. Only a couple of studies have been made on ELF and other studies have been made at extremely high level. The effects are not well documented so I am not disagreeing with you at all. There is a public health need to lay the issue to rest, and unlike the case with Agent Orange, which is fraught with methodologic difficulties, a prospective VDT study, not a case control or a retrospective study, would be comparatively easy to do given the huge numbers of women of child-bearing age that work at VDTs. Even a freshman epidemiology class could design such a study. Once you have such a widespread prevalence of an exposure in the population, doing the study is not very difficult. More than 15 million people work at VDTs now. By 1990 more than half of the workforce will spend more than half of their time in some type of VDT-related work, according to Arthur D. Little. A study of the effects of VDTs is not methodologically challenging, but it needs to be done to reassure the public.

THOMAS SHEPARD: I'd like to comment about the VDT screens. As most of you already know, the study of spontaneous abortion is a complete morass because normally we expect nearly 60% of all conceptions to end in spontaneous abortion. So the more you ask, the higher the rate is. Also, it is very difficult to research spontaneous abortions right now.

JEANNE STELLMAN: That is why you would have to enroll subjects prospectively.

Congenital Cytomegalovirus Infection: Prospects for Prevention[a]

ROBERT F. PASS, CECELIA HUTTO,[b] SERGIO STAGNO,
WILLIAM J. BRITT, AND CHARLES A. ALFORD

Departments of Pediatrics and Microbiology
University of Alabama at Birmingham
University Station
Birmingham, Alabama 35294

The remarkable success of the rubella immunization program in eliminating rubella epidemics and in nearly eliminating congenital rubella in the United States stands in sharp contrast to our problems with congenital cytomegalovirus (CMV) infection. Only around 20 cases of congenital rubella syndrome are reported to the CDC each year now, and the total number of cases in this country has been estimated at approximately 100 per year.[1] Congenital CMV infection is present in around 1% of live births or roughly 36,000 infants per year.[2] If only 10 to 20 percent of the infected infants suffer sequelae, then simple arithmetic reveals that between 3,600 and 7,200 children per year in the United States will have handicaps due to congenital CMV infection.[3] Since CMV infection is endemic, we can expect this number to be affected each year. Prevention of congenital CMV infection is clearly an important public health goal.

Recognition that CMV could be transmitted to the fetus when maternal infection occurred more than a year before conception pointed out the complexity of the host virus relationship and dampened enthusiasm for use of the live virus vaccine approach that had been so successful with rubella.[4] If maternal immunity from natural infection could not prevent transmission of virus to the fetus, there was little reason to think that immunity induced by an attenuated vaccine virus could. Furthermore, there was concern that vaccine virus might even reactivate and spread to the fetus. Although questions concerning the merits and potential risks of live CMV as an immunizing agent remain, knowledge gained in recent years suggests that an effective means of stimulating maternal immunity could result in fewer damaging congenital CMV infections.

PRIMARY VERSUS RECURRENT MATERNAL INFECTION

Congenital CMV infection has been considered due to *recurrent* maternal infection when there is preconceptional maternal immunity, in contrast with initial acquisition of virus during pregnancy, *primary* maternal infection. Evidence for transmission of CMV to the fetus with recurrent maternal infection includes reports of consecutive siblings with congenital infection, longitudinal studies of

[a] This work was supported in part by Grants 5 M01 RR32, HD 10699, and HD 17966 from the National Institutes of Health.

[b] Current address: Department of Pediatrics, University of Miami School of Medicine, Box 016960, Miami, FL 33101.

women with laboratory evidence of infection before pregnancy and observation of a high congenital infection rate in an African population with almost universal acquisition of CMV before puberty.[6-8] Review of published data on rates of maternal seropositivity and of congenital infection have shown that populations with high rates of immunity among women of childbearing age also have high rates of congenital infection. Conversely, populations in which a large proportion of women are susceptible to CMV tend to have lower rates of congenital CMV infection.[3] These data have been interpreted as further evidence for recurrent maternal CMV infection as a cause of congenital infection. When the great majority of women are seropositive before sexual maturity, it is likely that recurrent maternal infections account for most congenital infections. Since lower socioeconomic status is associated with acquisition of CMV earlier in life, congenital infection rates in the U.S. have been found to be higher among low-income groups (TABLE 1).[5,9] Although recurrent maternal infection probably accounts for more congenital CMV infections on a worldwide basis, for some populations, primary infection during pregnancy is more commonly the source of congenital infection. This appears to be true for middle- and upper-income groups in the U.S. with low rates of immunity among young women.[3,5,9] Of much greater significance than the number of congenital CMV infections attributable to primary or recurrent maternal infections is their relative importance as causes of damaging congenital CMV infections.

Primary infection can be proven by detecting maternal conversion from seronegative to seropositive during pregnancy or by detection of specific IgM antibody using an assay of proven sensitivity and specificity.[10] Recurrent maternal infection can be conclusively identified as the source of congenital infection only if the mother is seropositive or known to have shed virus before conception. Data based on mothers presumed to be in the recurrent group because they lacked IgM antibody to CMV early in pregnancy must be interpreted with caution. Obviously, categorizing maternal infections requires long-term, longitudinal study of large numbers of pregnant women and their offspring. Follow-up through subsequent pregnancies is usually required to identify congenital infections as due to a maternal recurrence, and longitudinal follow-up of infected infants for years is required to define sequelae. Preliminary results comparing infants born after primary maternal infection with those born after maternal recurrence are shown in TABLE 2.[5] Since the presence of symptoms at birth is a very strong predictor of CNS sequelae,[11] it appears that children born after primary infection are more likely to be disabled; the cord IgM and urine infectivity results suggest that they have more extensive infection. Ahlfors et al.[8] published results of a similar study that were

TABLE 1. Prevalence of Maternal Antibody to CMV and Incidence of Congenital Infection According to Socioeconomic Status (SES) in Two U.S. Cities

Location	Maternal Seropositivity	Congenital CMV
Houston, Texas[9]		
low SES	83%	1.2%
upper SES	50%	.6%
Birmingham, Alabama[5]		
low SES	82%	1.6%
upper SES	55%	.6%

TABLE 2. Severity of Congenital CMV Infections Due to Primary Maternal Infection Compared with Those Due to a Recurrent Maternal Infection[a]

Evidence of Infection in Newborns	Type of Maternal Infection	
	Primary	Recurrent
Symptomatic newborns	5/33 (15%)	0/27 (0)
Cord IgM $\geq$ 20 mg/dl	19/30 (63%)	4/19 (21%)
Urine infectivity[b]	3.4	2.3

[a] From Stagno, et al.[5]
[b] $\log_{10}$ TCID$_{50}$ of 0.2 ml urine.

interpreted as evidence for a significant role for recurrent maternal infection as a cause of damaging congenital infection; however, only one child with damage convincingly linked to CMV was born to a mother with established preconceptional immunity. Although maternal immunity does not prevent transmission of virus to the fetus, it does appear to decrease the likelihood of central nervous system damage; immunization of seronegative women could potentially be used to prevent damaging congenital CMV infection.

SOURCES OF PRIMARY MATERNAL INFECTION

Risk factors for primary maternal CMV infection during pregnancy may help identify populations that would benefit most from immunization and that could be used to assess vaccine efficacy. The sources for maternal CMV infections are not known and are likely to be multiple. Horizontal transmission of CMV has been demonstrated with blood products and organ transplantation.[12] Increased prevalence of antibody to CMV in sexually active groups and in pregnant women with histories of multiple sex partners or sexually transmitted diseases suggests that CMV can be acquired through sexual contact.[13–15] Another potentially important source of CMV for mothers is young children shedding CMV. Children who have congenital infection or acquire CMV early in life usually shed virus for years in saliva or urine.[12] Evidence for transmission of CMV from children to family members caring for them is summarized in TABLE 3. In addition to these data, Dworsky et al.[19] noted that seroconversions were twice as frequent during second pregnancies, and Stagno et al.[20] found a significant association between maternal seroconversion and the presence of a child in the home. In a longitudinal family study, Taber et al.[21] found that CMV infection in parents was often preceeded by

TABLE 3. Transmission of CMV from Young Children to Parents or Family Members

Study	Evidence for Child to Adult Transmission
Spector[16]	Mother seroconverted after exposure to newborn; strains similar by restriction enzymes.
Yeager[17]	Seroconversion within one year in 7/15 susceptible mothers of infants with hospital-nursery-acquired CMV.
Dworsky[18]	Seroconversion in aunt of congenitally infected newborn; CMV strains similar by restriction enzymes.

seroconversion in a young child. Although there is increasing evidence that children transmit CMV to adults in the home environment, it is not clear that hospital workers providing care for children are at increased risk, since results of published studies are not in agreement.[19,22-24] Day-care workers undoubtedly have frequent contact with children excreting CMV; whether they are at increased risk for acquisition of CMV has not yet been determined.[25]

In conclusion, there is evidence that maternal immunity prevents damaging fetal CMV infections, and therefore, vaccine to prevent maternal CMV infections during pregnancy could be of benefit to society. The observations indicating that CMV can be transmitted from child to mother or other caregiver in the home environment suggest that we could identify circumstances carrying high risk for acquisition of CMV. Such high-risk groups of women would be ideal for testing candidate vaccines. What type of vaccine is most likely to be safe and effective is speculative at this point in time. Further knowledge of the host response to specific viral polypeptides and glycoproteins after natural infection will be important in assessing the immune response generated by any candidate vaccine. It will also be important to devise a means for assessing vaccine efficacy in protecting recipients from natural infection.

REFERENCES

1. BART, K. J., W. A. ORENSTEIN, S. R. PREBLUD, *et al.* 1985. Pediatr. Infect. Dis. **4:** 14–21.
2. ALFORD, C. A., S. STAGNO & R. F. PASS. 1980. Natural history of perinatal cytomegaloviral infection. *In* Perinatal Infections. Ciba Foundation Symposium. Elsevier. New York. pp. 125–147.
3. STAGNO, S., R. F. PASS, M. E. DWORSKY & C. A. ALFORD. 1982. Clin. Obstet. Gynecol. **25:** 563–576.
4. STAGNO, S., D. W. REYNOLDS, E-S HUANG, *et al.* 1977. N. Engl. J. Med. **296:** 1254–1290.
5. STAGNO, S., R. F. PASS, M. E. DWORSKY, *et al.* 1982. N. Engl. J. Med. **306:** 945–949.
6. YEAGER, A. S., H. P. MARTIN & J. A. STEWART. 1977. Clin. Pediatr. **16:** 455–458.
7. SCHOPFER, K., E. LAUBER & U. KRECH. 1978. Arch. Dis. Child. **53:** 536–539.
8. AHLFORS, K., S. A. IVARSSON, S. HARRIS, *et al.* 1984. Scand. J. Infect. Dis. **16:** 129–137.
9. MONTGOMERY, J. R., E. O. MASON, A. P. WILLIAMSON, *et al.* 1980. South. Med. J. **73:** 590–593.
10. GRIFFITHS, P. D., S. STAGNO, R. F. PASS, *et al.* 1982. J. Infect. Dis. **145,** 647–653.
11. PASS, R. F., S. STAGNO, G. J. MYERS & C. A. ALFORD. 1980. Pediatrics **66:** 758–762.
12. PASS, R. F. 1985. J. Infect. Dis. In press.
13. JORDAN, M. D., W. E. ROUSSEAU, G. R. NOBLE, *et al.* 1973. N. Engl. J. Med. **288:** 932–934.
14. DREW, W. L., L. MINTZ, R. C. MINER, *et al.* 1981. J. Infect. Dis. **143:** 188–192.
15. Chandler, S. H., E. R. Alexander & K. K. Holmes. 1985. J. Infect Dis. In press.
16. SPECTOR, S. A. & D. H. SPECTOR. 1982. Pediatr. Infect. Dis. **1:** 405–409.
17. YEAGER, A. S. 1983. Pediatr. Infect. Dis. **2:** 295–297.
18. DWORSKY, M. E., A. LAKEMAN & S. STAGNO. 1984. Pediatr. Infect. Dis. **3:** 236–238.
19. DWORSKY, M. E., K. WELCH, G. CASSADY & S. STAGNO. 1983. N. Engl. J. Med. **309:** 950–953.
20. STAGNO, S., G. CLOUD, R. F. PASS, *et al.* 1984. J. Med. Virol. **13:** 347–353.
21. TABER, L. H., A. L. FRANK, M. D. YOW & A. BAGLEY. 1985. J. Infect. Dis. **151:** 948–952.
22. YEAGER, A. S. 1975. J. Clin. Microbiol. **2:** 448–452.
23. FRIEDMAN, H. M., M. R. LEWIS, D. M. MENEROFSKY & S. A. PLOTKIN. 1984. Pediatr. Infect. Dis. **3:** 233–235.

24. AHLFORS, K., S.-A. IVARSSON, T. JOHNSON & K. RENMARKER. 1981. Acta Paediatr. Scand. **70:** 819–823.
25. HUTTO, C., R. RICKS, M. GARVIE & R. F. PASS. 1985. Pediatr. Infect. Dis. **4:** 149–152.

DISCUSSION

THOMAS SHEPARD: Would you exclude a seronegative woman from a day-care center?

ROBERT PASS: A year ago if someone asked that question, I would have said, "No, I wouldn't; we just don't know what the risk is." Now, to be responsible, you would have to tell that person that there is a risk of getting CMV. The risk to caregivers seems to be greater than it would be if they didn't work in a day-care center. That risk can probably be controlled by careful handwashing and by taking care of children four years of age or older, not only because those children are less likely to shed virus, but also because the kind of interaction of the teacher is different with older children.

PART III.   MOLECULAR APPROACHES TO HUMAN GENETICS AND DEVELOPMENTAL DISABILITIES

Introduction

DAVID SOIFER

Department of Molecular Biology
Institute for Basic Research in Developmental Disabilities
New York State Office of Mental Retardation &
Developmental Disabilities
Staten Island, New York 10314

Dr. Zena Stein has spoken about the role of hypothyroidism in the epidemiology of mental retardation world-wide. In my own laboratory, and in several others here in the United States, in France, and in Israel, we have become interested in the molecular basis for thyroid hormone action in the developing nervous system. At least one action of thyroid hormone is to regulate the synthesis of a family of microtubule-associated proteins, the tau proteins, at critical times in neural differentiation. Dr. Stein indicated that in iodine-deficient populations, simple replacement therapy is not an adequate remedy to prevent or minimize mental retardation: Presumably, the maternal thyroid glands are no longer adequate to respond to increased dietary iodine. It would be useful to know the mechanism for thyroid regulation of tau protein gene expression—how do you turn these genes on and off? With that information it may be feasible to intervene in cases of maternal hypothyroidism, to prevent the consequent defficiencies in brain development.

These kinds of questions are part of the raison d'être for meetings such as this one. By bringing epidemiologists, clinicians, and other health-care providers together with cell and molecular biologists, it has been possible for many of us to become aware of problems of mutual interest that were hidden because of the gulfs that separate our disciplines. I hope that we are able to find more of these areas of common interest and new areas of discussion for all of us.

In addition to the kinds of research to be discussed in this section, research involving the use of genetic probes for mapping specific sites on chromosomes and the use of recombinant DNA and other kinds of molecular probes to study cell surface changes, there are a variety of other concerns where molecular biology is appropriate to the study of developmental disabilities. These range from examination of primary transcripts in individual cells and populations of cells to the study of translation products and the modification of proteins in various tissues of developmentally disabled individuals. It is a wide-open and exciting area of research for many investigators. What is most important is that we learn to speak each other's languages so that we can take maximum advantage of the flow of useful information from the research laboratory and apply the latest laboratory tools to dealing with clinical problems as they are redefined and clarified.

The Fragile X Syndrome[a]

W. TED BROWN, EDMUND C. JENKINS,
MICHAEL S. KRAWCZUN, KRYSTYNA WISNIEWSKI,
RAUL RUDELLI, IRA L. COHEN, GENE FISCH,
ENID WOLF-SCHEIN, CHARLES MIEZEJESKI,
AND CARL DOBKIN

Institute for Basic Research in Developmental Disabilities
New York State Office of Mental Retardation &
Developmental Disabilities
Staten Island, New York 10314

INTRODUCTION

The discovery of the Fragile X (fra(X)) syndrome represents a major advance in our understanding of mental retardation. Males who possess a marker on the X chromosome at position q27, referred to as the fra(X) chromosome, have a distinct inherited form of mental retardation known as the fra(X) syndrome. Over the last several years it has been recognized that this X-linked syndrome is the most common hereditary form of mental retardation. There are about 25% more mentally retarded males than females in the population. This excess was first noted by Penrose[1] in 1938 and has been since verified in numerous surveys. It has been suggested by several authors that the explanation for this excess may be due to the presence of undiagnosed forms of X-linked mental retardation.[2-4] There are more than 50 X-linked disorders that are associated with mental retardation, but their individual frequencies are rare.[5] The fra(X) syndrome is different in that it appears to be quite common. The prevalence of males affected with the fra(X) syndrome is estimated to be about 1 in 1350, whereas only about 1 in 2,033 females are so afflicted.[6] This disorder is second only to Down's syndrome (trisomy 21) as a genetic cause of mental retardation. With the advent of methods for detecting fra(X) in blood and amniotic fluid cells, population screening and prenatal diagnosis of the syndrome have now become possible. New methods for testing for genetic conditions using DNA probes and restriction fragment length polymorphisms (RFLPs) are being applied to fra(X). They hold the promise of improved testing and increased basic understanding of the syndrome. In the following, we review the clinical features and current studies of the fra(X) syndrome.

HISTORY OF FRAGILE X

The first description of a family in which there was a marker X chromosome in association with X-linked mental retardation was given by Lubs[7] in 1969. In this family there were four male members identified who had the marker and were retarded, whereas four female carriers who had one normal and one marker

[a] This work was supported in part by Grant MH 38201 from the National Institutes of Health.

129

chromosome were normal. The marker was described as "a small satellite separated from the main long arm of the chromosome by a constriction" and was similar to that shown in FIGURE 1. Although a marker X chromosome was also reported by Escalante[8] in 1971 in a family with three retarded brothers, the importance of Lubs' initial observation was not realized until 1976. At that time, two reports[9,10] demonstrated the marker in 25 affected males and one affected female among ten families. In retrospect, this long delay in recognizing fra(X) appears to have been due to the introduction in the early 1970s of improved methods for growing cells in the laboratory.

In 1977, Sutherland[11,12] discovered that a cell culture medium deficient in folic acid was necessary to detect the marker chromosome. He observed that the fragile site on the marker X chromosome could only be seen in a folate-deficient medium, a medium that was no longer in general use. This type of medium was in general use in the late 1960s and was the kind that had been used by Lubs. After the discovery of the need to use folate-deficient media, additional families with apparent X-linked retardation were tested and the fra(X) chromosome was soon identified in many.

In the late 1970s and early 1980s, several studies reported that about 40% of families with two or more retarded brothers would show the fra(X) chromosome when tested.[13-23] Accordingly, if 40% of the 25% excess of male mental retardation is due to the fra(X) syndrome, then about 10% of retarded males may have the fra(X) chromosome. Cytogenetic surveys[24-29] of institutionalized males have given prevalence estimates ranging from about 2% to 9%. Bloomquist et al.[30,31] conducted a prevalence study of all retarded males in a rural Swedish population and found that about 1 in 1,500 males had the fra(X) syndrome. A large epidemiologic study by Webb et al.[6] in England found a prevalence of 1 in 1,350. Thus, fra(X) is very common and most cases probably have not been tested yet and recognized.

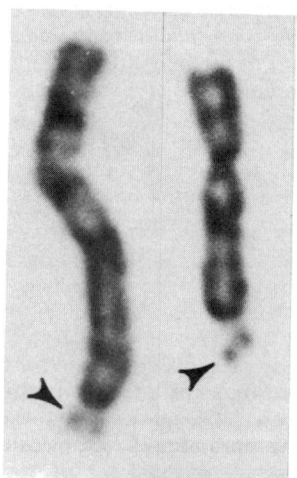

FIGURE 1. Fragile X chromosomes. A stretching or breaking is observed in band q27.3.

CLINICAL FEATURES OF AFFECTED MALES

Compared with persons having many other genetic and chromosomal syndromes, males affected with fra(X) are frequently fairly normal in physical appearance. This probably accounts for the fact that the syndrome was not recognized as a distinct entity until after the fra(X) chromosome was discovered. Males with the fra(X) syndrome may show subtle facial features. FIGURE 2 shows two male brothers with the fra(X) syndrome. Physical features variably present include a large head circumference, a prominent forehead, a narrow mid-face diameter, narrow inter-eye distances, a high arched palate, some facial asymmetry, and prominent and long ears. Hyperextensible joints, mitral valve prolapse, scoliosis, somewhat prominent thumbs, and flat feet may also be present. However, a distinctive feature of most adult fra(X) males is enlarged testicular volume, known as macroorchidism. Normal adult males have a mean testicular volume of about 17 ml with an upper limit of about 25 ml, whereas adult fra(X) males usually have a volume of more than 25 ml and frequently 50 to 100 ml.

Before the era of fra(X) testing, several families were identified that showed the inheritance of X-linked mental retardation associated with macroorchidism.[32,33] When these families were later retested, most were found to have the fra(X) chromosome. Within a group of institutionalized males who showed macroorchidism, we found that 80% were positive for fra(X).[34] Although others have

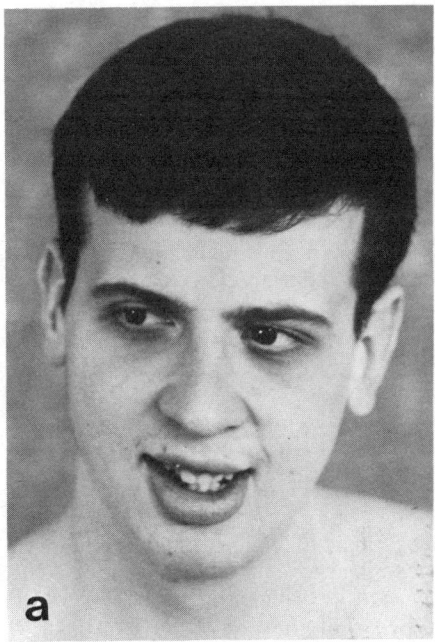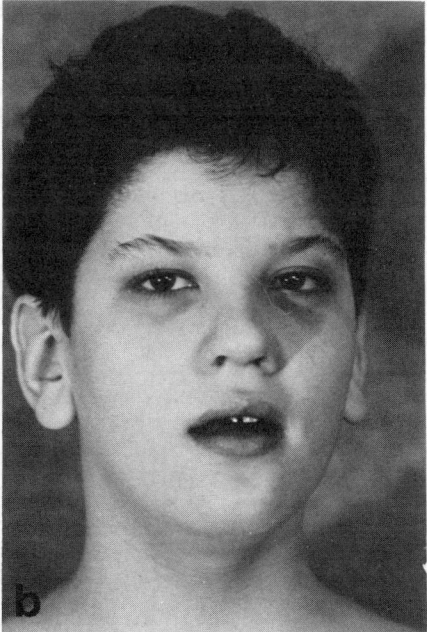

FIGURE 2. Two brothers with the fragile X chromosome. **(a)** 17-year-old brother; **(b)** 12-year old brother.

reported somewhat lower frequencies of mentally retarded males with ma-croorchidism to also have fra(X), macroorchidism appears to be the single most useful physical indicator of males who are likely to have fra(X).[35,36]

Fra(X) has been reported in most ethnic groups,[5] but an increased frequency in a specific ethnic group has not been established so far. However, among the population we have tested for fra(X), we have found a higher than expected number of Jewish persons positive for fra(X).[37] Of 44 probands who were diag-nosed as fra(X)-positive, 16 were of Jewish descent (39%). When fra(X)-negative clients are used as a case control, the proportion (17/57) of fra(X)-positive Jewish clients was significantly greater than the proportion (27/177) of fra(X)-positive non-Jewish clients ($\chi^2 = 4.89$; $p < 0.03$). Sources of possible ascertainment bias were eliminated as factors. Specifically, these were: the population of Jewish people residing in the catchment area (16%), the population of developmentally disabled Jewish people in the catchment area served by New York State (17%), the proportion of Jewish people admitted to the clinic for diagnosis (17%), and the proportion of Jewish people assessed by our division within the clinic (24%). This suggests there may be an increased risk of fra(X) among Jewish people.

COGNITIVE AND NEUROLOGICAL ABNORMALITIES

In our experience the IQ of the majority of males is in the moderate to severely retarded range. Some persons are profoundly retarded and some are in the mild range. A few males with learning disabilities and IQs above 70 have also been reported by Hagerman et al.[38] Neurologic evaluations suggest that a static central nervous system process is present without focal lateralizing signs in the majority of males.[39] Impairment of fine motor coordination and the presence of soft neuro-logic findings are common. Fra(X) males frequently show speech delay and a paucity of language. In addition, repetitive speech patterns may be present. In the higher functioning fra(X) males, speech abnormalities described as cluttering have been reported.[40] Hyperactivity among fra(X) males also appears common. Ap-proximately one-fourth of fra(X) males have a history of seizures which are usu-ally transient or are well controlled with anticonvulsant medication. The elec-troencephalogic and neuroradiologic findings are generally nonspecific.[39]

Limited neuropathologic information is available. We have studied one fra(X) male who died at age sixty-two.[41] Immaturity of dendritic spines and abnormali-ties in synaptic contacts were identified as the main differences in brain anatomy in this patient.

THE ASSOCIATION OF FRA(X) WITH AUTISM

Autistic persons are usually male and a 4 : 1 male to female ratio is common. In several early studies, fra(X) males were described as autistic or psychotic. One autistic fra(X) male was recorded by Turner et al.[13]; two identical male twins with autism were observed by Proops and Webb[42]; and Mattei et al. reported two cases with "infantile psychosis."[43] In these early studies, the significance of the associ-ation between fra(X) and autism was not recognized.

In 1982 we identified five autistic males with fra(X) and emphasized the proba-ble significant association of fra(X) with autism.[44,45] Our finding of an association of fra(X) with autism has now also been confirmed by many others. Meryash *et al.*[46] reported a case of a fra(X) male with autism. Levitas *et al.*[47] found that 6 of 10 males with fra(X) had autism and described the psychological and behavioral profiles seen in these males. Brondum-Nielsen[48] reported that 20 of 27 fra(X) males showed psychiatric and behavioral problems in addition to mental retardation and that one had been diagnosed as autistic at an early age.

We have now screened 183 autistic males and found 24 (13.1%) positive for fra(X).[49] The overall frequency of finding fra(X)-positive males among 12 surveys of autistic males was about 12%.[49] We identified a total of 150 fra(X)-positive males in 55 families; twenty-four of these fra(X) males (17.3%) had a diagnosis of autism. In addition, some of the older males had been hyperactive, having had severe behavioral disturbances as children, and had been institutionalized, but had not been diagnosed as autistic. By current diagnostic criteria,[50] they might have also been categorized as autistic. We recommend that all autistic males be tested for fra(X) since it appears to be the single most common specific biomedical cause of autism.

THE CARRIER FEMALE

About one-half of female carriers show fra(X) expression. However, many normal carriers are cytogenetically negative. Approximately one-third of female carriers are mentally impaired and of these about 80% show fra(X) expression. Turner *et al.*[15] found that in a school for the mildly retarded, 6 of 78 girls with IQs in the range of 55–75 and normal appearance also possessed the fra(X) chromosome. Among their female relatives, 18 were found to be carriers: of these, six showed some degree of mental impairment. In their epidemologic study, Webb *et al.*[6] found the prevalence of mentally impaired females to be 1 in 2,033. Although the level of mental impairment in females is generally in the mild range, there are reports of several families that have severely retarded females. Webb *et al.*[51] also noted three female carriers of fra(X) in one family whose level of retardation ranged from the moderate to the mildly retarded level, along with two other female carriers who had borderline mental function. We have also identified several fra(X)-positive carriers who showed severe to profound retardation. Thus, in addition to its association with mental retardation among males, a significant frequency of female mental retardation appears to be due to the fra(X) chromosome.

The fact that some carrier females are affected appears consistent with random inactivation of one of the two X chromosomes. There appears to be a correspondence between late replication of the normal X chromosome (reflecting X inactivation) in carrier females and the presence of mental impairment.[5] A report of two female identical twin carriers indicated that one was severely retarded whereas the other was of high intelligence. If the active chromosome was the fragile one in the majority of the affected sister's brain cells, then brain function may have been affected, giving a result similar to that of fra(X) males.

We have observed a characteristic profile of cognitive defect in some carriers.[53] Our study of seven females revealed an increased verbal and decreased performance score on the Wechsler IQ test, a relative lower subtest score on arithmetic, digit span, block design, and object assembly.[54] A similar profile in eight carriers has also been reported by Kemper *et al.*[55] Thus, the fra(X) mutation may predispose affected females to a characteristic cognitive profile deficit.

GENETICS OF INHERITANCE AND NONPENETRANCE IN THE MALE

X-linked inheritance implies that the mother randomly passes on one of her two X chromosomes to each of her children. The father passes on his X chromosome to his daughter and his Y chromosome to his sons. If a mother is a carrier of fra(X), then half of her children will receive the fra(X) and half will not. Thus, half of her daughters on average would be carriers and half of her sons would be expected to have the fra(X) syndrome.

Within families where there is a pattern of X-linked mental retardation, the strength of association between the presence of the fra(X) marker and mental retardation is very significant.[56] Thus, when a male fetus or infant is detected that is found to be fra(X)-positive, the risk is high that this individual will be mentally retarded. However, many families have now been identified where apparently normal males were transmitting the faulty chromosome.[57-66] These males were negative on testing and were of normal intelligence, but passed a fra(X) chromosome to their daughters. Their daughters were nearly always fra(X)-negative, but frequently had retarded sons. Hence, these males, the grandfathers, are referred to as nonpenetrant since they showed no expression of either the fra(X) chromosome or mental impairment. Two such pedigrees that we have studied[67] are shown in FIGURES 3 and 4.

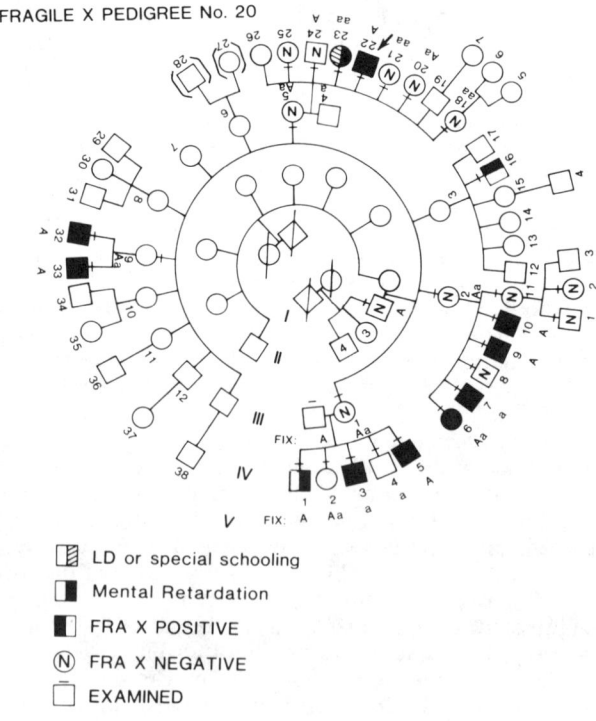

FRAGILE X PEDIGREE No. 20

◨ LD or special schooling

■ Mental Retardation

■ FRA X POSITIVE

Ⓝ FRA X NEGATIVE

□ EXAMINED

FIGURE 3. Pedigree of a fragile X family with 12 mentally impaired first cousins. The grandfather was an apparently normal transmitting male. This family shows loose linkage with the factor IX probe.

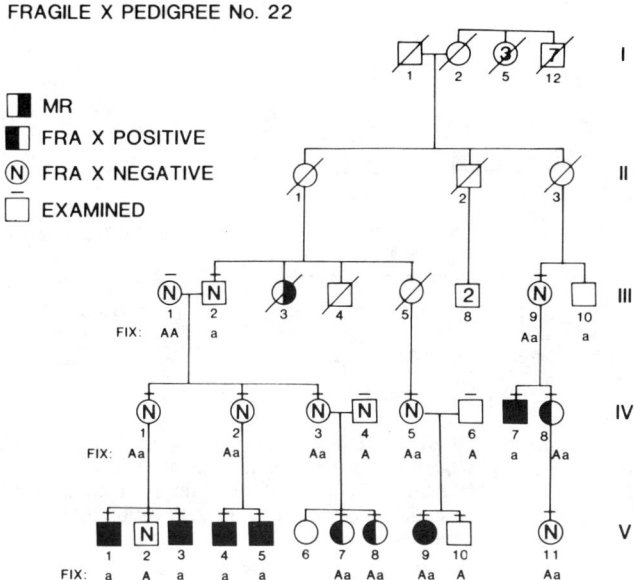

FIGURE 4. Pedigree of a fragile X family showing a transmitting male (III-2). This family shows tight linkage between factor IX and fragile X.

In 1984 and 1985, Sherman et al.[68,69] analyzed a large number of fra(X) pedigrees and observed unusual features about the inheritance of the fra(X) syndrome. They found a 20% deficit of fra(X)-positive males, which indicated that penetrance was about 80%. Instead of the expected 50 : 50 ratio of affected to normal brothers, they found a ratio of 40 : 60, respectively. The sibships of NP males showed a penetrance of only 18% among their brothers. That is to say, 18% of the expected 50% or 9% of their brothers were mentally impaired. Thus, the sons of the mothers of NP males are much less likely to be affected than are the sons of their daughters. This observation has come to be referred to as the "Sherman Paradox," as illustrated in FIGURE 5.[5] It suggests that very unusual genetic inheritance patterns exist in fra(X).

The mental status of the carrier mother appeared related to the expression in her offspring,[68,69] as illustrated in FIGURE 6. Instead of the 40 : 60 ratio, if the mother herself was mentally impaired, then the expected 50 : 50 ratio was indeed found; i.e., half of her sons were affected. Likewise, about twice as many of her daughters were affected, i.e., 37% instead of 18%. The fraction of sibships with one affected male was not increased above what would be expected if all mothers of these sons were carriers.[68] This suggested that all new mutations would have to occur at the time of the mother's conception or in a previous generation.

NEW MUTATION IN FRAGILE X

Fra(X)-positive males have fathered children, but because they are retarded they usually do not marry or reproduce. As a result, their genes are lost from the

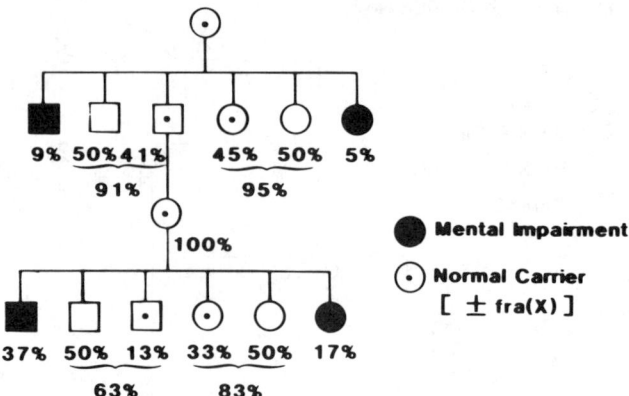

FIGURE 5. The Sherman Paradox ("The sons of mothers of transmitting [nonpenetrant] males are much less likely to be affected than are the sons of their daughters."). Thus, this schematic pedigree illustrates the low likelihood of the siblings of a nonpenetrant male being affected in comparison to his grandchildren. Such different ratios of expression have not been reported in any other genetic disease.

gene pool. A state of genetic equilibrium for the prevalence condition in the population can be assumed if the number of cases neither increases nor decreases. For an equilibrium to be maintained, the new mutation rate must equal the rate of loss of their genes from the population. On the basis of surveys of a large number of fra(X) families by Sherman *et al.*,[68,69] it does not appear that sporadic cases of fra(X) males occur. This implies that no fra(X) males are new mutations, but rather that all have mothers who are carriers. This study suggested that all new

Overall Expression of Mental Impairment
In Fragile X Pedigres

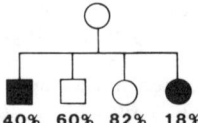

Offspring of Affected Mothers:

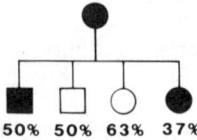

FIGURE 6. The offspring of affected fra(X) mothers have a higher rate of expression than do the offspring of unaffected carriers. A 20% deficit of affected males exists in these pedigrees.

mutations appear to be derived from sperm and none from eggs. Sherman et al. calculated an extraordinarily high new mutation rate for fra(X). They estimated the new mutation rate for sperm to be 1 in 1,389. This new mutation rate is higher by nearly 10-fold than that known for any other known genetic disease. This suggests there may be something very extraordinary about the nature of the fra(X) mutation to cause it to occur at such a high rate and lead to such a high prevalence of the condition in the population.

PRENATAL DIAGNOSIS

The prenatal detection of the fra(X) chromosome is an important step in the prevention of the syndrome. Obtained from amniotic fluid, fetal amniocytes are similar to fibroblasts in growth properties. After Sutherland[11,12] discovered that folic acid and thymidine concentrations need to be reduced in culture medium in order to induce the fra(X) chromosome, Glover[70] and Tommerup et al.[71] reported that 5-fluorodeoxyuridine (FUdR), a specific inhibitor of thymidylate synthetase, was effective in inducing fragile site expression in lymphocytes and also fibroblasts. By means of the FUdR method for fra (X) detection in fibroblasts, it is now possible to detect the fra(X) site in amniocytes.

We first demonstrated the feasibility of prenatal identification of fra(X) in amniotic fluid cells in 1981,[72] and these findings were later confirmed in a prospective diagnosis.[73] We have now identified three male and five female fetuses positive for fra(X) after prenatal diagnosis in 25 studies.[74,75] Two male cases were examined after elective termination; they were found to have macroorchidism, demonstrating that the syndrome is expressed prenatally (FIG. 7).[76] Neuropathologic findings suggested brain immaturity as well. On a world-wide basis there have been about 150 trials and 31 fetuses have been shown to have fra(X).[5] Although still experimental, the use of amniotic fluid cell cultures and chorionic villus sampling combined with DNA marker has allowed the prenatal diagnosis of fra(X) to become a reliable test. This allows for primary prevention of the mental retardation associated with the fra(X) syndrome.

ATTEMPTS AT FOLIC ACID THERAPY

To date there has not been any deficiency of folic acid or related metabolites demonstrated in any fra(X) subjects. However, since fragile sites can be induced by culturing in folate-deficient medium, it was suggested that administration of excess folic acid to fra(X) subjects be tested to determine if it might affect behavior in these subjects. Initially, Lejeune et al.[77] used folic acid to treat a 10-year-old fra(X)-positive male, who was described as mentally retarded, agitated, and aggressive, and "spectacular amelioration of his behavioral problems" was noted. After 8 days, the treatment was stopped for 15 days and his poor behavior returned. Lejeune[78] subsequently reported treating 15 additional males with folic acid therapy and noted behavioral improvements and decreased fra(X) frequencies.

However, Lejeune's studies were uncontrolled and a placebo effect was likely to be present. Therefore, we undertook a therapeutic trial of folic acid in a controlled double-blind study.[79,80] Two fra(X) brothers (FIG. 2) (one was 19 and the other 12 years of age) were given 8 days of placebo followed by 8 days of folic acid

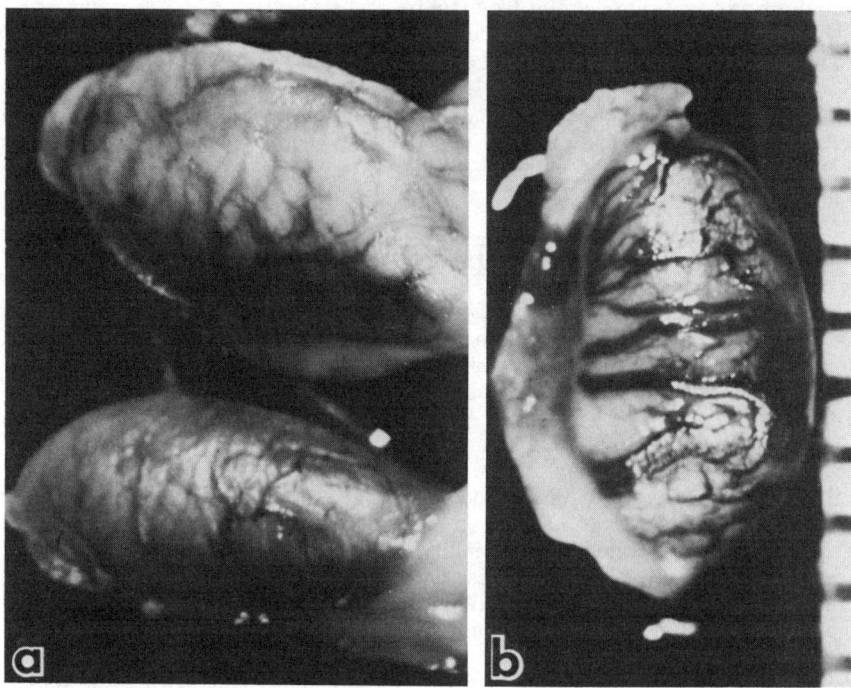

FIGURE 7. **(a)** Fragile X fetal testis shown on *top* showing macroorchidism present during the fetal period compared with age-matched control on *bottom*. **(b)** Marked congestion of vessels were noted (scale in mm). (From Jenkins *et al.*[74] Reprinted by permission from the *American Journal of Medical Genetics*.)

or 8 days of folic acid followed by 8 days of placebo. Fra(X) frequencies were determined twice a week. Serial Stanford-Binet and Peabody IQ testing showed no significant change in intelligence. However, the parents correctly estimated which boy was receiving therapy. They based their assessment on the boys' visuomotor and verbal performances, as well as increased ability to concentrate.

After the controlled trial, the subjects were put on high oral folic acid (1,000 mg/day) for an additional five months. A decrease in fra(X) frequency was observed when low folic acid medium was used for fra(X) detection. This effect persisted even after washing of the cells prior to culture. No significant reduction was seen using the FUdR method of fra(X) induction. Serial IQ testing (Stanford-Binet and Peabody) showed no significant changes. The parents' perception of their boys' improved behavior and ability to concentrate persisted when high-dose oral therapy was given. However, we observed no dramatic differences. We initiated another controlled trial with five subjects aged 5 to 25 years, using either 250 mg/day of folic acid or placebo for a 3-month period, followed by two 3-month periods of crossover (ABA design) to either drug or placebo. No change in IQ and no consistent behavioral changes related to therapy were noted.[81] In addition to our study, Carpenter *et al.*[82] reported a trial with four subjects, one of whom appeared to show some improvement. Hagerman *et al.*[83] undertook a double-blind trial with 25 subjects. No significant improvement on standard IQ and other psychologic tests were found for the group as a whole. However, some of the younger, prepubertal males showed apparent improvement in behavior as well as in psychologic test scores. Gustavson *et al.*[84] saw some improvement occurring only in young subjects. Hogge *et al.*[84] administered folic acid to a pregnant mother found to be carrying a fra(X)-positive male fetus. After delivery, the infant was said to show apparent developmental delay, indicating a probable lack of effect of this therapy. Rosenblatt *et al.*[86] in a controlled study of two 14-year-old identical twin brothers reported no effect of folic acid.

In summary, although we have observed no dramatic behavioral changes with folic acid therapy, improvement may have occurred in some individuals, but overall it appeared that its effects were inconsistent. Additional controlled trials are needed to determine whether the syndrome can be treated in this manner and deviant behavior significantly ameliorated.

THE FRAGILE X CHROMOSOME—BIOCHEMICAL BASIS FOR INDUCTION

The nature of the biochemical mutation underlying the fragile site on the X chromosome is unknown, but appears to involve the availability of thymidine. A critical step necessary for fragile site induction appears to be the inhibition of thymidine synthesis, as illustrated in FIGURE 8. If folate-deficient medium is employed, then the folic acid derivative that is necessary for thymidylate synthetase activity is decreased and thymidine monophosphate is not synthesized. The same inhibition occurs in the presence of folic acid when thymidylate synthetase

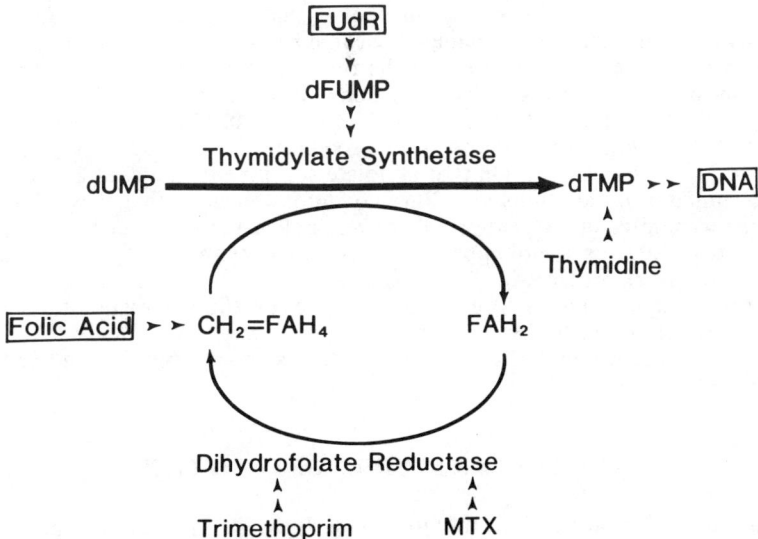

FIGURE 8. Metabolic pathway involved with fragile X expression. A critical step appears to be the inhibition of deoxythymidine monophosphate (dTMP) synthesis.

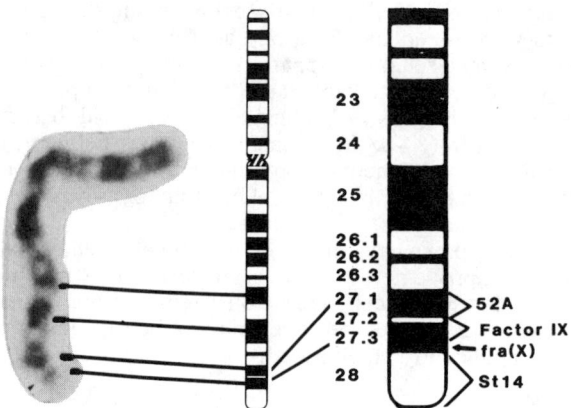

FIGURE 9. The fragile X chromosomes showing fragile sites at band Xq27.3. A portion of the dark staining band 27.3 appears above and below the fragile site. (From Krawczun *et al.*[93] Reprinted by permission from *Human Genetics*.)

is inhibited by the use of FUdR. Methotrexate, trimethoprim, fluorodeoxycytidine, and trifluorothymidine also block thymidine synthesis and can be used as alternatives to induce fragile site expression.[11,12,87–90] Sutherland has also found that a great excess of thymidine will induce fra(X) sites.[91] This excess appears to inhibit the synthesis of the enzyme, thymidylate synthelase, which may reduce the availability of cytidine. Thus, the induction of the fra(X) site appears in part to be dependent on thymidine availability.

The fra(X) site was first sublocalized to band Xq27.3 by Brookwell and Turner[92] and established by Krawczun, Jenkins and Brown,[93] as shown in FIGURE 9. When viewed under the scanning electron microscope, it appears to resemble an isochromatid gap.[94] In addition to the fragile site at Xq27, 12 other heritable folate-sensitive fragile sites with chromosomal locations (2q11, 2q13, 6p23, 7p11, 8q22, 9p21, 9q32, 10q23, 11q23, 12q13, 16p21, and 20p11), a nonfolate-sensitive site at 16q22, and a bromo-deoxyuridine-requiring site at 10q25 have also been described.[90,94] The observation that all folate-sensitive fragile sites appear to respond similarly to variations in culture conditions suggests that they all share a similar mechanism of expression. However, unlike the fra(X) site, most of the other autosomal sites do not appear to be associated with developmental disabilities. There are also FUdR-induced and spontaneous "constitutive" fragile sites that occur in all individuals. The FUdR-sensitive constitutive fragile sites occur at bands 1p31, 3p14 and 16q23; "common" fragile sites occur at bands 2q31, 3p14, 6q26, 7q32, 16q23, and Xp22.[70] It appears that some constitutive and common fragile sites are identical.

MOLECULAR STUDIES OF FRAGILE X

The molecular nature of the mutation underlying fra(X) is not known. There is no metabolic abnormality that has been associated with fra(X), which suggests a particular abnormal gene or a gene product. Because we lack knowledge about what specific gene might be involved, indirect approaches have been employed using DNA recombinant techniques to map the location of the mutation. Near the site of the fra(X) locus on the X chromosome are other segments of the X chromosome known as DNA probes, which have been isolated and found to show identifiable variations between persons. The variations are known as DNA polymorphisms and are inherited. These variations are detected in laboratory tests of blood samples by the use of restriction enzymes and appear as variations in fragment lengths. Hence, they are also known as restriction fragment length polymorphisms (RFLPs). The RFLPs have no relationship to the fra(X) locus, but are just nearby neighbors on the X chromosome. By identifying the nearby DNA RFLPs, their inheritance can be traced within a family. If DNA polymorphisms are very close to the fra(X) locus, then every time a given polymorphism is inherited, it is very likely that the fragile site on the X chromosome is also inherited, even though the fragile site may not be physically visible or expressed. It is analogous to tracing the inheritance of fingerprints of regions of the X chromosome which are adjacent to the fra(X).

Several genes are known to be located on either side of the fra(X) locus, as illustrated in FIGURE 10. These include the gene that is missing in hemophilia A, called factor VIII, and the gene that is missing in hemophilia B, called factor IX. DNA probes for factor IX and factor VIII are known to be polymorphic and have associated RFLPs. They have been used to trace the inheritance of the fra(X) locus in families. There are other DNA segments that are polymorphic and that do not encode genes. They are simply random DNA RFLPs. One that has been found to be polymorphic and close to factor IX is called 52A. Likewise, at the location of factor VIII is another RFLP known as ST14. The locations of several of these probes are illustrated in FIGURE 10.

Most polymorphisms exist in one of two possible variations which are identified by numbers or letters, for example, factor IX is either A or a, and 52A is either 1 or 2. Since females have two X chromosomes, a carrier mother may have both types of a given polymorphism, i.e., Aa or 12. If she has both types of a polymorphism she is termed a heterozygote for that polymorphism; otherwise she is a homozygote. Her sons, affected and unaffected, receive one or the other of her two X chromosomes and hence one or the other of the two types of the polymorphism. Thus, the inheritance of a given type of polymorphism can be correlated with the inheritance of the fra(X) locus. It is essential for the mother to be a heterozygote to be informative for the inheritance of a given polymorphism. The frequency of heterozygotes in females for either factor IX or 52a is about 50%. The DNA probe, ST14, has been found to be highly polymorphic and exists in forms 1 though 9, as well as types a or b. More than 90% of the time, a woman is heterozygous for ST14 and its inheritance is informative. The chromosomal locations of 52A and factor IX are known to be above or proximal to the fra(X) locus. Factor VIII and St14 are below it and near the end of the chromosome. These four probes bracket the site of the fra(X) locus. By tracing the inheritance of these RFLPs within families, potential carriers can be tested for the inheritance of the fra(X) chromosome, whether or not it is expressed cytogenetically.

Depending upon the distance on the chromosome between the probes and the fra(X) locus, there is a certain probability of recombination occurring with each offspring. Genetic distances are measured in terms of likelihood of recombination. Data from studies of large normal pedigrees indicate that the genetic distance between 52A and factor IX is about 10% recombination, between factor IX and St14 about 28% recombination, and between St14 and factor VIII about 1.7% recombination, as illustrated in FIGURE 8.[96] The fra(X) locus lies somewhere in between factor IX and St14.

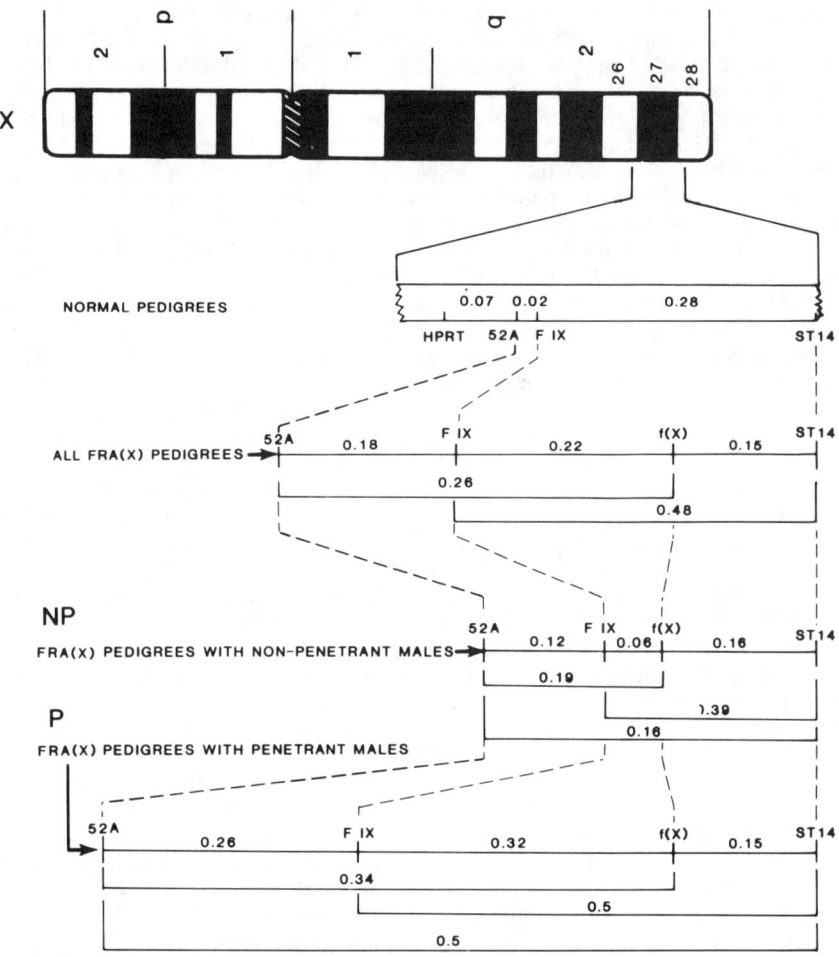

FIGURE 10. Recombination of fractions around the fragile X locus for DNA probes 52A, F9 and St14. Data from normal pedigrees is from Drayna *et al.*[95] Seven families in which NP males were identified showed less recombination than 30 pedigrees where such males were not identified. (From Brown *et al.*[97] Reprinted by permission from the *American Journal of Medical Genetics.*)

We have observed that some families show very little recombination between factor IX and fra(X). In other families there is a high rate of recombination. This suggests genetic heterogeneity. Initially, we studied eight fra(X) families, and suggested that there were two types of fra(X) families and that genetic heterogeneity was present.[67] This appeared to be related to the presence or absence of a nonpenetrant male within the family. We subsequently studied 16 fra(X) pedigrees for the inheritance of factor IX, 52a, and ST14 RFLPs and combined our information with that published for 16 other fra(X) pedigrees.[97] The distance between fra(X) and St14 appeared to be about 15 centimorgans. Based on the combined information, we concluded that in families where nonpenetrant males had been identified, a lower rate of recombination (close to 6%) was likely than that found in other fra(X) families (close to 32%) where only penetrant males had been identified. This is illustrated in FIGURE 10.

Although the distance between factor IX and the 52A probe in normal pedigrees was previously found to be about 1.7%,[96] in fra(X) pedigrees overall our analysis indicated it was about 18% (FIG. 10). This was a significant difference. It suggested the fra(X) locus may cause a higher rate of recombination in that region of the chromosome.

On the basis of further analysis, it now appears to us that most large fra(X) pedigrees have males that are likely to be nonpenetrants. This can be shown by the use of the probes combined with linkage analysis. For example, analysis of the pedigrees shown in FIGURES 3 and 4 has indicated that the grandfathers were most likely transmitting nonpenetrant males. Thus, whether a family is categorized as penetrant or nonpenetrant is unlikely to be directly related to the heterogeneity observed. We have identified several other large families with nonpenetrant males. Some show tight linkage, whereas others show loose linkage between factor IX and fra(X). This continues to suggest the presence of genetic heterogeneity. Genetic heterogeneity could explain in part the high new mutation rate and may suggest that clinical differences exist between families.

This method of testing for fra(X) by DNA linkage analysis has several important potential uses. First, it complements cytogenetics. With it we can determine whether a daughter or a sister of a fra(X) male is likely to have inherited the fra(X) chromosome, even if she is cytogenetically negative. Second, it can be useful for prenatal diagnosis. It can increase the accuracy of prediction of fra(X) when amniotic fluid or chorionic villus samples are examined. Third, it is useful for predicting which of the parents of the carrier mother may have transmitted the fra(X) chromosome. We have observed that frequently it may have been inherited from her father, even though he was of normal intelligence and cytogenetically negative. It is possible, in many situations, that the grandfather was a transmitting, nonpenetrant male, in which the expression of the chromosome was in some way suppressed and not expressed. It is also possible that many normal brothers may be transmitting males. This can now be investigated by DNA linkage analysis. The use of molecular markers combined with cytogenetics allows for increased information and improved genetic counseling to be given to families.

FUTURE RESEARCH ISSUES

Since the nature of the fra(X) mutation is not known, molecular studies are needed to isolate and characterize the mutation. One approach to this problem would be to isolate large pieces of DNA such as can now be separated by special types of gel eletrophoresis, as illustrated in FIGURE 11. After this separation, it may be possible to isolate the segment of the X chromosome that contains the fragile site. This may be identified by finding that two probes that are known to be on either side of the fra(X) location are both present in the isolated piece of DNA. Then molecular cloning approaches can be used to distinguish that which is different about the fra(X) site as compared to the same sequence from normal individuals.

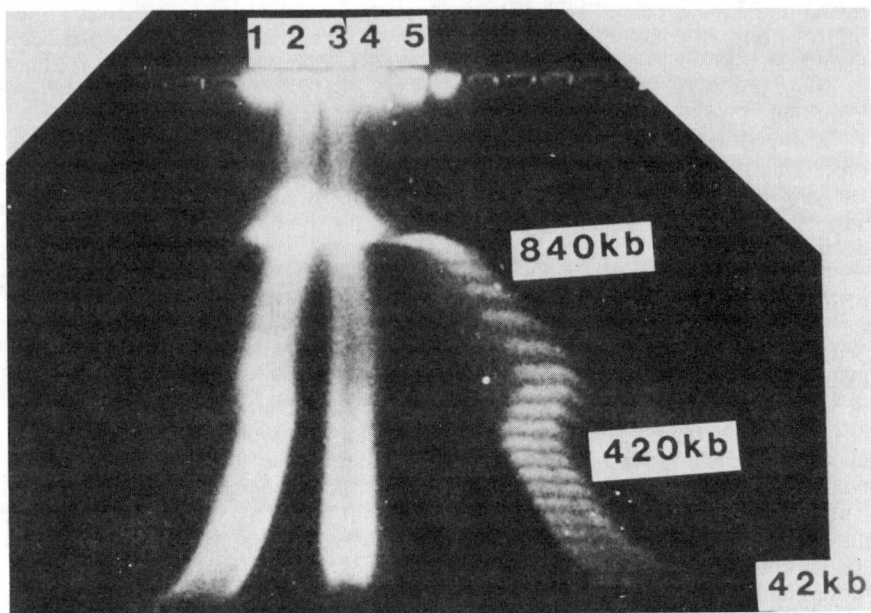

FIGURE 11. Pulsed field-gradient agarose gel electrophoresis of fra(X) cell line DNA with phage concatemer "ladder." *Lanes 1 and 4:* undigested fra(X) DNA. *Lane 2:* fra(X) digested with restriction enzymes *Not*I, which recognizes a 12-bp sequence. *Lane 3:* fra(X) DNA digested with enzyme *Kpn*I, which recognizes a 4-bp sequence. *Lane 5:* size marker "ladder" of 42-kb lambda phage concatemer. Twenty marker bands can be seen on the ethidium-stained gel, showing resolution between 5×10^4 to 9×10^5 base pairs. The ethidium staining also shows that the average size *Not*I fragment is much larger than the average *Kpn*I size.

Although the nature of the molecular basis of the fra(X) mutation is unknown, several hypothesis have been presented. One such hypothesis regarding fra(X) inheritance suggests that a recombination or amplification of a DNA sequence occurs.[97] There may be a DNA region rich in pyrimidines, such as thymidine, which is prone to undergo unequal crossing over during meiosis. Unequal crossing over could lead to rearrangement and amplification or deletion of sequences that may be important in affecting the expression of nearby genes. If a sequence were amplified this might lead to a region that would be visible as a cytogenetic lesion. The hypothesis is attractive from the standpoint that it may explain some of the puzzling features of fra(X). For example, the apparent high mutation rate could be due to the fact that there is a high frequency of rearrangements of a common repetitive-type DNA sequence. The size of the amplified region could be variable and this could result in its being nonvisible in transmitting males. This could also be reflected in the differences in the expression that is seen in different males with the fra(X) syndrome. The observation of tight linkage in some families and not in others may relate to the likelihood of crossing over, which could be due to the length of repetitive sequences. If normal individuals have a small pyrimidine-rich repetitive sequence on the X-chromosome, this might explain the occasional observation of a low frequency of apparent fra(X) chromosome seen in normal individuals such as we have reported.[98]

The fact that some males, the NP males, are carriers and do not express the mutation, may also suggest that there are environmental factors that can modulate the expression of the fragile site. Once the exact nature of the mutation is known, it is possible that a specific means of treating the condition can be found.

In summary, it appears that the fra(X) is present in from 5 to 10% of mentally retarded individuals. The populations of the mentally retarded and their families deserve to have accurate testing done to identify the affected individuals and those family members who are at risk of transmitting the fra(X) chromosome to their children. This will allow for primary prevention through prenatal diagnosis. It will also allow for secondary prevention through the introduction of early intervention programs to assure the best possible learning environment for fra(X) males in order to minimize their developmental disabilities. It is clear that for the fra(X) syndrome, genetic counseling and prenatal diagnosis can make a major difference in the prevention of mental retardation.

ACKNOWLEDGMENTS

We thank Frances Sanfillipo for secretarial assistance and Richard Weed for photographic assistance.

REFERENCES

1. PENROSE, L. S. 1938. A clinical and genetic study of 1,280 cases of mental defect (special report series no. 299). Medical Research Council. London, England.
2. LEHRKE, R. 1972. A theory of X-linkage of major intellectual traits. Am. J. Ment. Defic. **76:** 611–619.
3. TURNER, G. & B. TURNER. 1974. X-linked mental retardation. J. Med. Genet. **11:** 109–113.
4. HERBST, D. S. & J. R. MILLER. 1980. Nonspecific X-linked mental retardation. II. The frequency in British Columbia. Am. J. Med. Genet. **7:** 461–469.
5. TURNER, G., J. M. OPITZ, W. T. BROWN, K. E. DAVIES, P. A. JACOBS, E. C. JENKINS, M. MIKKELSEN, M. W. PARTINGTON & G. R. SUTHERLAND. 1986. Conference report: Second International Workshop on the Fragile X and on X-linked Mental Retardation. Am. J. Med. Genet. **23:** 11–67.
6. WEBB, T. P., S. E. BUNDEY, A. I. THAKE & J. TODD. 1986. Population incidence and segregation ratios in the Martin-Bell syndrome. Am. J. Med. Genet. **23:** 573–580.
7. LUBS, H. A. 1969. A marker X chromosome. Am. J. Hum. Genet. **21:** 231–244.
8. ESCALANTE, J. A., H. GRUNSPUN & O. FROSA-PESSOA. 1971. Severe sex-linked mental retardation. J. Genet. Hum. **19:** 137–140.
9. GIRAUD, F., S. AYME, J. F. MATTEI & M. G. MATTEI. 1976. Constitutional chromosomal breakage. Hum. Genet. **34:** 125–136.
10. HARVEY, J., C. JUDGE & S. WIENER. 1977. Familial X-linked mental retardation with an X chromosome abnormality. J. Med. Genet. **14:** 46–50.
11. SUTHERLAND, G. R. 1977. Fragile sites on human chromosomes: Demonstration of their dependence on the type of tissue culture medium. Science **197:** 265–266.
12. SUTHERLAND, G. R. 1977. Marker X chromosomes and mental retardation. New Engl. J. Med. **296:** 1415.
13. TURNER, G., R. TILL & A. DANIEL. 1978. Marker X chromosomes, mental retardation and macroorchidism. N. Engl. J. Med. **299:** 1472.
14. JACOBS, P. A. 1979. More on marker X chromosomes, mental retardation and macroorchidism. N. Engl. J. Med. **300:** 737–738.

15. TURNER, G., R. BROOKWELL, A. DANIEL, M. SELIKOWITZ & M. ZILIBOWITZ. 1980. Heterozygous expression of X-linked mental retardation and X-chromosome marker fra(X) (q27). New Engl. J. Med. **303:** 662–664.
16. TURNER, G., A. DANIEL & M. FROST. 1980. X-linked mental retardation, macroorchidism, and the Xq27 fragile site. J. Pediat. **96:** 837–841.
17. TURNER, G. & J. M. OPITZ. 1980. Editorial comment: X-linked mental retardation. Am. J. Med. Genet. **7:** 407–415.
18. HOWARD-PEEBLES, P. N. & G. R. STODDARD. 1980. Familial X-linked mental retardation with a marker X chromosome and its relationship to macro-orchidism. Clin. Genet. **17:** 125–128.
19. HERBST, D. S. 1980. Nonspecific X-linked mental retardation I. A review with information from 24 new families. Am. J. Med. Genet. **7:** 497–501.
20. HERBST, D. S. & J. R. MILLER. 1980. Nonspecific X-Linked mental retardation. II. The frequency in British Columbia. Am. J. Med. Genet. **7:** 461–469.
21. SHAPIRO, L. R., P. L. WILMOT, M. D. KUHR, E. LILIENTHAL & L. C. HIGGS. 1982. Genetic counseling for normal parents with two or more retarded children: A diagnostic dilemma. *In* Clinical Genetics: Problems In Diagnosis And Counseling. A. M. Willey, T. P. Carter, S. Kelly & I. Porter, Eds.: 147–152. Academic Press. New York.
22. FISHBURN, J., G. TURNER, A. DANIEL & R. BROOKWELL. 1983. The diagnosis and frequency of X-linked conditions in a cohort of moderately retarded males with affected brothers. Amer. J. Med. Genet. **14:** 713–724.
23. FRYNS, J. P. & H. VAN DEN BERGHE. 1982. X-linked mental retardation and fragile (xq27) site. Clin. Genet. **23:** 203.
24. CARPENTER, N. J., L. G. LEICHTMAN & B. SAY. 1982. Fragile X-linked mental retardation. A survey of 65 patients with mental retardation of unknown origin. Am. J. Dis. Child. **136:** 392–398.
25. BRONDUM-NIELSEN, K., N. TOMMERUP, H. V. DYGGVE & C. SCHOU. 1982. Macroorchidism and fragile X in mentally retarded males. Clinical, cytogenetic and some hormonal investigations in mentally retarded males, including two with the fragile site at XQ28, fra(X)(Q28). Hum. Genet. **61:** 113–117.
26. LINNA, S. L., S. SIMILA, E. HARO & R. HERVA. 1984. Prevalence of fragile X-chromosome. Lancet **i:** 220–221.
27. WEBB, T., A. THAKE, J. TODD & S. BUNDEY. 1984. Prevalence of fragile X-chromosome. Lancet **i:** 220.
28. JANCAR, J. 1984. Prevalence of fragile X-chromosome. Lancet **i:** 220.
29. RHOADES, F. A. 1984. The fragile (X) syndrome in Hawaii: A summary of clinical experience. Am. J. Hum. Genet. **17:** 209–214.
30. BLOOMQUIST, H. K., K. H. GUSTAVSON, I. NORDENSON & A. SWEINS. 1982. Fragile site X chromosomes and X-linked mental retardation in severely retarded boys in a northern Swedish county. A prevalence study. Clin. Genet. **60:** 278–280.
31. BLOOMQUIST, H. K., M. BOHAM, S. O. EDVINSSON, C. GILLBERG, K. H. GUSTAVSON, G. HOLMGREN & J. WAHLSTROM. 1984. Frequency of the fragile X syndrome in infantile autism. Clin. Genet. **27:** 113–117.
32. TURNER, G. & B. TURNER. 1974. X-linked mental retardation. J. Med. Genet. **11:** 109–113.
33. CANTU, J. M., H. E. SCAGLIA, M. GONZALEZ-DIDDI, P. HERNANDEZ-JAUREGUKI, M. E. MORENO, J. GINER, A. ALCANTAR & G. PEREZ-PALACIOS. 1978. Inherited congenital normofunctional testicular hyperplasia and mental deficiency. Hum. Genet. **41:** 331–339.
34. BROWN, W. T., P. M. MEZZACAPPA & E. C. JENKINS. 1981. Screening for fragile X syndrome by testicular size measurement. Lancet **ii:** 1055.
35. HOWARD-PEEBLES, P. N. 1983. Screening of mentally retarded males for macroorchidism and the fragile X chromosome. Am. J. Med. Genet. **15:** 631–635.
36. HUTTON, L., R. N. RANKIN & J. POZSONYI. 1985. High resolution ultrasound of macro-orchidism in mental retardation. J. Clin. Ultrasound **13:** 19–22.
37. FISCH, G. S., W. T. BROWN, I. L. COHEN & E. C. JENKINS. 1985. An increased risk of fragile X among Jewish people? Am. J. Hum. Genet. **37:** A196.

38. HAGERMAN, R. J., M. KEMPER & M. HUDSON. 1985. Learning disabilities and attentional problems in boys with the fragile X syndrome. Am. J. Dis. Child. **139:** 674–678.
39. WISNIEWSKI, K. C., J. H. FRENCH, S. FERNANDO, W. T. BROWN, E. C. JENKINS, E. FRIEDMAN, A. L. HILL & C. M. MIEZEJESKI. 1985. The fragile X syndrome: Associated neurological abnormalities and developmental disabilities. Ann. Neurol. **18:** 665–669.
40. HANSON, D. M., A. W. JACKSON, III & R. J. HAGERMAN. 1986. Speech disturbances (cluttering) in mildly impaired males with the Martin-Bell/fragile X Syndrome. Am. J. Med. Genet. **23:** 195–206.
41. RUDELLI, R. D., W. T. BROWN, K. WISNIEWSKI, E. C. JENKINS, M. LAURE-KAMIONOWSKA & F. CONNELL. 1985. Adult fragile X syndrome. Clinico-neuropathologic findings. Acta Neuropathol. **67:** 289–295.
42. PROOPS, R. & T. WEBB. 1981. The "fragile" X chromosome in the Martin-Bell-Renpenning syndrome and in males with other forms of familial mental retardation. J. Med. Genet. **18:** 366–373.
43. MATTEI, M. G., J. F. MATTEI, C. AUMERAS, M. AUGER & F. GIRAUD. 1981. X-linked mental retardation with the fragile X. A Study of 15 families. Hum. Genet. **59:** 281–289.
44. BROWN, W. T., E. FRIEDMAN, E. C. JENKINS, J. BROOKS, K. WISNIEWSKI, S. RAGUTHU & J. H. FRENCH. 1982. Association of fragile X with autism. Lancet **i:** 100.
45. BROWN, W. T., E. C. JENKINS, E. FRIEDMAN, J. BROOKS, K. WISNIEWSKI, S. RAGUTHU & J. H. FRENCH. 1982. Autism is associated with the fragile X syndrome. J. Aut. Devel. Disabil. **12:** 303–307.
46. MERYASH, D. L., L. S. SZYMANSKI & P. S. GERALD. 1982. Infantile autism associated with the fragile-X syndrome. J. Aut. Dev. Disord. **12:** 295–301.
47. LEVITAS, A., R. J. HAGERMAN, M. BRADEN, B. RIMLAND & P. McBOGG. 1983. Autism and the fragile X syndrome. J. Dev. Behav. Pediatr. **4:** 151–158.
48. BRONDUM-NIELSEN, K. 1983. Diagnosis of the fragile X syndrome (Martin-Bell syndrome). Clinical findings in 27 males with the fragile site at Xq28. J. Ment. Defic. Res. **27:** 211–226.
49. BROWN, W. T., E. C. JENKINS, I. L. COHEN, G. S. FISCH, E. G. WOLF-SHEIN, A. GROSS, L. WATERHOUSE, D. FEIN, A. MASON-BROTHERS, E. RITVO, B. A. RUTTENBERG, W. BUCKLEY & S. CASTELLS. 1986. Fragile X and autism: A multicenter survey. Am. J. Med. Genet. **23:** 334–352.
50. AMERICAN PSYCHIATRIC ASSOCIATION. 1980. Diagnostic and Statistical Manual (DSM) III. Washington, DC.
51. WEBB, G. C., J. L. HALLIDAY, D. B. PITT, C. G. JUDGE & M. LEVERSHA. 1982. Fragile (X) (q27) sites in a pedigree with female carriers showing mild to severe mental retardation. J. Med. Genet. **19:** 44–48.
52. TUCKERMAN, E., T. WEBB & S. E. BUNDEY. 1985. Frequency and replication status of the fragile X, fra(X)(q27-28), in a pair of monozygotic twins of markedly differing intelligence. J. Med. Genet. **22:** 85–91.
53. MIEZEJESKI, C. M., E. C. JENKINS, L. A. HILL, K. WISNIEWSKI, J. H. FRENCH & W. T. BROWN. 1986. A profile of cognitive deficit in females from fragile X families. Neuropsychologia. In press.
54. MIEZEJESKI, C. M., E. C. JENKINS, A. L. HILL, K. WISNIEWSKI & W. T. BROWN. 1984. Verbal vs nonverbal ability, fragile X syndrome, and heterozygous carriers. Am. J. Hum. Genet. **36:** 227–229.
55. KEMPER, M. B., R. J. HAGERMAN, R. S. AHMAD & R. MARINER. 1986. Cognitive profiles and the spectrum of clinical manifestations in heterozygous fra(X) females. Am. J. Med. Genet. **23:** 139–156.
56. SILVERMAN, W., R. LUBIN, E. C. JENKINS & W. T. BROWN. 1983. Quantifying the strength of association between fra(X) chromosome marker presence and mental retardation. Clin. Genet. **23:** 436–440.
57. MARTIN, J. P. & J. BELL. 1943. A pedigree of mental defects showing sex linkage, J. Neurol. Neurosurg. Psychiatry **6:** 154–157.
58. DUNN, H. G., H. RENPENNING, J. W. GERARD, J. R. MILLER, T. TABATA & S.

FEDEROFF. 1963. Mental retardation as a sex-linked defect. Am. J. Ment. Defic. **67:** 827–848.

59. FRYNS, J. P. & H. VAN DEN BERGHE. 1982. Transmission of fragile (X)(Q27) from normal male(s). Hum. Genet. **61:** 262–263.

60. GARDNER, R. J. M., R. D. SMART, J. M. CORNELL, L. M. MERCKEL & P. BREIGHTON. 1983. The fragile X chromosome in a large Indian kindred. Clin. Genet. **23:** 311–317.

61. JACOBS, P. A., M. MAYER, J. MATSUURA, F. RHOADES & S. C. YEE. 1983. A cytogenetic study of a population of mentally retarded males with special reference to the marker (X) syndrome. Hum. Genet. **63:** 139–148.

62. VAN ROY, B. C., M. C. DESMEDT, R. H. RAES, J. E. DUMON & J. G. LEROY. 1983. Fragile X trait in a large kindred: Transmission also through normal males. J. Med. Genet. **20:** 286–289.

63. CAMERINO, G., M. G. MATTEI, J. F. MATTEI, M. JAYE & J. L. MANDEL. 1983. Close linkage of fragile X linked mental retardation syndrome to haemophilia B and transmission through a normal male. Nature **306:** 701–707.

64. SIMOLA, K. O. J. 1984. X-linked mental retardation with the marker X chromosome— A clinical and cytogenetic study. Thesis, University Of Helsinki.

65. FROSTER-ISKENIUS, U., A. SCHULTZE & E. SCHWINGER. 1984. Transmission of the marker X syndrome trait by unaffected males: Conclusion from studies of large families. Hum. Genet. **67:** 419–427.

66. RHOADES, F. A., A. C. OGLESBY, M. MAYER & P. A. JACOBS. 1982. Marker X syndrome in an oriental family with probable transmission by a normal male. Am. J. Med. Genet. **12:** 205–217.

67. BROWN, W. T., A. C. GROSS, C. B. CHAN & E. C. JENKINS. 1985. Genetic heterogeneity in the fragile X syndrome. Hum. Genet. **71:** 11–18.

68. SHERMAN, S. L., N. E. MORTON, P. A. JACOBS & G. TURNER. 1984. The marker (X) syndrome; A cytogenetic and genetic analysis. Ann. Hum. Genet. **48:** 21–37.

69. SHERMAN, S. L., P. A. JACOBS, N. E. MORTON, U. FROSTER-ISKENIUS, P. N. HOWARD-PEEBLES, K. B. NIELSEN, M. W. PARTINGTON, G. R. SUTHERLAND, G. TURNER & M. WILSON. 1985. Further segregation analysis of the fragile X syndrome with special reference to transmitting males. Hum. Genet. **69:** 289–299.

70. GLOVER, T. W. 1981. FUdR induction of the X chromosome fragile site: Evidence for the mechanism of folic acid and thymidine inhibition. Am. J. Hum. Genet. **33:** 234–242.

71. TOMMERUP, N., H. POULSEN & K. BRONDUM-NIELSEN. 1981. 5-fluoro-2-deoxyuridine induction of the fragile site on X28 associated with X linked mental retardation. J. Med. Genet. **18:** 374–376.

72. JENKINS, E. C., W. T. BROWN, C. J. DUNCAN, J. BROOKS, M. BEN-YISHAY, F. M. GIORDANO & H. M. NITOWSKY. 1981. Feasibility of fragile X chromosome prenatal diagnosis demonstrated. Lancet **i:** 1291.

73. SHAPIRO, L. R., P. L. WILMOT, P. BRENHOLZ, A. LEFF, M. MARTINO, M. J. MAHONEY & J. C. HOBBINS. 1982. Prenatal diagnosis of the fragile X chromosome. Lancet **i:** 99–100.

74. JENKINS, E. C., W. T. BROWN, J. BROOKS, C. J. DUNCAN, R. D. RUDELLI & H. M. WISNIEWSKI. 1984. Experience with prenatal fragile X detection. Am. J. Med. Genet. **17:** 215–239.

75. JENKINS, E. C., W. T. BROWN, M. G. WILSON, M. S. LIN, O. S. ALFI, E. R. WASSMAN, J. BROOKS, C. J. DUNCAN, A. MASIA & M. S. KRAWCZUN. 1986. The prenatal detection of the fragile X chromosome: Review of recent experience. Am. J. Med. Genet. **23:** 297–312.

76. RUDELLI, R. D., E. C. JENKINS, K. WISNIEWSKI, R. MORETZ, J. BRYNE & W. T. BROWN. 1983. Testicular size in fetal fragile X syndrome. Lancet **i:** 1221–1222.

77. LEJEUNE, J. 1982. Is the fragile X syndrome amenable to treatment? Lancet **i:** 273–274.

78. LEJEUNE, J. 1981. Metabolisme des monocarbones et syndrome de l'X fragile. Bull. Acad. Nat. Med. **165:** 1197–1206.

79. BROWN, W. T., E. C. JENKINS, J. FRIEDMAN, J. BROOKS, C. J. DUNCAN, I. COHEN, L.

HILL, K. WISNIEWSKI & J. FRENCH. 1982. A controlled trial of folic acid therapy in Fragile X individuals. Am. J. Hum. Genet. **34:** 82A.

80. BROWN, W. T., E. C. JENKINS, E. FRIEDMAN, J. BROOKS, I. L. COHEN, C. J. DUNCAN, A. L. HILL, M. N. MALIK, V. MORRIS, E. WOLF, K. WISNIEWSKI & J. H. FRENCH. 1984. Folic acid therapy in the fragile X syndrome. Am. J. Med. Genet. **17:** 289–297.

81. BROWN, W. T., I. L. COHEN, G. S. FISCH, E. G. WOLF-SHEIN, V. JENKINS, M. N. MALIK & E. C. JENKINS. 1986. High-dose folic acid treatment of fragile (X) males. Am. J. Med. Genet. **23:** 263–271.

82. CARPENTER, N. J., D. H. BARBER, M. JONES, W. LONDLEY & C. CARR. 1983. Controlled six-month study of oral folic acid therapy in boys with fragile X-linked mental retardation. Am. J. Hum. Genet. **35:** 82A 243.

83. HAGERMAN, R., P. MCBOGG, A. LEVITAS, L. MCGAVRAN, A. SMITH, B. BERRY, M. BRADEN, K. VAN HOUSEN, K. NEWALL & I. MATUS. 1983. Folic acid treatment of the fragile-X syndrome. Am. J. Hum. Genet. **35**(6): 92A: 274.

84. GUSTAVSON, K. H., K. DAHLBOM, A. FLOOD, G. HOLMGREN, H. K. BLOMQUIST AND G. SANNER. 1985. Effect of folic acid treatment in the fragile X syndrome. Clin. Genet. **21:** 463–467.

85. HOGGE, W. A., S. A. SCHRONBERG, T. W. GLOVER, F. HECHT & M. S. GOLBUS. 1984. Prenatal diagnosis of fragile (X) syndrome. Obstet. Gynecol. **63:** 19S–21S.

86. ROSENBLATT, D. S., E. A. DUCHENES, F. V. HELLSTROM, M. S. GOLICK, J. J. VEKEMANS, S. F. ZEESMAN & E. ANDERMANN. 1985. Folic acid blinded trial in identical twins with fragile X syndrome. Am. J. Hum. Genet. **37:** 543–552.

87. SUTHERLAND, G. R. 1979. Heritable fragile sites on human chromosomes I. Factors affecting expression in lymphocyte culture. Am. J. Hum. Gent. **31:** 125–135.

88. LEJEUNE, J. N., N. LEGRAND, J. LAFOURCADE, M. O. RETHORE, O. RAOUL & C. MAUNOURY. 1982. The fragile X effect of trimethoprime treatment [in French]. Ann. Genet. **25:** 149–151.

89. JACKY, P. B. & G. R. SUTHERLAND. 1983. Thymidylate synthetase inhibitors and fragile site expression in lymphocytes. Am. J. Hum. Genet. **35:** 1276–1283.

90. SUTHERLAND, G. R. & F. HECHT. 1985. Fragile sites on human chromosomes. A. G. Motulsky, P. S. Harper & M. Bobrow, Eds. Oxford University Press. New York.

91. SUTHERLAND, G. R., E. BANKER & A. FRATINI. 1985. Excess thymidine induces folate-sensitive fragile sites. Am. J. Med. Genet. **22:** 433–443.

92. BROOKWELL, R. & G. TURNER. 1983. High-resolution banding and the locus of the Xq fragile site. Hum. Genet. **63:** 77.

93. KRAWCZUN, M. S., E. C. JENKINS & W. T. BROWN. 1985. Analysis of the fragile-X chromosome: Localization and detection of the fragile site in high-resolution preparations. Hum. Genet. **69:** 209–211.

94. HARRISON, C. J., E. M. JACK, T. D. ALLEN & R. HARRIS. 1983. The fragile X: A scanning electron microscope study. J. Med. Genet. **20:** 280–285.

95. SUTHERLAND, G. R., P. B. JACKY, E. BAKER & A. MANUEL. 1983. Heritable fragile sites on human chromosomes. X. New folate-sensitive fragile sites: 6p23, 9q32, and 11q23. Am. J. Hum. Genet. **35:** 432–437.

96. DRAYNA, D. & R. WHITE. 1985. The genetic linkage map of the human X chromosome. Science **230:** 753–758.

97. BROWN, W. T., A. G. GROSS, C. B. CHAN & E. C. JENKINS. 1986. DNA linkage studies in the fragile X syndrome suggests heterogeneity. Am. J. Med. Genet. **23:** 643–664.

97. NUSSBAUM, R. L., S. D. AIRHART & D. H. LEDBETTER. 1986. Recombination and amplification of pyrimidine-rich sequences may be responsible for initiation and progression of the Xq27 fragile site: An hypothesis. Am. J. Med. Genet. **23:** 715–721.

98. JENKINS, E. C., W. T. BROWN, J. BROOKS, C. J. DUNCAN, M. M. SANZ, W. B. SILVERMAN, K. P. LELE, A. MASIA, E. KATZ, R. A. LUBIN & S. L. NOLIN. 1986. Low frequencies of apparently fragile X chromosomes in normal control cultures: A possible explanation. Exp. Cell Biol. **54:** 40–48.

DISCUSSION

JOSEPH FRENCH: What will it mean if one gets a tightly linked probe or probes, if indeed there are two different genotypes, phenotypic similarity or practical applications?

TED BROWN: Having a probe that is tightly linked allows us to test for carriers in families; for example, in the family that had tight linkage, the sisters of the males who were affected were at risk of having fragile X and being carriers. We could not tell them definitely that they were not carriers just by the absence of fragile X cells in the blood. By having a restriction-fragment-length polymorphism that was tightly associated with the marker, we could tell them with a very high confidence level that they were not carriers. We were able to offer that kind of counseling, so having tightly linked markers allows for carrier testing. It also allows for prenatal testing using the DNA markers; thus it has immediate clinical application.

JOSEPH FRENCH: Apropos of that, do you have any experience where a carrier identification by the usual karyotypic analysis has failed, but with your probes it has been successful?

TED BROWN: In one family, for example, all three sisters were negative cytogenetically, but they were obligate carriers. This was one case where the DNA probe was associated with the fragile site. One of the granddaughters who was cytogenetically negative lacked the associated polymorphism and thus was counseled; she was not a carrier.

Now, interestingly, it looks like all daughters of nonpenetrant males are almost always negative cytogenetically. There are very few exceptions. There is something unusual going on in the transmission from the nonpenetrant grandfather to his daughters.

QUESTION: What is the frequency of the different alleles of factor IX polymorphism?

TED BROWN: It is about 70%-30%. Seventy percent for the more common, and 30% for the less common. That means that about 50% of women are heterozygous for the polymorphism, thus it is useful in about half of the families. We tested 23 families and found that about half of carriers were heterozygous. There are two other new polymorphisms that we are now using for the factor IX probe. This will increase the frequency of informative families to about 70%.

Application of Molecular and Somatic Cell Genetics to the Study of Chromosome 21 [a]

MARTHA LIAO LAW [b,c]

[b]Mental Retardation Research Center
B. F. Stolinsky Laboratory
Department of Pediatrics
University of Colorado Health Sciences Center
Denver, Colorado 80262

MARGARET VAN KEUREN [c]

[c]Eleanor Roosevelt Institute for Cancer Research
Denver, Colorado 80262

Down's syndrome (trisomy 21), the most common viable aneuploidy and most frequently identified cause of mental retardation, occurs in 1 in 600 to 1 in 1,000 live births[1] in the United States. Although the chromosomal abnormality for the syndrome has been known for more than twenty years,[2] the biochemical and molecular basis governing the expression of the phenotype is poorly understood. Down's syndrome patients have increased incidence of other clinical conditions such as congenital heart defects, infectious disease (especially pneumonia), endocrine dysfunction, and leukemia.[3–5] Also, a condition that closely simulates Alzheimer's disease is seen in 100% of Down's syndrome individuals over the age of 40.[6,7] Therefore an understanding of Down's syndrome at the biochemical and molecular level is important because it will provide us with clues in elucidating the mechanisms underlying the development of these other diseases as well.

Because Down's syndrome is characterized by the presence of an extra copy of all or part of chromosome 21, it is reasonable to suggest that the syndrome is caused by an excess of normal genetic material. The expression of such additional genetic information on chromosome 21 may also be responsible for the increased incidence of such disorders as leukemia, infection, endocrine dysfunction, and premature aging. One approach to the study of this problem is to search for genes that are located on or controlled by chromosome 21 and to explore the structure and function of these genes and the regulation of their expression. The goal is to identify which of these genes are involved in the pathology associated with Down's syndrome.

In our institute, the study of chromosome 21 is facilitated by employing both somatic cell genetics and recombinant DNA technologies. While the first method involves construction of cell hybrids containing human chromosome 21 from various sources including patients with Down's syndrome, the latter leads to the isolation of cloned genes and DNA fragments from chromosome 21. By combining the two techniques, we have been able to assign genes and DNA fragments to

[a] This manuscript is contribution number 535 from the Eleanor Roosevelt Institute for Cancer Research. It was aided by Grant HD17449 from the National Institutes of Health.

regions of chromosome 21 including the region q22, which has been shown by cytogenetic analyses in various laboratories[8-11] to be responsible for Down's syndrome.

Most of these cell hybrids were generated in the laboratories of Carol Jones and David Patterson by fusing an auxotrophic mutant, either Ade⁻C or Ade⁻G,[12,13] of a Chinese hamster ovary cell line CHO-K1 with human cells from individuals with normal or abnormal chromosome 21. Hybrids were isolated in selective medium such that only those containing the gene complementing the Ade⁻C or Ade⁻G deficiency could survive. Since both of these genes, which code for enzymes of purine synthesis, have been assigned to human chromosome 21,[14,15] the hybrids must contain this chromosome. Cytogenetic analyses were used to determine the human chromosomal content. In the case of defining different deletions of chromosome 21 when the breakpoints in these deletions are very close to each other, even the most advanced cytogenetic techniques involving prometaphase chromosome banding were unable to distinguish one from another. Therefore, there is a need to develop new methods that will allow us to identify and characterize fine structure of chromosome 21 and we hope that from such a fine-structure map of the chromosome, we can detect precisely what genes and DNA sequences are in the region q22. On the other hand, the use of a certain class of cloned repetitive human DNA to construct fine-structure maps of human chromosomes have been shown to be particularly efficacious in regional mapping of human chromosomes such as 12[16] and 11[17,18]. The work presented here describes a further application of such a strategy to the regional mapping of human chromosome 21 and demonstrates its usefulness in defining chromosomal abnormalities not detectable by cytogenetic analysis.

The rationale behind this approach is that a repetitive sequence of a few thousand copies occurs in multiple sites in the human genome instead of a single site as in the case of a unique sequence. If one can distinguish these sites from each other by using a repeat sequence as a probe, then this sequence can be used as a genetic marker for defining multiple sites in the human genome. This method will increase considerably the number of landmarks along each chromosome and hence will facilitate the identification of translocations, deletions, and other chromosomal rearrangements.

One such repetitive sequence that can differentiate one site from another in the human genome was isolated from a λ-phage DNA library of human chromosome 12 constructed by one of us (MLL). It is a 2.2-kb EcoR1 fragment that detects multiple bands by Southern blot analysis of digested DNA from a hamster/human hybrid containing a single human chromosome. The bands detected by hybridization are human specific and exhibit patterns that appear to be distinct for each chromosome;[16] hence, if a hamster/human hybrid containing the same chromosome but with a deletion is analyzed, the absence of one or more of these bands will define such a deletion on the chromosome. In situ hybridization on metaphase chromosomes revealed that this sequence was present in the euchromatin region of all human chromosomes except Y.[19] Characteristics of such a repetitive sequence had enabled us to regionally assign the 2.2-kb fragment and five other related sequences on human chromosome 12[16] and 26 DNA fragments to four regions on human chromosome 11.[18] Furthermore, we were able to demonstrate the use of this repetitive sequence probe for the detection of chromosomal deletion associated with genetic disease by confirming a deletion on chromosome 12 of a mentally retarded patient.[20]

In analyzing a series of hamster/human cell hybrids containing human chromosome 21, we used as probes both the 2.2-kb sequence and a recombinant DNA clone containing a related sequence from chromosome 21, designated 21#1. 21#1 was isolated from a chromosome-21 DNA library[21] by virtue of its hybridization to the 2.2-kb sequence. Restriction mapping and Southern blot hybridization showed that an internal 0.58-kb *Pvu*II fragment of 21#1 was the segment hybridizing strongly to the 2.2-kb probe. This 0.58-kb fragment was found to be a more useful probe than the 2.2-kb fragment for regional mapping, as the hybridization bands detected by Southern blot analysis are much more distinct.

FIGURE 1 shows a Southern blot of *Eco*R1-digested DNA of three independent chromosome 21 cell hybrids and probed with the 0.58-kb fragment. A hamster/human hybrid, 72532X-6, containing an intact chromosome 21 as the only human chromosome was provided by Carol Jones.[22] Hybrid 2Fu[r]1 contains only the long arm of chromosome 21 that is translocated to a hamster chromosome.[23,24] Since the centromere is missing, the breakpoint is in the band q11.1.[25] Hybrid R2-10 contains a ring chromosome 21 as the only human chromosome. The human parental cells, GM6137,[26] were provided by G. Stetten and S. Antorarakis. The breakpoints at p11.2 and q22.3 had joined to form the ring chromosome.[25,26]

In the figure, the black dots designate some of the DNA fragments that are common to all three hybrids and hence can be located to the region of chromosome 21 that is common to all the hybrids, namely q11.1 to q22.3. The DNA fragment indicated by a black solid square is present in 2Fu[r]1 and 72532X-6 but not in R2-10, hence it must be a marker in the region on the long arm that was deleted in R2-10, namely q22.3 to qter. The open circle represents a DNA fragment that is present in R2-10, indicating that this fragment probably resides in the region joining the two breakpoints at p11.2 and q22.3 to form the ring structure. This particular fragment should contain DNA sequences from both the short and long arms. The open squares represent some of the DNA fragments that were observed in 2Fu[r]1 only. Although it had only one breakpoint at the centromere, results here seemed to imply that this breakpoint was more complicated than expected. Perhaps the organization of DNA at the centromere is not linear as in the rest of the chromosome.

Another plausible explanation for the bands that were different from one cell to another could be restriction-fragment-length polymorphisms (RFLPs) present on the chromosome 21 of different individuals. In order to test this possibility, DNA from four somatic cell hybrids, each containing a chromosome 21 from a different source, were used to hybridize to the 0.58-kb probe. These hybrids were WAVR4dF94a, a mouse/human chromosome-21 hybrid provided by F. Ruddle; SCC16-5, a mouse/human chromosome-21 hybrid, a gift from D. R. Cox; Thy B1-33-6-1, a mouse/human chromosome-21 hybrid from C. Bostock and 72532X-6. Southern blot hybridization results (FIG. 2) showed that the band patterns were very similar among these four cell lines with the exception of 72532X-6. This cell line showed three prominent hybridization bands that were different from the others. These were marked with an asterisk (*) in FIGURE 1 and FIGURE 2 and were not included for consideration in the regional assignment of these related DNA fragments on chromosome 21. Our result also implies that RFLPs of chromosome 21 from different individuals are not readily detectable by probing *Eco*R1 digests with the 0.58-kb fragment as a probe and that most of the extra bands observed may be due to chromosomal abnormalities, that is, deletions, translocations, or other rearrangements.

Another set of hybrids was constructed by introducing chromosomal breakage using high doses of radiation followed by cell fusion[27] and was characterized by isozyme and cytogenetic analyses for the presence of a whole or part of chromosome 21. Our Southern blot analysis of these hybrids using the 0.58-kb probe

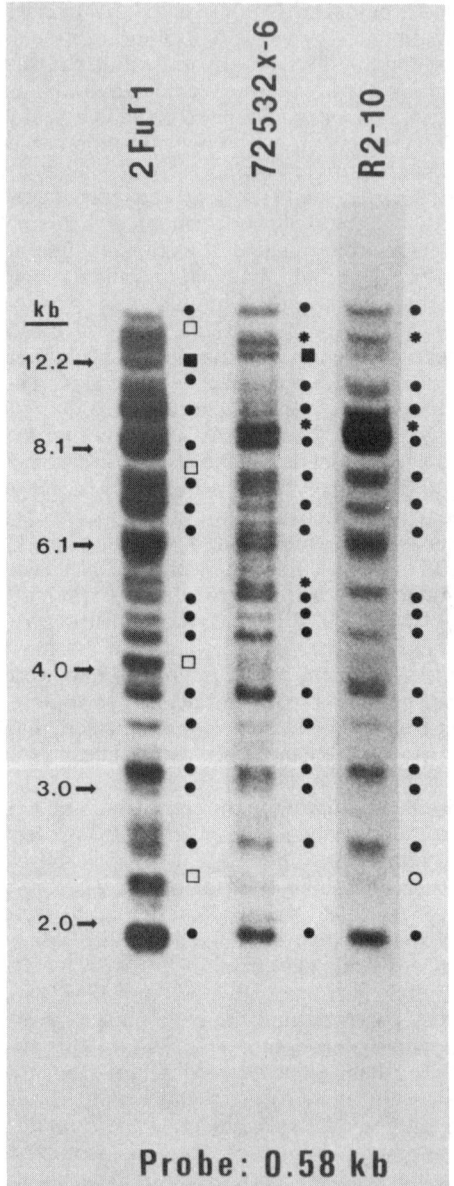

FIGURE 1. Southern blot of *Eco*R1-digested DNA of three hamster/human hybrids containing different parts of human chromosome 21 and hybridized to a ^{32}P-labeled 0.58-kb fragment.

showed some interesting results that were not observed by the previous methods. When compared to that of the parental cell line, the multiple-band patterns of these hybrids revealed that one of these hybrids had lost a large amount of human material, while another one, which by cytogenetic analysis was identical to the parent, had lost a specific DNA fragment.[28] This experiment is another example

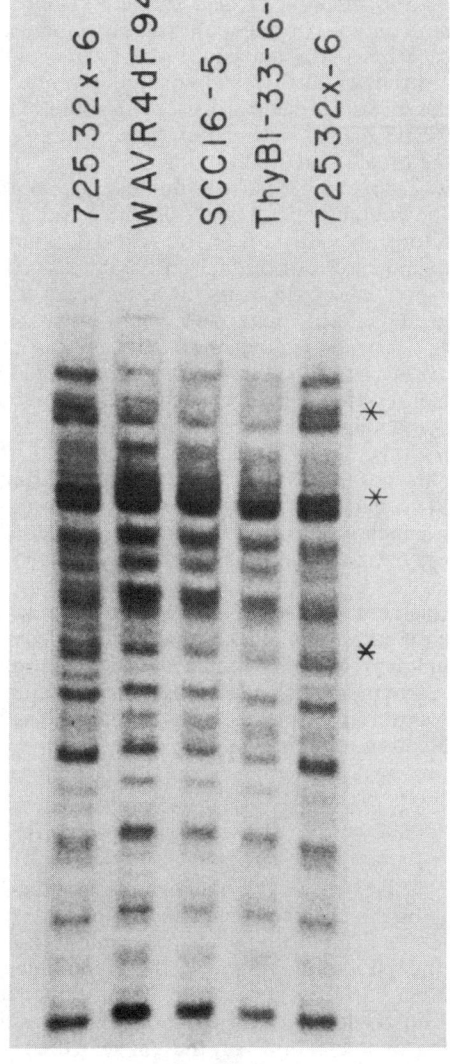

FIGURE 2. Southern blot of *Eco*R1-digested DNA of four rodent/human hybrids containing different human chromosome 21 and hybridized to a ^{32}P-labeled 0.58-kb fragment.

that shows the high resolution of our method in identifying chromosomal deletions in somatic cell hybrids.

Preliminary results were obtained in identifying chromosomal translocation in a somatic cell hybrid constructed between a Chinese hamster ovary cell mutant and myeloblasts with a translocation t(8;21)(q22; q22) from a patient with acute myelogenous leukemia (AML).[29] On the Southern blot probed with the 0.58-kb fragment, we were able to identify a band in this hybrid that was not present in a hybrid with a single chromosome 8[30] and a hybrid with a single chromosome 21.[31]

Although this extra band may still result from a RFLP in this patient, a more exciting interpretation is that this particular band could contain a DNA fragment that occurred at or near the site of rearrangement. Since this translocation is a nonrandom chromosome rearrangement seen almost exclusively in the M2 subtype of AML, and since the C-*mos* oncogene (8q22) located near the site of rearrangement on chromosome 8 is not translocated, the fragment identified by the 0.58-kb repeat probe could potentially be important in the etiologic study of the t(8;21) M2 subtype of AML. Moreover, chromosome band 21q22, the site of the breakpoint on chromosome 21 involved in this translocation, is a region that, when trisomic, leads to the development of Down's syndrome. Since it has been shown that patients with Down's syndrome have an increased risk of leukemia,[32] future investigation in the regulation and expression of DNA sequences in the region 21q22, or on their effect on expression of other genes may provide us with useful information on the molecular genetics of Down's syndrome as well as leukemia.

It is thus important that these DNA fragments defining deletions and translocations be cloned into plasmid or phage vectors so that they may serve as useful markers for genetic and molecular biology studies. In this study, the fragment containing the exact sequence to the 0.58-kb probe plus its flanking sequences on both the 3' and 5' ends was identified, and it was shown to be absent in one of the chromosome-21 hybrids. Using high-stringency washing after hybridization (0.1% SDS, 0.1 XSCC,[d] 65°C for 2 to 10 hours), we were able to show that the DNA fragment containing the 0.58-kb sequence was unique to chromosome 21 just as the original 2.2-kb probe was unique to chromosome 12.[28] Therefore, the advantage of using a repetitive sequence such as the 2.2-kb fragment and the 0.58-kb fragment as described above, rather than a unique sequence is that it can serve as a repetitive sequence defining multiple sites in the genome and also function as a unique sequence defining a particular site. For the detection of deletions and other rearrangements, especially in small chromosomes such as 21, it is the former property that makes it very efficient in the initial assignment of a chromosome location.

SUMMARY

An extra copy of human chromosome 21 has been known for over twenty years to be the chromosomal abnormality in Down's syndrome; however, the biochemical and molecular basis governing expression of the phenotype is still poorly understood. Using the methods of somatic cell and molecular genetics, we have been studying genes and DNA sequences on chromosome 21 by constructing hamster/human hybrids containing a whole or partial chromosome 21 and assigning their locations on the chromosome. In particular, a family of repetitive sequences, some having only a few thousand copies in the human genome, have been used as cloned DNA markers to define deletions in these somatic cell hybrids. We have shown that this approach can significantly improve the resolution of fine chromosomal structures over the conventional cytogenetic analysis. The rationale behind this approach is the observation that a repetitive sequence probe often forms multiple bands after hybridizing to a Southern blot of digested hybrid

---

[d] SDS stands for sodium dodecyl sulfate and 20X SCC contains 3 M sodium chloride and 0.3 M sodium citrate, pH 7.2.

DNA, and the band pattern appears to be unique for each human chromosome. Therefore, each band (sequence) can be assigned to a particular region of human chromosome 21 by comparing the band patterns from hybrids containing different portions of the chromosome. Results presented here showed that a 0.58-kb repetitive sequence probe can be used to identify deletions, translocations, and other more complicated rearrangements of chromosome 21 seen in patients with abnormalities of this chromosome. The advantage of using such a repetitive sequence probe over a unique sequence is that it can serve both as a repetitive sequence defining multiple sites (multiple bands on a Southern blot) in the genome and at the same time serve as a unique sequence defining a particular site (individual band). For the detection of deletions and other rearrangements, especially in small chromosomes such as 21, it is the former property that makes it very efficient in the initial assignment of a chromosome location.

ACKNOWLEDGMENTS

The technical assistance of Mr. Ralph Berger is greatly appreciated.

REFERENCES

1. ADAMS, M. M., J. D. ERICKSON, P. M. LAYDE & G. P. OAKLEY. 1981. Down syndrome: Recent trends in the United States. J. Am. Med. Assoc. **246:** 758–760.
2. LEJEUNE, J. M., M. GAUTIER & R. TURPIN. 1959. Etude des chromosomes somatiques de neuf enfants mongoliens. C. R. Acad. Sci. **248:** 1721–1722.
3. MIKKELSEN, M. 1981. New aspects of a well-known syndrome (Down Syndrome-Mongolism). Eur. J. Pediatr. **136:** 5–7.
4. SCOGGIN, C. J. & D. PATTERSON. 1982. Down's syndrome as a model disease. Arch. Int. Med. **142:** 462–464.
5. ROWLEY, J. D. 1981. Down syndrome and acute leukemia: Increased risk may be due to trisomy 21. Lancet, **ii** 1020–1022.
6. GLENNER, G. G. 1983. Banbury report 15: Biological aspects of Alzheimer's disease. Cold Spring Harbor Symp. pp. 137–144.
7. GLENNER, G. G. & C. W. WONG. 1984. Alzheimer's disease and Down's syndrome: sharing of a unique cerebrovascular amyloid fibril protein. Biochem. Biophys. Res. Commun. **122:** 1131–1135.
8. NEIBUHR, E. 1974. Down syndrome, the possibility of a pathogenetic segment on chromosome 21. Hum. Genet. **21:** 99–101.
9. CERVENKA, J., R. J. GORLIN & G. R. DJAVADI. 1977. Down syndrome due to partial trisomy 21q. Clin. Genet. **11:** 119–121.
10. HAGEMEIJER, A. & E. M. E. SMIT. 1977. Partial trisomy 21, further evidence that trisomy of band 21q22 is essential for Down phenotype. Hum. Genet. **38:** 15–23.
11. BRADLEY, C. M., D. M. COX, D. PATTERSON & A. ROBINSON. 1982. Gene dosage effect for phosphoribosylglycineamide synthetase (GARS) in a patient with Down syndrome and non-Robertsonian t(21:21) translocation. Pediatr. Res. **16:** 190A.
12. IRWIN, M., D. C. OATES & D. PATTERSON. 1979. Biochemical genetics of Chinese hamster cell mutants with deviant purine metabolism, isolation and characterization of a mutant deficient in the activity of phosphoribosylaminoimidazole synthetase. Somatic Cell Genet. **5:** 203–216.
13. OATES, D. C. & D. PATTERSON. 1977. Biochemical genetics of Chinese hamster cell mutants with deviant purine metabolism characterization of Chinese hamster cell mutants defective in phosphoribosylpyrophosphate amidotransferase and phosphonbosylglycine-amide synthetase and an examination of alternatives to the first step of purine biosynthesis. Somatic Cell Genet. **3:** 561–577.

14. PATTERSON, D., S. GRAW & C. JONES. 1981. Demonstration by somatic cell genetics of coordinate regulation of genes for two enzymes of purine synthesis assigned to human chromosome 21. Proc. Natl. Acad. Sci. USA **78:** 405–409.
15. MOORE, E. E., C. JONES, F. T. KAO, D. C. OATES. 1977. Synteny between glycineamide ribonucleotide synthetase and superoxide dismutase (soluble). Am. J. Hum. Genet. **29:** 389–396.
16. LAW, M. L., J. N. DAVIDSON & F. T. KAO. 1982. Isolation of the human repetitive sequence and its application to regional chromosome mapping. Proc. Natl. Acad. Sci. USA **79:** 7390–7394.
17. GUSELLA, J. F., C. JONES, F. T. KAO, D. HOUSMAN & T. T. PUCK. 1982. Genetic fine-structure mapping in human chromosome 11 by use of repetitive DNA sequences. Proc. Natl. Acad. Sci. USA **79:** 7804–7808.
18. LAW, M. L., F. T. KAO, C. JONES & T. T. PUCK. 1984. Use of cloned repetitive DNA sequences for mapping human chromosomes. Cytogenet. Cell Genet. **37:** 519.
19. FUNDERBURK, S. J. & M. LAW. 1982. Chromosome mapping of a repetitive DNA sequence via *in situ* hybridization. Am. J. Hum. Genet. **34:** 160A.
20. FUNDERBURK, S. J., R. S. SPARKES, I. KILSAK & M. L. LAW. 1984. Chromosome deletion mapping of interspersed low-copy repetitive DNA. Am. J. Hum. Genet. **36:** 769–776.
21. KRUMLAUF, R., M. JEANPIERRE & B. D. YOUNG. 1982. Construction and characterization of genomic libraries from specific human chromosomes. Proc. Natl. Acad. Sci. USA **79:** 2971–2975.
22. MILLER, Y. E., C. A. JONES, C. H. SCOGGIN, D. PATTERSON. A chromosome-21-associated cell surface antigen present on fetal brain. Trisomy **21:** In press.
23. PATTERSON, D., C. JONES, H. MORSE, P. RUMSBY, Y. MILLER & R. DAVIS. 1983. Structural gene coding for multifunctional protein carrying orotate phosphoribosyltransferase and OMP decarboxylase activity is located on the long arm of human chromosome 3. Somatic Cell Genet. **9:** 359–374.
24. PATTERSON, D. & V. B. SCHANDLE. 1983. A comparison of Chinese hamster/human hybrid cells containing different fragments of chromosome 21 using cytogenetic, biochemical, and molecular approaches. *In* Banbury Report on Recombinant DNA Applications to Human Disease. C. T. Caskey & R. L. White, Eds. Vol. **14:** 215–223. Cold Spring Harbor Laboratory. Cold Spring Harbor.
25. VAN KEUREN, M. L., P. C. WATKINS, H. A. DRABKIN, E. W. JABS, J. F. GUSELLA & D. PATTERSON. 1986. Regional localization of DNA sequences on chromosome 21 using somatic cell hybrids. Am J. Hum. Genet. **38:** 793–804.
26. STETTEN, G., B. SIROKA, V. L. CARSON & C. D. BOEHM. 1984. Prenatal detection of an unstable ring 21 chromosome. Hum. Genet. **68:** 310–313.
27. GRAW, S. & D. PATTERSON. Unpublished data.
28. LAW, M. L., L. TUNG, M. VANKEUREN, S. GRAW & R. BERGER. Manuscript in preparation.
29. DRABKIN, H. A., M. DIAZ, C. M. BRADLEY, M. M. LEBEAU, J. D. ROWLEY & D. PATTERSON. 1985. Isolation and analysis of the 21q+ chromosome in the acute myelogeneous leukemia 8;21 translocation: Evidence that c-mos is not translocated. Proc. Natl. Acad. Sci. USA **82:** 464–468.
30. JONES, C., D. PATTERSON & F. T. KAO. 1981. Assignment of the gene coding for phosphoribosylglycineamide formyltransferase to human chromosome 14. Somatic Cell Genet. **7:** 399–409.
31. LAW, M. L. & H. A. DRABKIN, R. BERGER & D. PATTERSON. Unpublished data.
32. EVANS, A. I. K. & J. K. STEWARD. 1972. Down syndrome and leukemia. Lancet **ii:** 1322.

DISCUSSION

DAVID KURNIT: In the first sequence you showed significant conservation to mouse DNA. Is it transcribed? Usually when you see conservation from man to mouse, such sequences are transcribed.

MARTHA LAW: None of them hybridize to mouse. Do you mean the mouse hybrids?

DAVID KURNIT: You showed A9 with a significant number of bands.

MARTHA LAW: A9 has nothing to do with the cell line of mouse called A9; it is a hybrid containing deleted 12. All of my repeat sequence bands show no cross hybridization to hamster or to mouse; however, the repeat sequence is transcribed.

TED BROWN: I was interested in your interpretation of the ring chromosome, the one that is present in the hybrid R2-10. Did you say the repeat sequence is present at the junction that is interrupted in the ring chromosome?

MARTHA LAW: There is a DNA band that is different in that particular hybrid, and I think that probably contains a DNA fragment at the junction of the two breakpoints that join together to form the ring, and that is why we see a band that is different. I am predicting that it contains DNA sequence from the short arm and from the long arm. The ultimate proof will be to isolate these sequences and then to clone them.

EDMUND JENKINS: At what level of band resolution do you work with cytogenetically? To say that we really cannot distinguish between one line and another while there is certainly deletion in one and not the other is a little puzzling.

MARTHA LAW: We definitely can see the deletions by cytogenetic analysis in the chromosome 12 hybrids. We can also see the deleted chromosome 12 from the patient cytogenetically. I am talking about the 21-deletion hybrids, which are very complicated. According to Sharon Graw, who constructed these hybrids, it was difficult sometimes to identify on chromosome 21 exactly what part was missing. Also we were using a chromosome-breaking agent x-ray at different doses, and at high dose a lot of complicated translocations, deletions, and other rearrangements can happen. It is sometimes impossible to identify these rearrangements cytogenetically.

Trisomy 16 Mice: Neural, Morphological, and Immunological Studies[a]

STEVEN E. KORNGUTH,[b] EDWARD T. BERSU,[c]
ROBERT AUERBACH,[d] HANNA M. SOBKOWICZ,[b]
HENRY S. SCHUTTA,[b] AND GRAYSON L. SCOTT[c]

[b]*Department of Neurology*
[c]*Department of Anatomy*
[d]*Department of Zoology*
University of Wisconsin-Madison
Madison, Wisconsin 53706

INTRODUCTION

The occurrence of trisomy (i.e, the presence of an additional chromosome in the genome) is associated with distinct and deleterious effects on the development and cellular processes of affected individuals. Efforts are now directed toward understanding the mechanisms by which the trisomic condition causes adverse effects. Trisomy for autosome number 21, Down's syndrome, is of particular interest because of its high frequency (1 in 700 live births), the associated mental retardation, and the observation that individuals with the condition survive to the fourth and fifth decades. The Down syndrome has been the single most extensively investigated entity involving mental retardation since it was first identified in the middle of the 19th century.[1,2]

The discovery of the chromosomal basis of Down's syndrome in 1959[3] provided the initial impetus to relate specific phenotypic effects with cellular processes that are altered by the presence of the additional genetic material. Such studies have included: (1) detailed characterizations of Down's syndrome morphologic phenotype;[4,5] (2) identification of gene dosage phenomena for genes located on chromosome 21 (e.g., superoxide dismutase I); and (3) observations of possible effects of trisomy on cell proliferation.[6]

An extension of this work is now possible using trisomic mice that can be generated by using males or females with appropriate Robertsonian translocation chromosomes.[7] The trisomy for murine autosome 16 is of interest because murine chromosome 16 contains several genes that are present in human chromosome 21. The shared genes include those coding for superoxide dismutase-1, the alpha and beta interferon receptors, and the response of cells to beta-adrenergic agonist stimulation.[8,9] Three of these genes code for cell surface receptor linked systems. As with the Down's syndrome, gene dosage effects also occur for the gene products of these loci in the trisomic mice. A difficulty with the trisomy 16 system is that none of these mice survive beyond term with breeding schemes that have been used so far.[10]

Our current investigations include an evaluation of the effects of the additional chromosome 16 on the expression of cell surface markers, (e.g., the major histo-

[a] The work described in this report has been supported in part by Grant No. HD03352 from the National Institutes of Health.

compatibility complex). Two hypotheses being examined include: (1) There is an imbalance in the expression of cell surface markers on cells in the trisomy 16 conceptus. This imbalance affects cell–cell interactions and normal organogenesis. (2) There is a compromised vascular development in the trisomy 16 fetus that causes a delay in the organogenesis/differentiation of selected organs. To examine these hypotheses, the placenta, vasculature, and the cerebellum are being studied in particular. These organs were selected because cell–cell interactions play an important role in their development and their development has been noted to be adversely affected by the trisomy 16 genome.

GENERATION OF TRISOMY 16 FETUSES AND NORMAL LITTERMATES

Trisomy 16 mice for our studies were produced by crossing all-acrocentric BALB/c females with males carrying two different Robertsonian translocation metacentric chromosomes. The metacentric chromosomes have autosome 16 in common and are designated as Rb(16.17)8Lub and RB(11.16)2H. The karyotype of these males consists of 36 acrocentric chromosomes and one each of the two metacentric chromosomes. The karyotype of the females is comprised of 40 acrocentric chromosomes.

The biologic basis for the above breeding scheme is based on the fact that the two translocation chromosomes can segregate together during meiosis.[7,11] This can result in aneusomic gametes, which if fertilized by normal gametes from an all-acrocentric animal, will produce trisomic embryos with 39 chromosomes including one each of the two metacentric chromosomes, for a total of 41 chromosome arms. Normal littermates produced from this cross have 39 chromosomes as well, but only one metacentic chromosome, for a total of 40 chromosome arms.

All viable animals that were produced in our studies were karyotyped from direct preparations of amniotic fluid or minced liver and intestine, using the standard technique of hypotonic treatment with 0.075 M KC1 and 3:1 methanol to acetic-acid fixation.[12,13] Air-dried slides were stained with 2% Aceto-Orcein and the metaphases evaluated for the chromosome number and translocation complement. Each trisomic animal was used in several of the studies reported here.

The frequency, viability, and developmental profile of any given trisomy is influenced significantly by the specific Robertsonian translocation chromosome involved and strain background.[14,15] TABLE 1 is a summary of the composition of litters that have been collected thus far from our studies. The frequency of viable trisomic conceptuses is between 25% and 30% for each day of fetal age examined; this was calculated as a percentage of the total number of viable fetuses within any given litter. Dead fetuses or resorptions were infrequent, which suggests that the majority of trisomy 16 conceptuses survive the gestational period, at least in the period from nine days to term (19 days postconception). The trisomy 16 mice do not survive beyond birth. The authors consider that the defects of the heart and circulatory system in the trisomy 16 fetuses are not compatible with extrauterine survival.

MORPHOLOGY OF THE FETUS AND PLACENTA

The trisomy 16 phenotype has been investigated at the morphologic and biochemical levels in several laboratories.[10,16,17] The affected conceptuses are de-

TABLE 1. Composition of Litters Collected from Rb(16.17)/(11.16) Male X BALB/c Female Crosses

Day of Gestation	Total No. of Crosses	Implantation Sites	Viable Fetuses		Fetal Wastage	
			Normal	Trisomy 16	Dead Fetuses	Resorp- tions
9	1	12	4	8	0	0
10	2	12	9	2	0	1
		13	5	8	0	0
11	4	12.2	6.5	4.2	0	1.5
12	2	11	7	4	1	0
		12	9	2	0	1
13	3	14.0	6.3	3.6	0	4.0
14	3	9.3	4.6	3.6	0	1.0
15	6	10.8	6.3	3.0	0.3	1.2
16	5	9.2	5.6	2.8	0	0.8
17	8	10.0	6.1	3.3	0.1	0.5
18	7	9.6	6.0	3.6	0	0.1
Newborn[c]	9	—	6.8	—	0.7	—

Mean Number per Female[a,b]

[a] Numbers of litters were not large enough to calculate standard deviations.

[b] In those cases where only two litters have been collected, numbers from both are included.

[c] Seven dead fetuses were collected from the nine dams shortly after the entire litters were born. Six of the neonates had open eyelids. Four of the seven animals showed a trisomy 16 karyotype; karyotype analysis was not possible on the remaining three neonates.

scribed as being "developmentally retarded" or hypoplastic. They are frequently edematous and the eyelids of many trisomies fail to close. They also have cardiovascular defects and hydronephrosis.

A detailed analysis of the progression of morphologic development is under way in our laboratory to determine whether the development of all organ systems is affected similarly by the trisomic condition. Studies have been done for murine trisomy 19 from nine and one-half gestational days to term.[12] The morphologic development of all organ systems in this trisomic condition showed a uniform pattern of delay throughout gestation. This is not the case for the trisomy 16 fetuses that have been examined thus far. For example, the external phenotypes of affected conceptuses from the same litter, or from different litters, show a considerable degree of variation. Some appear similar to normal littermates while others appear more similar to fetuses of the previous gestational day (FIG. 1). This variability is not seen in the normal littermates. In most cases the trisomic fetuses weighed less than their normal littermates, where the measurement was not confounded by any recognizable edema. At 18 days, for example, the mean weight of normal conceptuses is 1.126 g, while that of the trisomies is 0.881 g.

The placentas of the trisomic and normal animals are being examined in detail for two reasons: (1) The placentas of all trisomies examined to date show signs of abnormal development and (2) a compromised placenta can be expected to play a major role in the altered growth and development of the affected fetus. A major feature of the trisomy 16 placenta is that it is reduced in size (FIGS. 2 and 3). This reduction is due to a decreased size of the labyrinth (FIG. 3). Decreased size of the

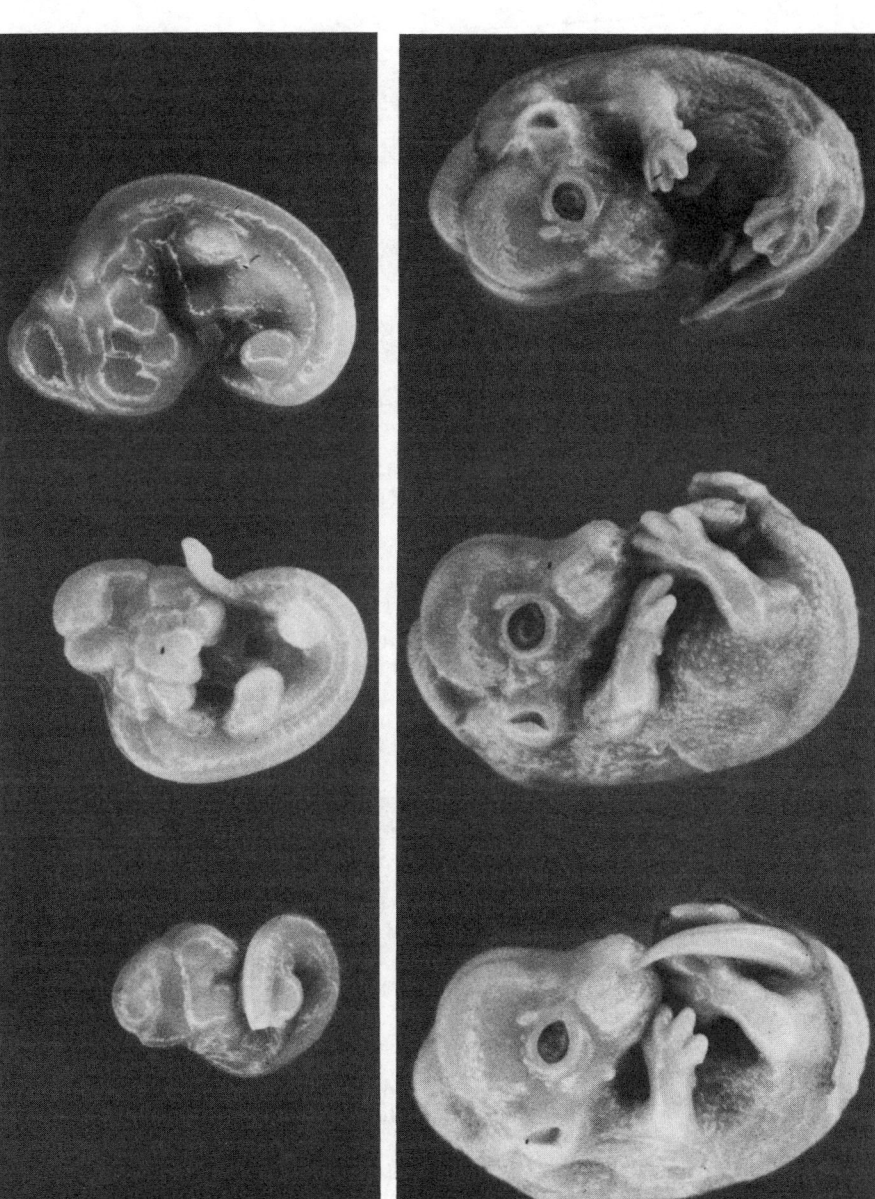

FIGURE 1. Variability in degree of morphologic development seen among trisomy 16 fetuses at 11 days (*top*) and 14 days (*bottom*). In each case, the normal fetus is in the center. Criteria for evaluating external phenotype include degree of limb, ear, eye, and tail development; appearance of branchial arches (11 days); and appearance and distribution of hair follicles and vibrissae (14 days).

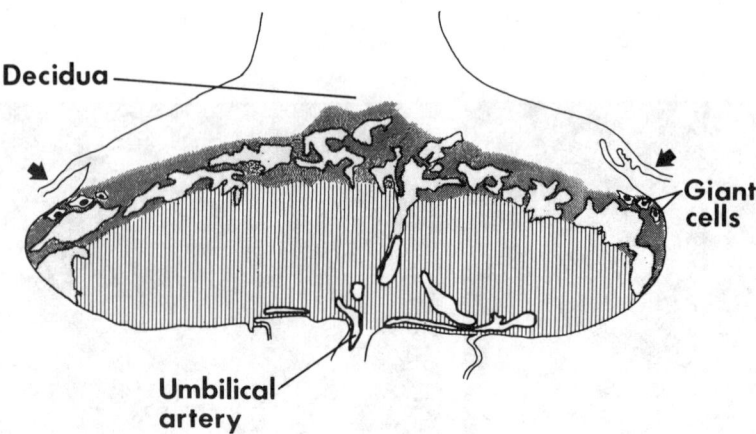

FIGURE 2. Diagram of a 15-day normal placenta, cut through center of placental disc (based on camera lucida drawing). The uterine wall (*large arrows*) is cut. *Vertical lines* show the labyrinth; *clear areas* within it are maternal sinusoids or fetal vessels. *Stippled area* is the basal zone. *Dark stipple* indicates areas populated by "glycogen cells," while *light stipple* indicates areas of "small trophoblast cells." (Terminology based on that of Davies and Glasser.[20]

labyrinth is also characteristic for murine trisomy 1[18] and trisomy 19 (Bersu, unpublished observations). In addition, Gearhart *et al.*[19] have reported a decreased surface-area ratio of fetal vasculature to that of the maternal blood sinuses in trisomy 16, based on a morphometric analysis of electron micrographs. These two elements are the major components that form the labyrinth.

There are several additonal features of the trisomy 16 placenta that make it easily distinguishable from the normal (Fig. 3). First, there are larger than normal numbers of small trophoblast cells within the labyrinth. Most of these are in the form of large clusters or strands of cells that persist to term. The strands extend from the basal zone of the placenta to the fetal surface. Delicate strands of these small cells are seen in 13- and 14-day normal placentas, where they surround the maternal blood sinuses ("septal trophoblast" of Davies and Glasser[20]); however, they are not normally present as large masses, and such masses are not seen in trisomy 19 placentas (Bersu, unpublished observations). Second, the basal zone, which is of trophoblast origin, is unusual in trisomy 16. This zone persists throughout gestation as a rather narrow layer composed of giant cells and small trophoblast cells. Normally, by 15 days, it differentiates into a broad band composed of clusters of small trophoblast cells and "glycogen cells,"[20] with only a small number of giant cells remaining at the edge of the placental disc.

The persistence of the islands of trophoblast cells within the labyrinth could reflect a failure of their usual transformation into components of the walls of the maternal blood sinuses. This could result from an alteration in the development of the fetal vascular bed. Failure of the basal zone to show any sign of differentiation into its more mature components is not understood at the present time; however, the observation indicates that the trisomy 16 genome significantly alters placental development, rather than just delaying its normal developmental progression.

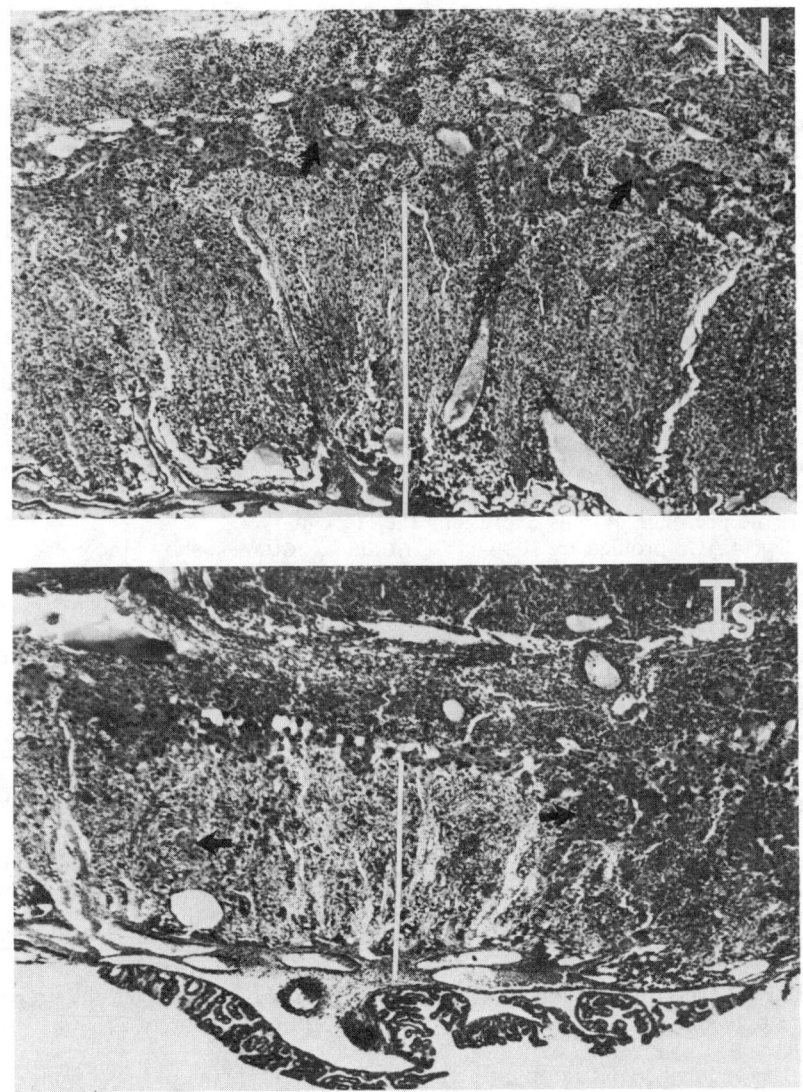

FIGURE 3.   Sections of placentas from 15-day normal (N) and trisomy 16 (Ts) fetuses, cut through center of placental disc. The *white vertical* line in each photomicrograph shows extent of labyrinth (see FIG. 2). The basal zone of the trisomic placenta consists almost exclusively of a narrow band of giant cells. *Arrows* indicate areas of small trophoblast cells in the basal zone of the normal placenta and in the labyrinth of the trisomic placenta. Original magnification, 40×; reduced to 60% of original size. Stained with hematoxylin.

HISTOCOMPATIBILITY ANTIGEN EXPRESSION ON PLACENTAL ENDOTHELIAL CELLS

The trisomy 16 fetuses produced in our laboratory carry one of the two number 17 autosomes in a Robertsonian translocation [Rb(16.17)8Lub]. Because the H-2 gene complex is located on chromosome 17 in mice, haplotyping is one means of identifying trisomic cells. Initial serological identification of the H-2 haplotype carried out by Dr. Chella David of the Mayo Clinic indicated that cells from the double metacentric males expressed the k haplotype at least at the H-2K end of the histocompatibility gene complex (manuscript in preparation). Because our trisomic embryos were generated using BALB/c dams (H-2d), it was possible to use an antibody to H-2K to distinguish embryo-derived cells from maternal cells in a mixed-cell suspension. This method was used to identify trisomic endothelial cells isolated from the placentas of 16-day-old mouse conceptuses. Collagenase was used to dissociate placental cells. After enrichment for endothelial cells by differential adhesion, the isolated cells were labeled with an anti-H-2k or anti-H-2b antiserum (NIH alloantisera). This was followed by labeling with a second antibody, fluoresceinated goat anti-mouse IgG (heavy and light chains) reagent. Labeled cells were then examined by flow cytometry (B-D FACS-IV) to determine the presence of cells expressing the H-2k marker.

Two FACS profiles are shown in FIGURE 4. FIGURE 4a shows the presence of H-2k positive cells in a suspension obtained from a trisomic placenta stained with anti-H-2k antiserum. FIGURE 4b shows cells from the same suspension stained with anti-H-2b antiserum. After correction for background labeling, 14% of the placental endothelial cells were stained with the haplotype-specific antiserum. Studies are now in progress to develop this technique for use in the isolation of trisomic placental cells by cell sorting.

LIGHT MICROSCOPICAL AND ULTRASTRUCTURAL STUDIES ON THE NERVOUS SYSTEM

Gross inspection of the brains obtained from the trisomy 16 mice and the normal control littermates revealed a marked reduction in the breadth and width of the trisomy cerebella and a reduction in the size of the colliculi. These two brain regions appeared disproportionately more affected than other brain regions examined. The trisomy and control littermates were removed from the pregnant dams between the 14th and 18th day after conception. A total of 25 animals was examined.

The cellular organization in the most affected brain regions was examined in serially sectioned brains, cut at 12 μm and stained by Nissl or hematoxylin and eosin procedures (FIG. 5). The cerebella of the trisomy 16 mice were reduced in size to 50–60% that of normal littermates. The external granule cell layer appeared reduced in thickness. The number of presumptive Purkinje cells per section was reduced in the trisomy 16 fetuses. The diameters of individual granule and presumptive Purkinje cells were similar in the trisomy and control animals. On a light-microscopic level, the most significant difference between the trisomy and control animals, at a given age, was the reduction in size of the cerebella.

To determine whether the cellular interactions in the 18-day fetal cerebella of trisomy 16 mice differed from control littermates, parasagittal sections were processed for examination in the electron microscope. The presumptive Purkinje

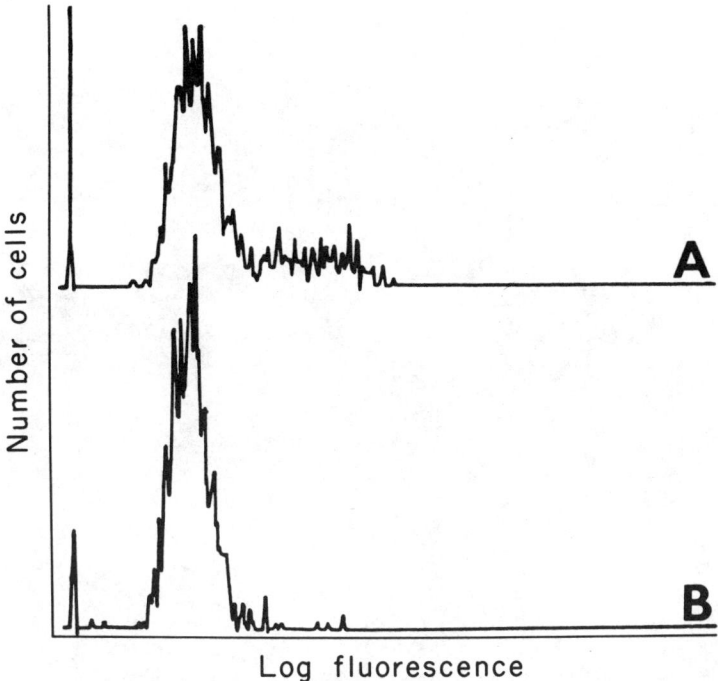

FIGURE 4. Histograms of endothelial cells obtained from a 16-day trisomy 16 placenta, using fluorescence-activated cell sorting. **(A)** Cells stained with anti-H-2k antiserum. **(B)** Cells stained with anti-H-2b.

cells appeared similar ultrastructurally in the trisomy and normal control animals (FIG. 6). The number of Purkinje cells entering into synaptic relationships with other cells was significantly reduced in the trisomy 16 fetuses. Montages were prepared of electron-micrographic images of the developing Purkinje cell layer. Each montage was of an area 1,400 square microns. Three trisomic fetuses from two litters and two control littermates were examined. The control cerebellar regions had an average of 18–25 identifiable synaptic contacts (vesicles, pre- and postsynaptic thickenings) per 1,400 square μm while these regions in the trisomy of the same litter had an average of six to eight identifiable contacts in the same area. A total area of at least 4,200 square microns was examined in cerebellar sections from each animal.

The endothelial cells of the vessels in the cerebella were examined because of: (1) the edema of the trisomy 16 animals; (2) the placental changes described above; and (3) because of an impression that the trisomy 16 fetuses have fewer vessels per unit area. The endothelial cells of the cerebellum had developed tight junctions in the trisomy and the control tissues. In the trisomy 16 vessels, however, there was an adhesion of leukocytes to the endothelial cells that was not seen in the controls (FIG. 7). The adhesion may be relevant to the edema because leukotrienes generated in the endothelial cells from arachidonic acid have been shown to increase both vascular permeability and leukocyte-endothelial cell adhesion.[21,22]

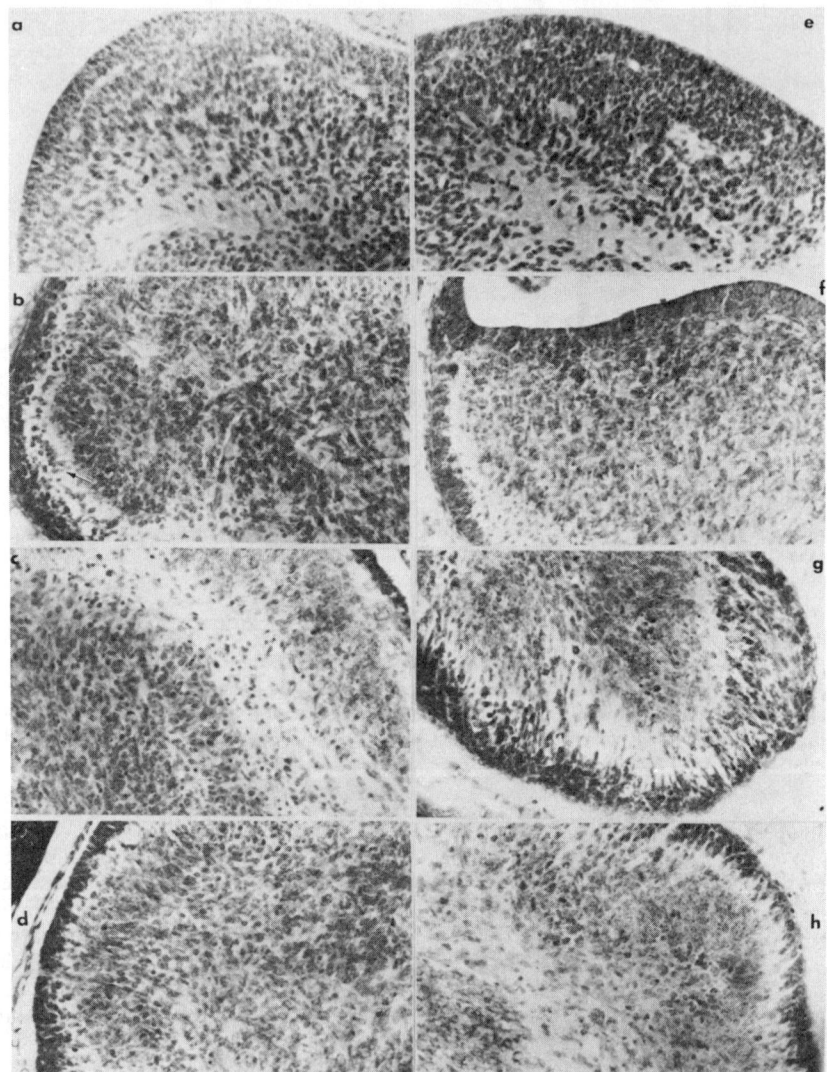

FIGURE 5. Photomicrographs of Nissl-stained sections (12 μm) of cerebella from normal control littermates **(a–d)** and trisomy 16 **(e–h)** mice at 14 days **(a,e)**, 15 days **(b,f)** and 18 days **(c,d,g,h)** after conception. Micrographs a–c and e–g are of parasagittal sections, while d and h are coronal sections. Many developing Purkinje cells can be recognized in the normal control sections at 15 days after conception (b, *arrow*). There is a paucity of these cells in the trisomy 16 brains at day 15. All micrographs are at an original magnification of 36×; reduced to 60% original size.

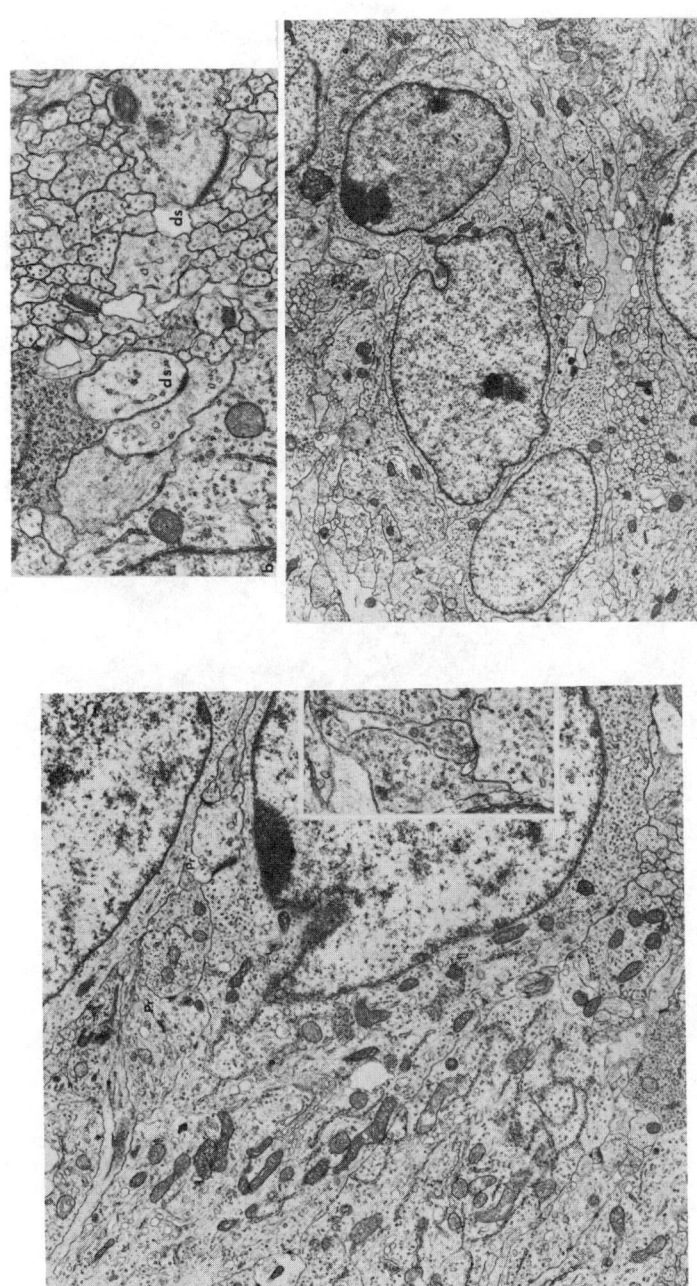

FIGURE 6. Electron micrographs of sections of cerebella from normal control (**a**) and trisomy 16 (**b,c**) mice at 17 days after conception. The cells visualized in (**a**) and (**c**) include Purkinje cells. The section of normal control contains synaptic endings on the soma of the Purkinje cell (a and *inset*) and the presynaptic ending is identified (Pr). The trisomy 16 cerebellum (b) possesses desmosomal-like contacts (ds) and these contacts frequently precede the appearance of typical synaptic contacts. Original magnifications: (a) 8,800×; (a, insert) 18,300×; (b) 18,300×; (c) 6,600×. All are shown reduced to 45% of original size.

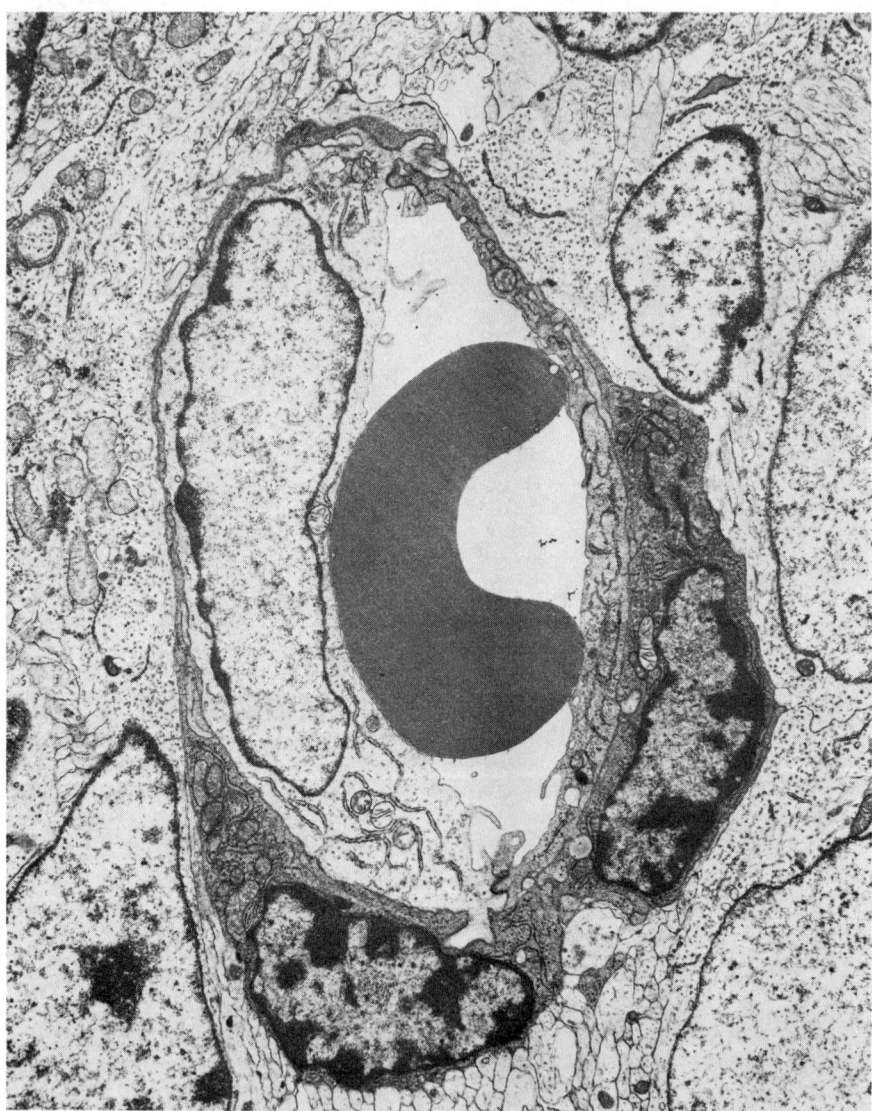

FIGURE 7.   Electron micrograph of a capillary in a trisomy 16 cerebellum at 17 days after conception. The capillary lumen contains a red blood cell and a white cell. The white cell appears to have formed junctional contacts with the capillary endothelial cell. Original magnification is 14,600×; reduced to 60% of original size.

TISSUE-CULTURE STUDIES OF THE NERVOUS SYSTEM

Because the trisomy 16 fetuses die at or before birth, tissue-culture techniques were used to investigate the survival and growth capacity of the tissues from the trisomy 16 mouse. Sixteen cultures were explanted. Seven of the cultures were excised from the cerebellum and nine from the tectum of a 17-day trisomy 16 fetus. Nine cultures of cerebellum and two cultures from tectum were excised from one normal littermate. The cultures were prepared as described previously.[23] All cultures were fixed for electron microscopy after 5 and 11 days *in vitro*.

All cultures from normal and trisomy fetuses developed outgrowth within the first 24 hours after explantation. The outgrowth zone increased with the time *in vitro* (FIG. 8). The first growth was formed by a multitude of fine fibers, apparently of neuronal origin, growing in a criss-cross manner, and forming a tight mesh around the explant. Within 48 hours, glial cells joined the fibrillar growth. Some of the glia were of astrocytic origin. In the outgrowth of cerebellar explants, small round cells with two or three short, branched processes were observed. These cells appear to correspond to granule cells. By the third day after explantation, some fibers formed bundles. The growing bundles displayed different directional preferences. Some grew radially, some encircled the explant, and some looped back at varying distances from the explant. The new growth from the normal and trisomy tissues was similar in extent and in directional preferences. Several days after explantation, nests of large cells that were presumably Purkinje neurons were seen within clear areas of the explanted cerebella of normal and trisomy fetuses.

Electron micrographs of two cerebellar explants supported the above observations on live tissues. At 11 days in culture, the growing neuronal and glial processes were present in both trisomy and control cultures. The growing processes, however, showed little if any tendency to interact. The growing processes did not yet form the glial cover that is observed in cultures of regenerating cerebella prepared from the postnatal and adult mouse. Purkinje (FIG. 9) and granule cells could be identified in both groups in culture. Their ultrastructure corresponded to that of cells in normal and trisomy fetal cerebella that were not cultured. Synaptic profiles were observed in cultures of both control and trisomic fetuses (FIG. 10). Numerous mitotic cells were observed in explants from the trisomic tissues.

MAJOR HISTOCOMPATABILITY ANTIGENS ON NEURAL CELLS OF TRISOMY AND CONTROL FETUSES

Our working hypothesis is that aberrant cell–cell interactions in the trisomy 16 mice and trisomy 21 human result from an imbalance of receptors on cells in the trisomy compared with normals. The altered response of the trisomy cells to beta-adrenergic agonists and the altered response to interferon indicate that many trisomy 16 cells may have a different receptor distribution on their cell surface than do control cells obtained from the same organ (Reed *et al.*[9] and Cox *et al.*[8]). The reported increase in the expression of interferon receptors (IFNR) on cells of trisomy 16 led to an examination of the level of H-2 markers on the nerve cells by immunohistochemical methods. Frozen sections of brain, including cerebella and colliculi, were cut at a thickness of 8 μm and then fixed in 80% ethanol. After the fixed sections were dried, they were incubated first with normal goat serum to

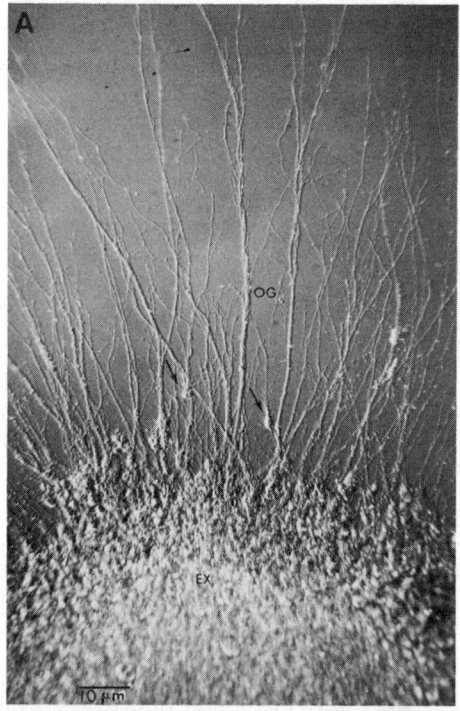

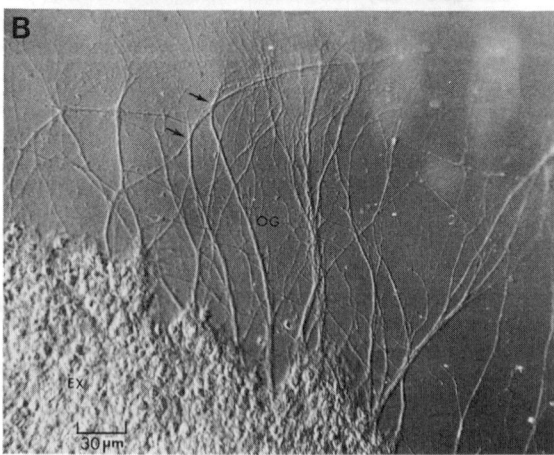

FIGURE 8. Fibrillar growth from cerebella of a trisomy 16 and normal control mouse, explanted at 17 days postconception. **(A)** Normal control, five days in culture. **(B)** Trisomy 16, ten days in culture. In both cases, note the criss-cross manner in which the fibers grow. Glial cells (*arrows*) are seen in the outgrowth zone. EX, explant. OG, outgrowth zone. Photographs were taken using Nomarski differential interference optics with a Leitz NPL Fluotar lens. Original magnification, 16×; reduced to 65% of original size.

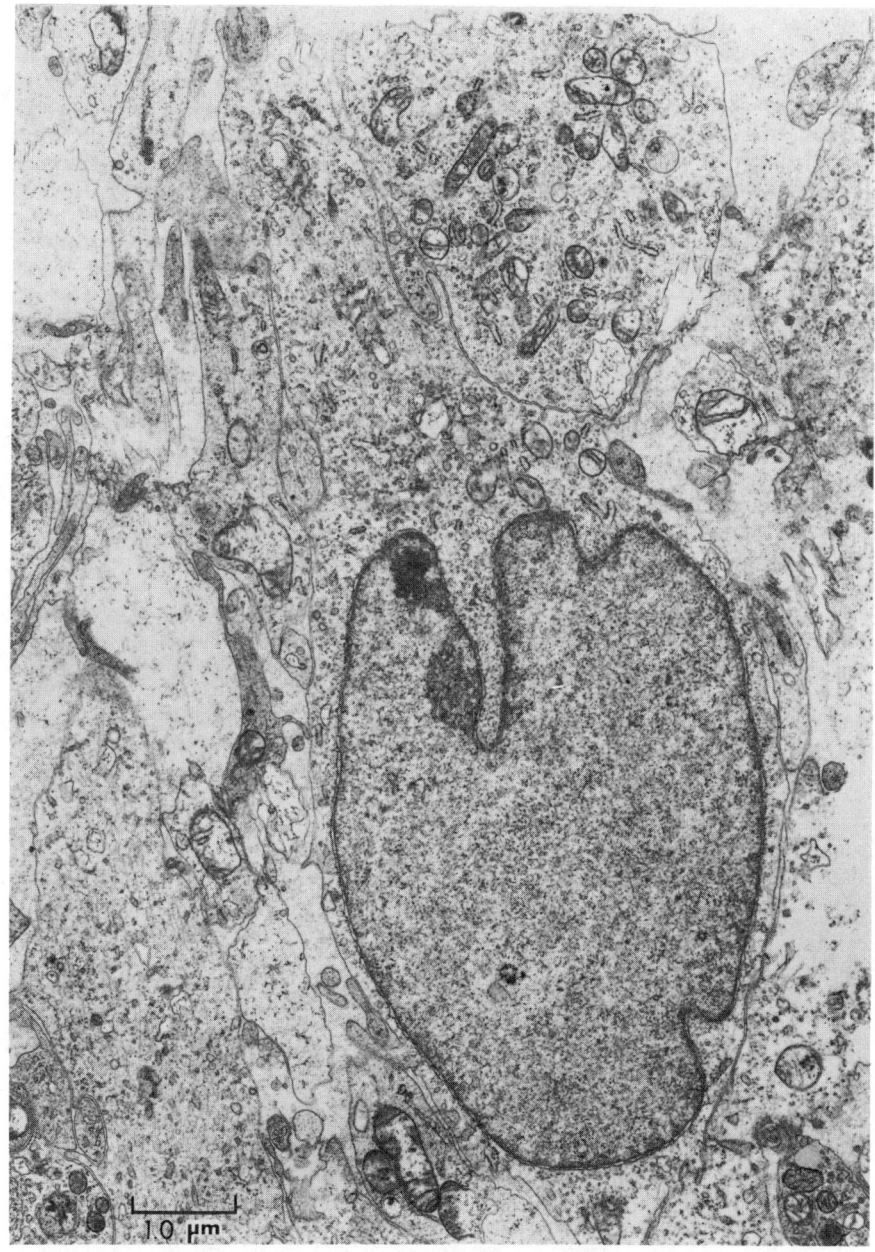

FIGURE 9. A young neuronal cell, presumably a Purkinje cell, in an 11-day culture from trisomy 16 mouse cerebellum. The invaginations of the nuclear membrane face the base of the dendrite. The cerebellum was explanted from a 17-day postconception fetus. Magnification, 14,000×.

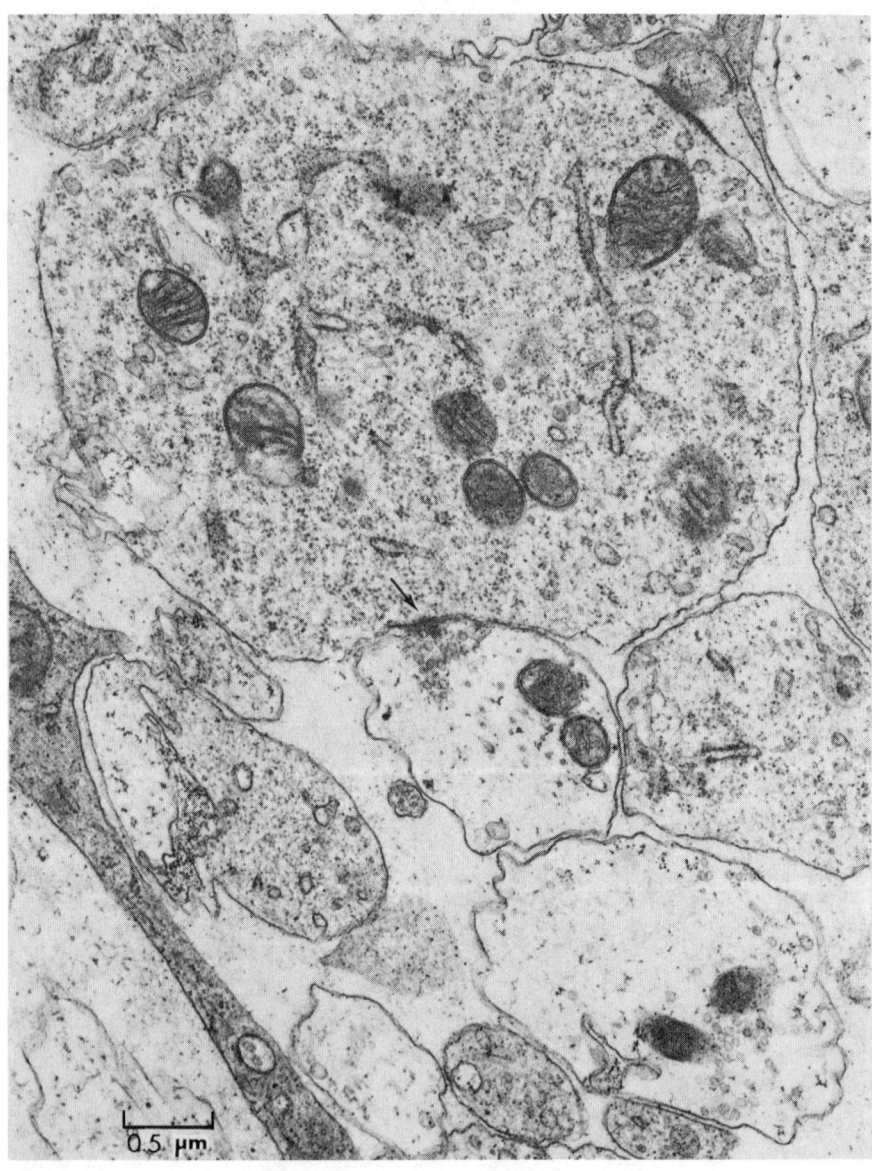

0.5 μm

FIGURE 10. A synaptic contact (*arrow*) in an 11-day culture from a fetal trisomy 16 cerebellum explanted at 17 days postconception. Magnification, 26,000×.

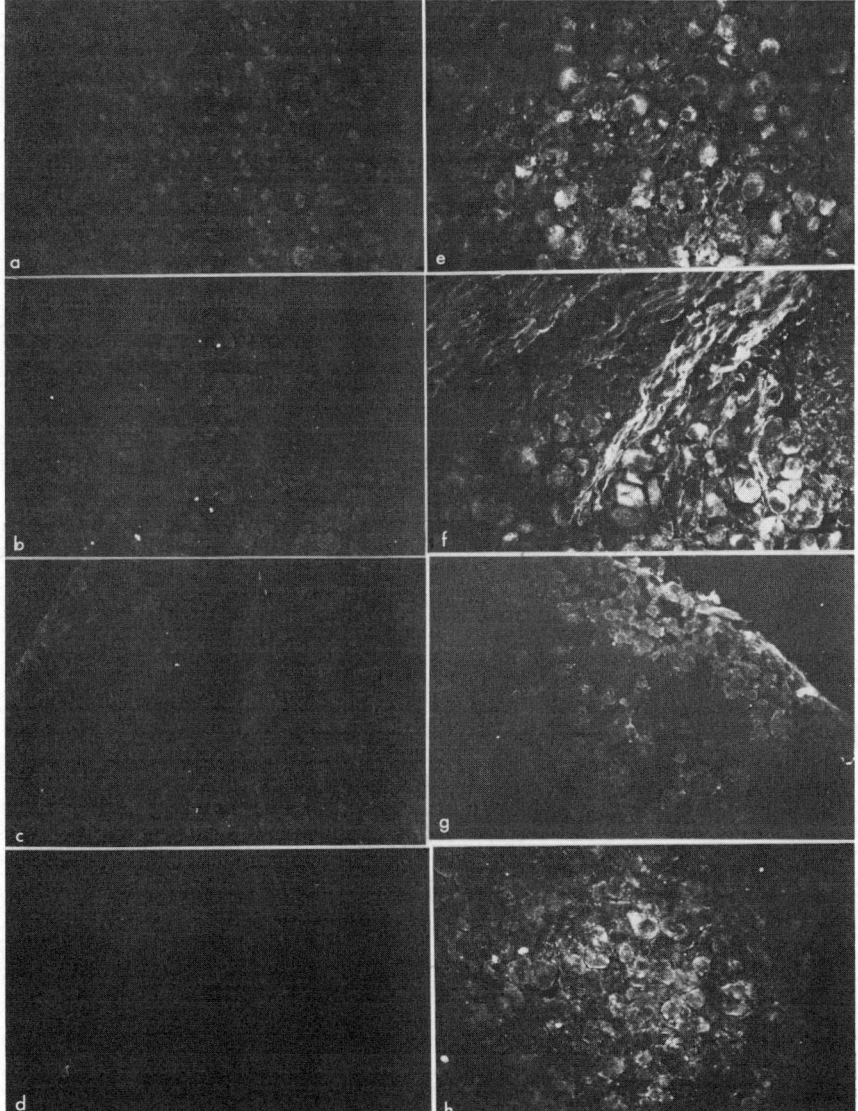

FIGURE 11. Sections of neural tissue of control **(a,b)** and trisomy 16 **(c–h)** mice after prior incubation with anti H-2K, H-2b antibodies and with fluorescein-labeled second antibody. Section **(a)** was colliculus incubated with anti-H-2b and then with second antibody; **(b)** was colliculus incubated with anti-H-2k and then with second antibody; **(c)** was colliculus incubated with anti-H-2b and second antibody; **(d)** was with second antibody alone; **(e,f)** were ganglia from trisomy 16 incubated with anti-H-2k and second antibody; **(g)** was cerebellum incubated with anti-H-2k and second antibody; **(h)** was colliculus incubated with anti-H-2k and second antibody. All photos were exposed to the ultraviolet source for the same time (3 min) and printed in the same manner. Original magnification is 36× for all photomicrographs. (Shown here reduced to 65% of original size.)

block nonspecific adsorption and then with alloantiserum (NIH D-25) that reacts with H-2Kk. Other sections were reacted with anti-H-2Kb, and a third group of sections were incubated with second antibody alone (fluorescein-labeled goat anti-mouse IgG). The third group of sections was not exposed to the mouse immunoglobulins.

The trisomy 16 cerebella and tectums of 17-day fetuses exhibited a higher degree of binding anti-H-2Kk antibodies than did the tissues from control littermates (FIG. 11). The H-2Kk markers were distributed on the majority of neurons in the trisomy animals. Incubation of the tissues with the anti-murine H-2Kb revealed a low level of reaction with both trisomy 16 and control littermates. Incubation of the tissues with second antibody alone (anti-mouse immunoglobulins) revealed no reactions.

CONCLUSIONS

(1) When double metacentric males, Rb(16.17)8Lub/Rb(11.16)2H, were bred against BALB/c females, a significant number of the resulting trisomy 16 fetuses survive to term or near term.

(2) Trisomy 16 fetuses show considerably greater variation in the degree of morphologic development than normal littermates. The trisomy 16 placenta is reduced in size and shows abnormal development of the labyrinth and basal zone compared with normal placentas.

(3) Cells derived from trisomy 16 placentas show a positive reaction with antibodies against the H-2k portion of the MHC. For this reason flow cytometry can distinguish the trisomic embryo-derived cells from maternal cells.

(4) The cerebella of the trisomy 16 fetuses appeared to be more affected than other brain regions when compared with control littermates. The number of synaptic contacts per unit area appeared reduced in trisomic cerebella. White blood cells were adherent to endothelial cells in the trisomy 16 fetuses at fetal day 18.

(5) The presence of an additional chromosome 16 does not appear to influence either the survival or capacity to grow of the cerebellar nerve cells *in vitro*, nor does it prevent formation of synaptic connections *in vitro*.

(6) The neural cells of the trisomy 16 mice have a higher level of H-2Kk on their cell surface than control littermates at the same fetal age. We may not conclude at present that this increased expression is related to the trisomy 16 phenotype. It is possible that the normal nonmetacentric chromosome 17 does not express H-2Kk. Alternatively the increased level of H-2Kk may indeed be a consequence of altered cell surface markers on the trisomy cells. (e.g. IFNR).

REFERENCES

1. SÉGUIN, E. 1846. Le Traitment Moral, l'Hygiène et l'Education des Idiots. Ballière, Paris.
2. DOWN, J. L. H. 1866. Observations on an ethnic classification of idiots. London Hospital Clinical Lectures and Reports. **3:** 259–262.
3. LEJEUNE, J., R. TURPIN & M. GAUTIER. 1959. Le mongolisme, premier exemple d'aberration autosomique humaine. Ann. Genet. **1:** 92–101.
4. BERSU, E. T. 1980. Anatomical analysis of the developmental effects of aneuploidy in man: The Down syndrome. Am. J. Med. Genet. **5:** 399–420.
5. SHAPIRO, B. L. 1983. Down syndrome—a disruption of homeostasis. Am. J. Med. Genet. **14:** 241–269.

6. HOEHN, H., M. SIMPSON, E. M. BRYANT, P. S. RABINOVICH, D. SALK & G. M. MARTIN. 1980. Effects of chromosome constitution on growth and longevity of human skin fibroblast cultures. Am. J. Med. Genet. **7:** 141–154.
7. GROPP, A. & H. WINKING. 1981. Robertsonian translocations: Cytology, meiosis, segregation patterns, and biological consequences of heterozygosity. Symp. Zool. Soc. Lond. **47:** 141–181.
8. COX, D. R., L. B. EPSTEIN & C. J. EPSTEIN. 1980. Genes coding for sensitivity to interferon (IfRec) and soluble superoxide dismutase (SOD-1) are linked in mouse and man and map to mouse chromosome 16. Proc. Natl. Acad. Sci. USA **77:** 2168–2172.
9. REED, W. D., M. L. OSTER-GRANITE, S. A. FISCHKOFF, P. J. RAAB & P. T. OZAND. 1983. Beta-adrenergic markers in cultured cells of human trisomy 21 and murine trisomy 16. Pediatr. Res. **17:** 141A.
10. MIYABARA, S., A. GROPP & H. WINKING. 1982. Trisomy 16 in the mouse fetus associated with generalized edema and cardiovascular and urinary tract anomalies. Teratology **25:** 369–380.
11. REDI, C. A., S. GARAGNA, B. HILSCHER & H. WINKING. 1985. The effects of some Robertsonian chromosome combinations on the seminiferous epithelium of the mouse. J. Embryol. Exp. Morph. **85:** 1–19.
12. BERSU, E.T. 1984. Morphologic development of the fetal trisomy 19 mouse. Teratology **29:** 117–129.
13. WHITE, B. J., J.-H. TJIO, L. C. VAN DE WATER & C. CRANDALL. 1972. Trisomy for the smallest autosome in the mouse and identification of the T1Wh chromosome. Cytogenetics **11:** 363–378.
14. GROPP, A. & G. GROHE. 1981. Strain background dependence of expression of chromosome triplication in the mouse embryo. Hereditas (abstract) **94:** 7.
15. WHITE, B. J., C. CRANDALL, E. RAVECHE & J.-H. TJIO. 1978. Laboratory mice carrying three pairs of Robertsonian translocations: Establishment of a strain and analysis of meiotic segregation. Cytogenet. Cell Genet. **21:** 113–138.
16. OSTER-GRANITE, M. L. & G. HATZIDIMITRIOU. 1985. Development of the hippocampal formation in trisomy 16 mice. Pediatr. Res. **19:** 328A.
17. SINGER, H., M. TIEMEYER, J. C. HEDREEN, J. GEARHART & J. T. COYLE. 1984. Morphologic and neurochemical studies of embryonic brain development in murine trisomy 16. Dev. Brain Res. **15:** 155–166.
18. CLAUSSEN, C. P. & U. ZIMMERMAN. 1977. Untersuchungen der Placenta bei fetaler Trisomie der Maus. Verh. Anat. Ges. **71:** 613–621.
19. GEARHART, J. D., M. L. OSTER-GRANITE & G. HATZIDIMITRIOU. 1985. Placental development in the trisomy 16 mouse. Pediatr. Res. **19:** 325A.
20. DAVIES, J. & S. R. GLASSER. 1968. Histological and fine structural observations on the placenta of the rat. Acta Anat. **69:** 542–608.
21. FORD-HUTCHINSON, A. W. 1985. Leukotrienes: Their formation and role as inflammatory mediators. Fed. Proc. **44:** 25–29.
22. STAUB, N. C., E. L. SCHULTZ, K. KOIKE & K. H. ALBERTINE. 1985. Effect of neutrophil migration induced by Leukotriene B$_4$ on protein permeability in sheep lung. Fed. Proc. **44:** 30–35.
23. SOBKOWICZ, H., B. BEREMAN & J. E. ROSE. 1975. Organotypic development of the organ of Corti in culture. J. Neurocytol. **4:** 543–572.
24. JOHNSTON, H. F., H. M. SOBKOWICZ, G. L. SCOTT & C. V. LEVENICK. 1984. Regeneration in the cerebellum of the postnatal and adult mouse in long-term cultures. Soc. Neurosci. Abstr. **10 (2):** 1024.

DISCUSSION

QUESTION: Since collagen affects the structure you are going to study, have you taken its effects into account in your culture system?

STEVEN KORNGUTH: At this stage of the study, we have not addressed which components in the culture medium permit the cultures to survive. The question that we are addressing is: Is the trisomy 16 cerebellum intrinsically unable to grow and survive beyond day term (i.e., day 20 of fetal life)? The answer is that these cerebella can grow 12–15 days *in vitro* when explanted at day 18 of fetal life and do very well in their growth. The cultures from the trisomy are not perceptibly different from the normal controls.

JOSEPH FRENCH: Have you done PET scanning or NMR, which would allow you to infer how rapidly metabolites are infused into brain and utilized?

STEVEN KORNGUTH: That is an excellent question. We have injected radiolabeled beads into the maternal blood to determine whether the delivery of blood to the placenta, from the maternal side, was similar in the trisomy 16 placenta and the normal control placenta. The answer is that the maternal sides of the placentas are similar. What isn't known is the delivery of the blood components from the fetal side of the placenta into the fetus. We are initiating a study to examine blood flow in the trisomy 16 and normal controls. These are single-exposure experiments. We have a new MRI system at Wisconsin; MRI imaging is, as you know, dependent on 2–3-millimeter slices. Since the entire animal's brain approximates 5 millimeters, the MRI is not well suited for these studies.

QUESTION: What happens to these animals after they are born? What is their life history?

EDWARD BERSU: We have seen the survival of two trisomy pups for a couple of hours. Most of the trisomy fetuses do have severe heart defects that would not be compatible with life at the time of birth. They are not of the variety, however, that would prohibit a normal gestation period.

STEVEN KORNGUTH: It should be pointed out that in terms of Down's syndrome conceptuses, the large majority do not survive at birth. The population of murine trisomy 16 fetuses we are examining is most comparable, we believe, to the most affected human trisomy 21 population.

EDMUND JENKINS: What is the feasibility of developing a partial trisomy 16 mouse that will survive?

DAVID KURNIT: David Cox is trying to do that, and basically he has made a construct that has less of chromosome 16, but he has yet to get one that is viable.

Molecular Dysmorphology: An Approach to Down's Syndrome

DAVID M. KURNIT

Genetics Division
The Children's Hospital
Boston, Massachusetts 02115

I wish to propose the following hypothesis to explain how gene dosage for genes encoded by chromosome 21 could yield birth defects in Down's syndrome: Increased dosage for genes on chromosome 21 increases the adhesiveness among trisomy 21 cells during embryogenesis, resulting in specific deficiencies of outgrowth of embryonic tissues that are later recognizable as characteristic congenital anomalies.

The fundamental question posed by Down's syndrome is how one can go from the karyotypic abnormality of trisomy 21 to the phenotypic abnormalities of Down's syndrome. Down's syndrome differs from more easily understood Mendelian syndromes: In Down's syndrome the alteration of genetic material is quantitative, not qualitative. Thus the genetic material is normal in Down's syndrome, and the error is "too much of a good thing."

Several years ago, I began to look at the cascade of abnormalities that one might expect from trisomy 21. I affirmed that dosage for genes on chromosome 21 occurred at the transcriptional level in fibroblasts from subjects with monosomy 21, normal subjects (disomy 21), and those with Down's syndrome (trisomy 21). Yoram Groner and colleagues have also affirmed the transcriptional basis for gene dosage using a cloned DNA probe from a gene, superoxide dismutase, encoded by chromosome 21. Several researchers, including Charles Epstein, David Cox, David Patterson, and W. Ted Brown, have shown that the transcriptional increase is reflected at the translational level, so that increased dosage for proteins encoded by genes on chromosome 21 is observed. To describe the concept of gene dosage and trisomy 21, I use the analogy of three factories turning out three fleets of cars, instead of the normal two factories turning out two fleets of cars. The next, more difficult question is: How does having too much of this good thing result in developmental anomalies?

As one looks at the list of anomalies associated with Down's syndrome, three points stand out: specificity, variability, and hypoplasia. The specificity is great enough so that individuals with Down's syndrome may be diagnosed by inspection, both externally (by the characteristic facies) and internally (by the characteristic endocardial cushion defects, the shape of the iliac bones, and propensity to specific defects such as duodenal obstruction). To illustrate the variability, only 40% of children with Down's syndrome have clinically significant congenital heart

a The data presented in this paper were published in 1985 in Volume 450 (pp. 191–204) of the *Annals of the New York Academy of Sciences* in an article entitled "Genetics of Congenital Heart Malformations: A Stochastic Model" by David M. Kurnit, John F. Aldridge, Rachael L. Neve, and Steven Matthysse. Included here are some of his more general comments on the problems of elucidating the molecular basis for the congenital malformations in Down's syndrome.

defects and 1% manifest overt duodenal obstruction. Edward Bersu and Drs. Epstein, Cox, Miyabara, and Gropp have found analogous specificity and variability looking at the rodent model for Down's syndrome, i.e., trisomy 16 in the mouse.

Christine Cronk in 1978 performed measurements that confirmed that persons with Down's syndrome are indeed shorter than average. Tom Kemper has pointed out that when one looks at differences in brain growth, the brain of a child with Down's syndrome is moderately small at birth, and lags even further behind in its growth rate during early childhood so that the microcephaly is more apparent in later childhood; Dr. Kemper speculates that the deficiency may reflect hypoplasia of a specific type of small neuron. Thus, both the entire body and the brain of persons with Down's syndrome are smaller than average, demonstrating that hypoplasia is an important part of the syndrome.

To make some sense out of the "laundry list" of abnormalities that we see, I would propose the following underlying rationales. (I must credit the late David Smith for many of these approaches). The microbrachycephaly in Down's syndrome reflects a brain with greater deficiency in anteroposterior outgrowth than bitemporal growth, apparently due to more marked deficiencies in frontal brain outgrowth. The characteristic facies of the Down syndrome is likely secondary to the relatively deficient frontal brain outgrowth. The most helpful way to look at a person with Down's syndrome is from the side: the lateral aspect reveals most dramatically the midface hypoplasia that results from lack of frontal brain outgrowth. This explains some of the other features that can be observed *en face,* such as the upslanting palpebral fissures, epicanthal folds, flat face, and short maxilla resulting in prominence of the relatively larger tongue; thus, a glance at the face is a glance at specific problems of deficient brain outgrowth during neurogenesis. The simian crease reflects short metacarpals at about 13 weeks of development, and the incurved fifth finger results from hypoplasia of the middle phalanx of that finger.

Taken together, the relating underlying theme is deficient outgrowth of specific tissues. The specificity negates simple theories hypothesizing that pancellular growth problems (e.g., those due to excess DNA interfering with DNA replication) are the cause. Not only did Holger Hoehn find that trisomy 21 cells propagate normally in culture, but we see other syndromes with excess DNA (e.g., trisomy 18 and trisomy 13) are distinct from trisomy 21 and from each other.

As suggested by other papers in this volume, one of the best places to look for intercellular specificity during development is at the cell surface. A number of molecules responsible for intercellular adhesion and migration during organogenesis have been localized to the cell surface. One can imagine that dosage for cell surface genes encoded by chromosome 21 would affect the way cells adhere, migrate, and otherwise interact during early development. This would predict that trisomy 21 cells are more adhesive than normal cells during development, and that models could be developed to explain why only a plurality of individuals with Down's syndrome manifest any of the particular defects associated with trisomy 21. In the previous paper published in the *Annals* (see footnote on page 179) that summarizes recent progress in my laboratory, affirmative evidence is presented for both of these points. First, in collaboration with Thomas Wright and Roslyn Orkin, fetal trisomy 21 fibroblasts from lung and heart, but not from skin, were found to aggregate more rapidly *in vitro* than were control fibroblasts. Second, in collaboration with Steven Matthysse and Jeff Aldridge, a simple model was presented using computer simulations of endocardial cushion outgrowth to explain how increased adhesiveness among endocardial cells can yield cardiac defects in a plurality of subjects. On the basis of these experimental and modeling approaches, we are now attempting to isolate genes on chromosome 21 that might mediate the increased adhesiveness we observed among trisomy 21 fibroblasts. In the long run, the availability of a rodent model for trisomy 21 suggests that it will be feasible to introduce the mouse homologues to genes on human chromosome 21 into mouse embryos, and to study how perturbations of the dosage for these genes affects morphogenesis.

Although we do not know the ultimate relevance of this model to the particulars of Down's syndrome, the elaboration of the model has heuristic value: It represents an example of approaching embryologic processes by "thinking like a cell," using the concept of relating a cluster of defects to their earliest identifiable single cause, making appropriate mathematical models, elaborating a unifying hypothesis, and then applying molecular methodologies to test the hypothesis.

DISCUSSION

STEVEN KORNGUTH: Was the temporal factor included in the modeling approach? For example, cell–cell adhesiveness may be expected to increase abruptly at particular stages of development corresponding to active periods of synaptogenesis.

DAVID KURNIT: We did not do that. A more elaborate simulation where we would try to model the entire AV canal would require a lot more iterations, and the simulations we performed were complex and time-consuming as they were.

EDWARD BERSU: Do you have any idea what types of cell surface phenomena are involved?

DAVID KURNIT: We are checking into the possibility that chromosome 21 encodes some type of glycosyl transferase. Increased glycosylation of cell surface markers could modify adhesiveness. We have looked at several transferases and we do not see differences as yet.

GENE FISCH: How many simulations did you run in any particular set of parameters?

DAVID KURNIT: Twenty-five to one hundred simulations.

GENE FISCH: It is not certain whether the pattern you are getting is a typical pattern of adhesion or the pattern that you were expecting. Were you satisfied with the results?

DAVID KURNIT: We ran sufficient numbers of simulations so that the statistical significance is satisfactory. We are satisfied.

GENE FISCH: There are all kinds of interesting parameters that could be introduced to the model, and these may affect the way the simulation came out. I am not arguing that your simulation is a bad one, but rather arguing that if you had used a larger number of simulations, say 1,000, you would be able to determine the statistical properties of the simulation.

DAVID KURNIT: We are continuing to refine the model.

MARTHA LAW: How do you get the monoclonal antibodies? Are they chromosome-21-specific?

DAVID KURNIT: They would appear to be. We injected C3H mice with the trisomy 21 fetal lung fibroblasts. Then we did fusions, and we screened the hybridomas for supernatants that reacted in an ELISA assay with mouse-human 21 hybrids and with the original trisomy 21 fibroblasts, but did not react with the mouse parent cell line.

DAVID SOIFER: This, I presume, is a lambda-gtll library?

DAVID KURNIT: Yes.

DAVID SOIFER: Were these genomic inserts? Or were these cDNA?

DAVID KURNIT: This was a cDNA library from the Down's syndrome fetal lung fibroblast cell line.

DAVID SOIFER: Did you generate your cDNA with a method that you selected for the poly(A) end or with a standard transcriptase/S1 nuclease?

DAVID KURNIT: We used oligo-dT priming and RNase H protocol.

TED BROWN: What is known about the genetics of the cell surface adhesion molecules? These molecules, which I believe are often glycoproteins, might be studied directly for gene dosage effects.

DAVID KURNIT: We do not know which, if any, of the identified cell surface adhesion molecules might be related to the phenomenon that we are seeing. There have been several such genes cloned; however, at present they are not available to us and have not been analyzed in relation to the gene dosage in Down's syndrome.

Introduction

DONALD A. SNIDER

*Institute for Basic Research in Developmental Disabilities
New York State Office of Mental Retardation &
Developmental Disabilities
Staten Island, New York 10314*

This session examines brain development and damage from the prenatal to postnatal periods. These authors examine the problems from different methodological perspectives and assess the potential for the understanding and prevention of mental retardation and developmental disabilities. Dr. Lucy Rorke of the Philadelphia Children's Hospital discusses how, over the past 25 years, the use of new technologies and services to prolong life in the prematurely born neonate have led to lower risk of permanent brain damage.

Dr. Alfred Brann, Jr., of Emory University Medical School in Atlanta presents data and a rationale for devoting more research to full-term as opposed to preterm births as the most feasible way to make further inroads in reducing brain disorders. An abstract of Dr. Brann's paper is included in these proceedings.

In an abstract of his presentation, Dr. Myron Winick outlines why both proportionate and disproportionate intrauterine growth retardation may have a final common pathway.

Dr. John Sturman reports on nutritional taurine and central nervous system development. Taurine is an amino acid that is found in abundant quantities in human breast milk but is missing from cow's-milk-based infant formulas. Research by Dr. Sturman and other scientists at the IBR led the U.S. Food and Drug Administration to allow taurine to be added to infant formula.

The Changing Pattern of Damage to the Perinatal Brain

LUCY BALIAN RORKE

Department of Neuropathology
The Children's Hospital of Philadelphia
Philadelphia, Pennsylvania 19104

Changes that have occurred in care and treatment of the neonate during the past quarter century have been remarkable indeed. Neonatology has developed as a complex and sophisticated subspecialty of pediatrics, and with it the introduction and increasing use of ultrasound, computerized tomographic (CT) scans, and now the magnetic resonance imaging (MRT) scan for diagnosis of intracranial lesions in the fetus and newborn. Moreover, establishment of neonatal intensive-care nurseries, use of complex monitoring equipment, and treatment of serious cardio-respiratory, metabolic, and gastrointestinal problems that are the common plague of the prematurely born, tiny infant have combined not only to increase the probability of survival, but to increase the likelihood that the life of that tiny creature will be relatively normal.

On the darker side though, both the death rate and probability of significant neurodevelopmental dysfunction increases as birth weight decreases.[1-3] Whereas those who die are of interest primarily to the pathologist, infants who survive may present a problem for themselves, their families, and society in general, especially if they are seriously disabled.

Although the major thrust of this symposium focuses on mental retardation and developmental disabilities, it has long been known that the neuropathologist has not contributed much to understanding the former, especially if children with chromosomal and metabolic defects are excluded.[4] On the other hand, children with developmental disabilities that may or may not include intellectual deficit often have an underlying brain or spinal cord lesion.

In view of the extraordinary expenditure of money and effort to prolong the life of the prematurely born neonate, we might well ask whether those who would otherwise have sustained permanent brain damage 25 years ago now have a lower risk of doing so. The epidemiologists among us will surely address this issue. From the vantage of the neuropathologist, it has become apparent that some lesions that were not uncommon 25 years ago are now seen only rarely or in altered form, whereas others now appear with increasing frequency. Still others continue to occur regularly.

The first group, namely brain lesions that have become relatively rare or whose manifestations have changed, includes kernicterus and meningitis. Babies with jaundice are currently rarely given exchange transfusions so that complications that might cause death consequent to this procedure have virtually been eliminated. This change has paralleled widespread use of "bili-lights" and has effectively kept bilirubin levels far below those values formerly common; thus, kernicterus, if now seen, is a pale counterpart indeed to the dramatic lesions encountered at the autopsy table in years past. The structure most commonly affected now is the thalamus, whereas damage was more typically localized in pallidum, hippocampus, and certain brain stem nuclei, that is, inferior colliculi, cranial nerve nuclei along the floor of the fourth ventricle, and the inferior olivary nuclei.[5]

Meningitis in the neonate, though fortunately not common, was formerly usually consequent to gram-negative organisms such as *E. coli.* In recent years, neonatal meningitis, though still rare, is more likely to be caused by *Streptococcus* Group B, an organism that has a more consistently devastating effect, as a severe vasculitis with vascular necrosis may produce subarachnoid hemorrhage and hemorrhagic necrosis of the brain along with the characteristic purulent exudate (FIGS. 1 and 2).

Four types of lesions continue to occur in spite of the best efforts of neonatologists: matrix-zone hemorrhage, periventricular leukomalacia, white-matter gliosis, and necrosis of thalamus with or without similar necrosis in striatum or lenticular nucleus.

Twenty-five years ago matrix-zone hemorrhages were chiefly of interest to pathologists and neurosurgeons, as it was thought that all infants who sustained such hemorrhages either died or developed hydrocephalus. Occasionally, an infant would come to necropsy for some lethal, non-central-nervous-system (CNS) disease, and brain examination would reveal a clinically silent, organized matrix-zone hemorrhage, not necessarily associated with hydrocephalus.

Increasing use of ultrasound and CT scan has reversed this situation, and pediatricians currently manifest intense interest in these lesions as it is now obvious that a considerable number of infants with such hemorrhages do not die. Towbin[6] has suggested a relationship between this lesion and mental retardation, but proof, to date, is lacking. Cells in the germinal matrix of the human brain consist primarily of spongioblasts that are destined to mature into astrocytes or

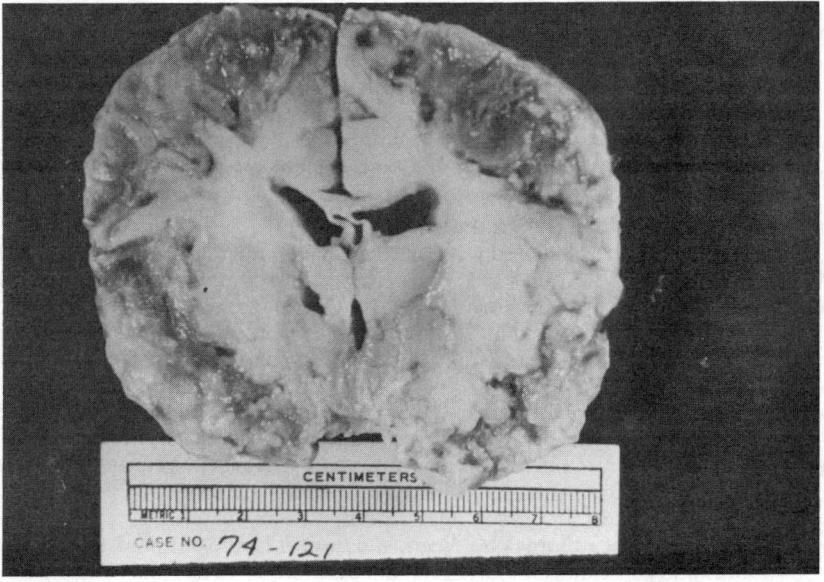

FIGURE 1. Coronal section of cerebrum from one-week-old infant showing hemorrhagic necrosis secondary to *Streptococcus* Group B meningitis and vasculitis.

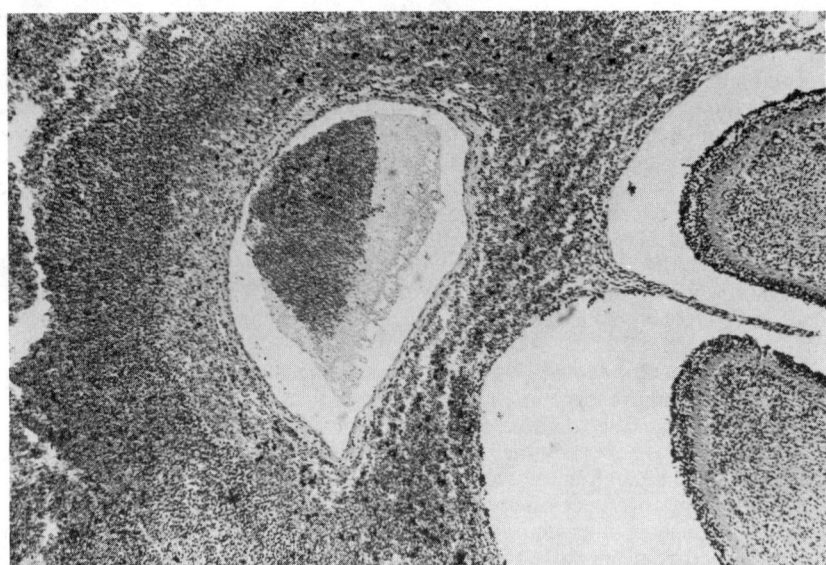

FIGURE 2. Acute vasculitis with subarachnoid hemorrhage adjacent to spinal cord in one-week-old infant with *Streptococcus* Group B meningitis (H & E; original magnification ×250, reproduced here at 60% of original size).

oligodendroglia. Neuroblasts from this structure provide the neuronal population of the pulvinar, nucleus lateralis posterior, and possibly other thalamic nuclei.[7] Since lesions in other portions of the brain are often associated with the matrix-zone hemorrhages, it may be difficult to determine what contribution each makes to the observed deficit in those without grade-four hemorrhage. On the other hand, it is conceivable that a subnormal population of neurons in the pulvinar-LP complex might contribute to some deficit in normal visual, speech, or integrative function as it is the main site of origin of thalamic fibers to the parietotemporo-occipital cortex. In addition, it receives afferents from several subcortical structures, suggesting a role in integrating mechanisms.[8]

Lesions commonly associated with matrix-zone hemorrhages include periventricular leukomalacia (PVLM) and white-matter gliosis. It has been suggested, but not proved, that white-matter gliosis may result from some nutritional deficiency or may be a less-exaggerated manifestation of hypoxia-ischemia than white-matter necrosis;[9,10] however, some investigators assert that it is a normal feature of the developing brain.[11] Whereas the white matter of the immature fetus contains many more glial cells than are normally present later in life, these are largely the so-called myelination glia and are morphologically different from those forms commonly found in white-matter gliosis, or perinatal telencephalic leukoencephalopathy, as it has been termed by Gilles and Murphy[9] (FIG. 3).

It is likely that a certain percentage of premature neonates with white-matter gliosis survive, but since there is no diagnostic test for it, the potential significance of this abnormality in isolation must remain speculative. Gilles and Murphy have suggested that such infants may manifest mild microcephaly and possibly some intellectual deficit later in life.[9]

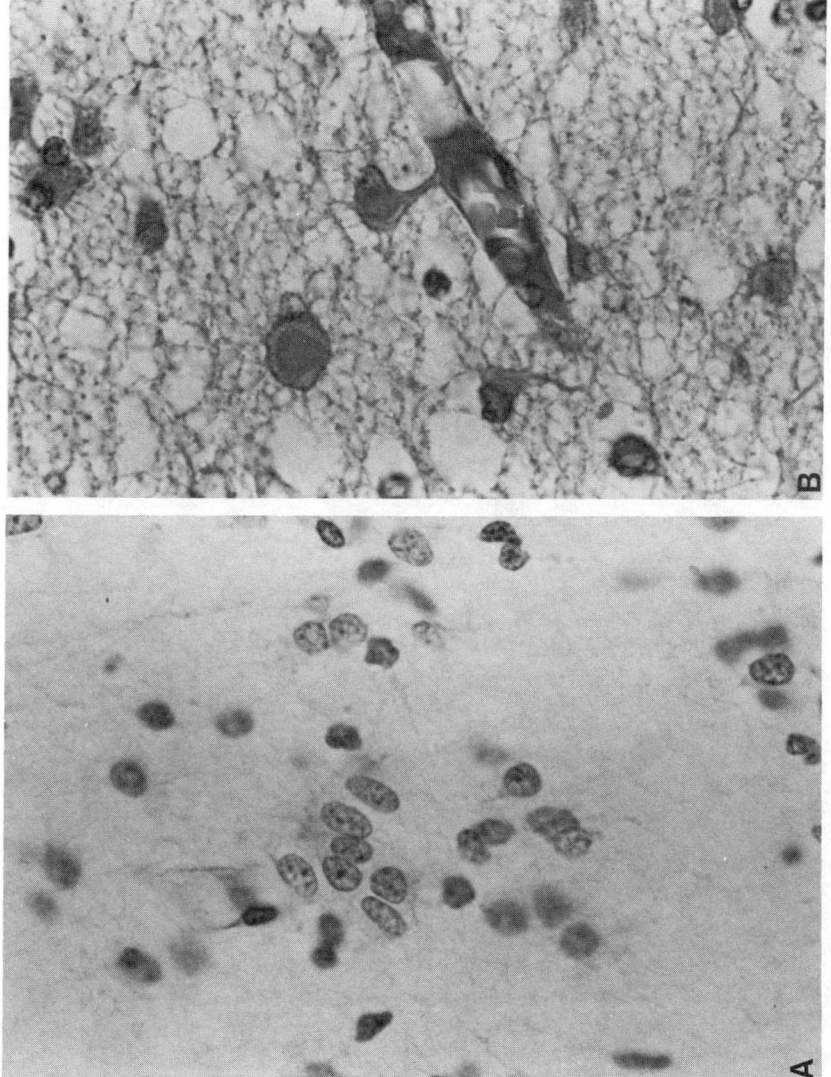

FIGURE 3. (A) Normal myelination glia in cerebral white matter of 2,250-gram neonate (luxol fast blue–cresyl violet; original magnification × 400, reproduced here at 60% of original size). (B) Pathological fiber-forming astrocytes in cerebral white matter of 1,740-gram neonate (H & E; original magnification ×400, reproduced here at 60% of original size).

Infants with PVLM are more likely to exhibit a fixed neurological deficit because of necrosis of fibers in the corona radiata at various cerebral levels. If this type of damage is minimal, it seems to occur more often adjacent to the occipital horns than at more anterior levels, and one might speculate on a possible relationship between these small lesions and the appearance of dyslexia later in life (FIG. 4). Although children whose only clinical problem is dyslexia rarely come to

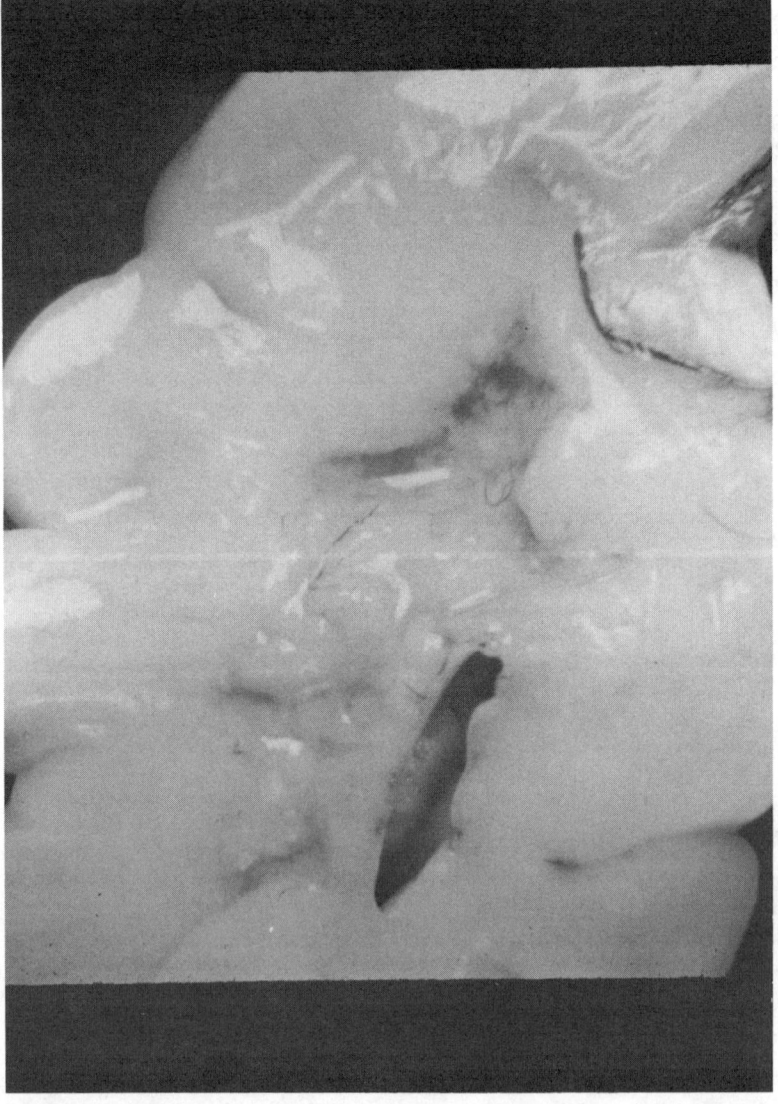

FIGURE 4. Acute necrosis of occipital periventricular white matter from brain of 1,325-gram, two-day-old neonate.

necropsy, identification of small white-matter lesions by MRI scanning now provides a simple technique for investigation of this hypothesis.

Necrosis of thalamus, most particularly the posterolateroventral nucleus, and occasionally other deep-grey masses, continues to occur at a low but steady rate. During the neonatal period, thalamic necrosis is commonly correlated with difficulties in sucking, and later in life infants and children with lesions in these regions typically display developmental retardation and movement disorders.[12]

Two major types of lesions appear to have increased in recent years: (1) a condition termed "hypotensive brain-stem necrosis" (HTBSN),[13] and (2) necrosis of basis pontis, often in association with cerebellar damage.

Hypotensive brain-stem necrosis is a term introduced by Gilles,[13] to describe a lesion characterized by necrosis of multiple tectal and tegmental nuclei consequent to cardiac arrest or shock. The cases that he described included both adults and children. Although the diagnostic term is somewhat misleading, as necrosis is generally not limited to the brain stem under these circumstances, the striking gross pathology sets the condition apart from others that may include damage to some brain-stem structures (FIG. 5).

HTBSN resembles the lesions produced by Windle et al.[14] and Miller and Myers[15] in experimental studies of asphyxia in monkeys.

Such lesions in human infants were rare 25 years ago, a phenomenon that was remarked upon from time to time, and not well understood. However, in the past five years it has become relatively frequent in my material at The Children's Hospital of Philadelphia, especially in infants who have had surgery for congenital heart disease or who have died after partial recovery from one episode of "near

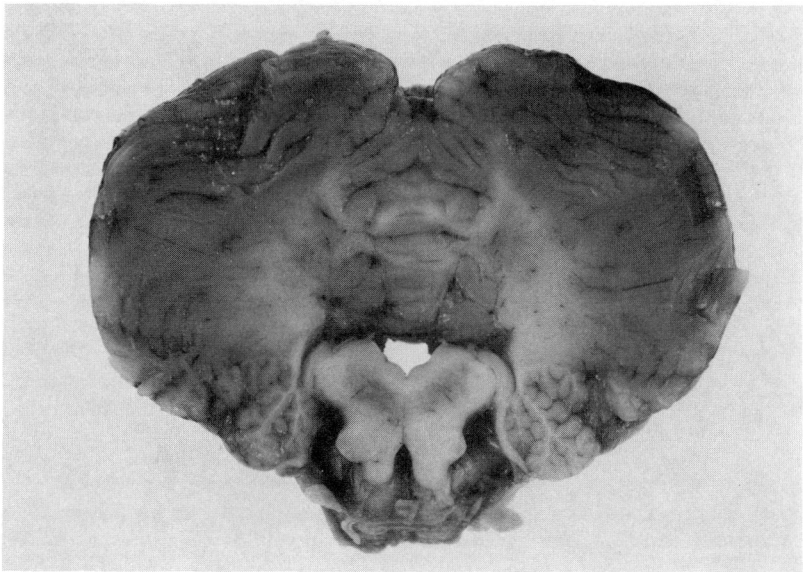

FIGURE 5. Transverse section of medulla and cerebellum of low-birth-weight (small for gestational age) infant with tetralogy of Fallot showing dusky discoloration of medullary tegmentum (and cerebellum) characteristic of "hypotensive brain-stem necrosis."

miss" sudden infant death syndrome (SIDS). As in the monkeys, however, damage to the nervous system is not confined to the brain stem but generally affects most nervous-system grey matter, including the spinal cord.

Finally, increased frequency of pontine and cerebellar necrosis also appears to be related to advanced technology and growth of the neonatal intensive-care unit. Selective necrosis of neurons in basis pontis was described by McAdams in 1967, who related the phenomenon to hypoxia consequent to pulmonary hemorrhage.[16] Friede independently associated necrosis of neurons in the basis pontis and subiculum, and introduced the term "pontosubicular necrosis" (PSN).[17] He suggested that pathogenesis was related to a specific selective vulnerability of neurons in those sites.

Our experience has not paralleled that of McAdams and Friede; that is, necrosis of basal pontine neurons is not necessarily associated with pulmonary hemorrhage but occurs with many types of cardiorespiratory disorders, and it is almost never combined with selective damage to subiculum with or without lesions elsewhere. We have noted an association between necrosis in basis pontis, or even frank infarction (primarily of the basis), and cerebellar necrosis (FIGS. 6, 7). This combination of lesions suggests that pathogenesis may be related to disturbances of blood flow through the vertebrobasilar system, rather than to a more global hypoxic phenomenon.[18]

Increased use of intubation for ventilation of infants with respiratory distress and hyperextension of the neck for this procedure may account for the appearance of this lesion. Gilles *et al.* reported instability of the atlanto-occipital joint in infants, and vulnerability of the vertebral arteries to constriction consequent to hyperextension.[19] These lesions in the infant brain stem and cerebellum, and

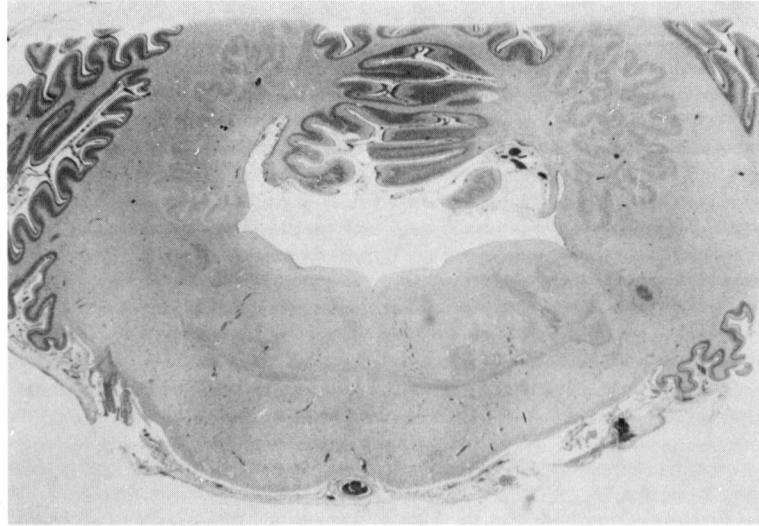

FIGURE 6. Section of midpons and cerebellum of 2,200-gram neonate who died three days after surgery for diaphragmatic hernia. Note marked reduction in size of basis pontis consequent to necrosis that includes the middle cerebellar peduncles and folia of the flocculi (H & E; original magnification ×15, reproduced here at 60% original size).

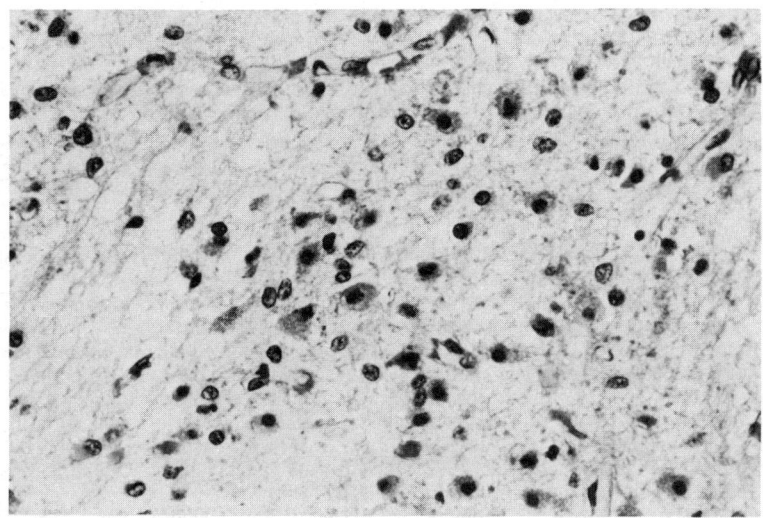

FIGURE 7. Photomicrograph of necrotic basis pontis shown in FIGURE 6 demonstrating necrosis of neurons and neuropil; a few astrocytic nuclei remain intact (H & E; original magnification ×250, reproduced here at 60% original size).

occasionally the occipital lobes, are basically similar to damage produced later in life consequent to hyperextension of the neck, sometimes in conjunction with chiropractic manipulations.

In summary, although advancing knowledge and use of sophisticated machines has led to increased survival of prematurely born infants and reduction or almost total disappearance of certain types of brain damage, prevention of a significant proportion continues to elude us, and treatment methods may even be contributing to a few types of lesions that were heretofore rare.

REFERENCES

1. MICHELSSON, K. & M. NORONEN. 1983. Neurological, psychological, and articulatory impairment in five-year-old children with a birth weight of 2,000 g or less. Eur. J. Pediatr. **141:** 96–100.
2. ROSS, G., A. N. KRAUSS & P. A. M. AULD. 1983. Growth achievement in low-birth-weight premature infants: Relationship to neurobehavioral outcome at one year. J. Pediatr. **103:** 105–108.
3. BUCKWALD, S., W. A. ZORN & E. A. EGAN. 1984. Mortality and follow-up data for neonates weighing 500 to 800 g at birth. Am. J. Dis. Child. **138:** 779–782.
4. PANETH, N. & R. I. STARK. 1983. Cerebral palsy and mental retardation in relation to indicators of perinatal asphyxia. Am. J. Obstet. Gynecol. **147:** 960–966.
5. HAYMAKER, W., C. MARGOLES, A. PENTSCHEW, H. JACOB, R. LINDENBERG, L. S. ARROYO, O. STOCHDORPH & D. STOWENS. 1961. Pathology of kernicterus and post-icteric encephalopathy. *In* Kernicterus and Its Importance in Cerebral Palsy. C. C. Thomas. Springfield, IL.
6. TOWBIN, A. 1969. Mental retardation due to germinal matrix infarction. Science **164:** 156–161.

192 ANNALS NEW YORK ACADEMY OF SCIENCES

7. Rakic, P. 1974. Embryonic development of the pulvinar-LP complex in man. *In* The Pulvinar-LP Complex. I. S. Cooper, M. Riklan & P. Rakic, Eds. C. C. Thomas. Springfield, IL.
8. Brodal, A. 1981. Neurological Anatomy in Relation to Clinical Medicine, Third Edition. Oxford University Press. Oxford, England. pp. 837–839.
9. Gilles, F. H. & S. F. Murphy. 1969. Perinatal telencephalic leukoencephalopathy. J. Neurol. Neurosurg. Psychiatry 32: 404–413.
10. Winick, M. 1976. Malnutrition and Brain Development. Oxford University Press. London, England.
11. Leech, R. W. & E. C. Alvord, Jr. 1974. Glial fatty metamorphosis: An abnormal response to premyelin glia frequently accompanying periventricular leukomalacia. Am. J. Pathol. 74: 603–612.
12. Rosales, R. K. & H. E. Riggs. 1962. Symmetrical thalamic degeneration in infants. J. Neuropathol. Exp. Neurol. 21: 372–376.
13. Gilles, F. H. 1969. Hypotensive brain-stem necrosis. Selective symmetrical necrosis of tegmental neuronal aggregates following cardiac arrest. Arch. Pathol. 88: 32–42.
14. Windle, W. F., H. N. Jacobson, M. I. R. de Ramirez de Arellano & C. M. Combs. 1962. Structural and functional sequelae of asphyxia neonatorum in monkeys (macaca mulatta). Res. Publ. Assoc. Res. Nerv. Ment. Dis. 39: 169–182.
15. Miller, J. R. & R. E. Myers. 1972. Neuropathology of systemic circulatory arrest in adult monkeys. Neurology (Minneapolis) 22: 888–904.
16. McAdams, A. J. 1967. Pulmonary hemorrhage in the newborn. Am. J. Dis. Child. 113: 255–262.
17. Friede, R. L. 1972. Pontosubicular lesions in perinatal anoxia. Arch. Pathol. 94: 343–354.
18. Rorke, L. B. 1982. Pathology of Perinatal Brain Injury. Raven Press. New York.
19. Gilles, F. H., M. Bina & A. Sotrel. 1979. Infantile atlanto-occipital instability. Am. J. Dis. Child. 133: 30–37.

Factors during Neonatal Life That Influence Brain Disorder[a]

ALFRED W. BRANN, JR.

Department of Pediatrics
Emory University Medical School
Atlanta, Georgia

The paper[1] presented here as an abstract described multiple factors affecting central nervous system (CNS) in the neonatal period that may influence the incidence of some types of long-term neurologic disability including static motor deficits, mental retardation, seizures, posthemorrhagic hydrocephalus, and microencephaly.

The paper particularly underscored the magnitude and importance of a problem that is not fully appreciated, intrapartum asphyxia affecting the full-term fetus/neonate. Three important factors contribute to this problem. First, more children with cerebral palsy are full-term infants than are preterm infants. Even though the incidence of CP is lower among full-term infants (3.38/1,000LB) than among preterm infants (90/1,000LB), this lower incidence is applied to a denominator that includes about 93% of the births in this country. Second, more asphyxiated full-term infants survive than do asphyxiated preterm infants. Third, there has been no significant reduction in the types of cerebral palsy seen in children who, as full-term neonates, had experienced intrapartum asphyxia. Thus, research efforts need to be directed in two areas: (1) at detecting women at low risk prenatally who will convert to a high risk during the intrapartum period, so that intervention can occur before suboptimal oxygenation harms the fetus, and (2) at detecting and treating early the neonate with hypoxic ischemic encephalopathy from intrauterine asphyxia.

REFERENCE

1. BRANN, ALFRED W. 1985. Factors during neonatal life that influence brain disorder. *In* Prenatal and Perinatal Factors Associated with Brain Disorders. J. M. Freeman, Ed. National Institute of Child Health and Human Development and National Institute of Neurological and Communicative Disorders and Stroke (NIH Publication No. 85–1149). Bethesda, MD.

[a] Dr. Brann's presentation was based on a chapter by the same title in *Prenatal and Perinatal Factors Associated with Brain Disorders.*[1]

Fetal Nutrition and Brain Growth

MYRON WINICK

Institute of Human Nutrition
Columbia University
New York, New York 10032

The development of the mammalian brain involves a series of sequential changes that include division, migration, and differentiation of neurons and glia, dendritic arborization, formation of synapses, synthesis and release of neural transmitters and neural hormones, and progressive increase in sensory motor and higher functions. All these changes, which begin before and extend beyond birth, may be affected by prenatal or postnatal nutrition.

In this presentation, analysis of two vastly different models of intrauterine growth retardation (IUGR) suggests that there may be a common final pathway through which intrauterine growth failure occurs in both models. This common final pathway may be related to failure of vascular system development, in particular of the placenta during gestation.

The first model that produces IUGR by ligation of the uterine artery near term (the so-called Wigglesworth procedure) results in disproportionate fetal growth failure. The brains of these IUGR animals are of normal size, with normal quantities of myelin and gangliosides; however, other organs, such as the liver, are much smaller than expected. The animals are free of behavioral abnormalities, and they are susceptible to hypoglycemia.

Maternal malnutrition (protein restriction) is the second most common method of producing IUGR. In this model, the size of the fetus is symmetrically reduced with the fetal brain being reduced in cell number to the same extent as the other organs. Myelination is reduced and the concentration of N-acetyl-neuraminic acid and gangliosides is also reduced. Pups show behavioral abnormalities that persist into adult life.

It is hypothesized that malnutrition operates by producing a chronic vascular insufficiency to the uterus and placenta. Maternal malnutrition during pregnancy results in a reduced blood volume expansion that leads to an inadequate increase in cardiac output, followed by a decreased placental blood flow. This decreased placental blood flow will result in a reduced placental size, and at the same time, in a reduced transfer of nutrients, the result being that of fetal growth retardation, although different than that produced by the acute vascular insufficiency of the first model.

Postnatally, those animals that have been subject to acute vascular clamping will not survive unless you supply them with exogenous glucose. They have completely depleted glycogen reserves, and they will not survive on their own. If supported, these animals do quite well postnatally without any behavioral deficits.

The animals subject to the other type of vascular insult, that is, the abnormal chronic decrease in blood volume and uterine blood flow, survive on their own, but they will show behavioral abnormalities later on.

That the above-mentioned models may be generalized to humans has been suggested by the kinds of different outcomes that one sees in populations where fetal malnutrition is rampant, versus populations in which smoking is one of the major causes of reduced fetal growth. Smoking produces an acute vascular

194

change with assymetrical growth retardation, whereas alcohol consumption during pregnancy fits into the chronic model and leads to symmetrical reduction.

DISCUSSION

CHARLES VORHEES: Dr. Winick, you said that alcohol fits into the chronic-induced, blood-flow reduction model. Could you comment on fetal growth failure in the models of binge drinking (the fetal alcohol facies) where animals or children are born with various kinds of overt defects?

MYRON WINICK: I did not say that the fetal alcohol syndrome was the result of the chronic model. I said that small doses of alcohol will induce fetal growth failure in the rat; whether accompanying malformations are seen is another issue. The fetal alcohol syndrome is really a question of direct embryonal alcohol toxicity. I have not addressed the pathogenesis of any congenital malformations; I tried to define models of fetal growth retardation. Obviously, the fetal alcohol syndrome produces malformed fetuses that also do not grow, so there seem to be, in these cases, both external and internal factors that simultaneously affect growth. I am restricting myself to those external factors that primarily affect fetal growth. Alcohol seems to do that when given in very small doses in the rat.

MARIA MICHEJDA: How do you know whether changes in blood volume precede those in blood flow or whether the reverse occurs? The new diagnostic techniques of Doppler measurements of umbilical blood flow in IUGR are very helpful. Would you consider this in your study?

MYRON WINICK: One approach might be to expand the blood volume artificially. By expanding the blood volume, one might be able to get normal nutrient transport, even in a malnourished situation. Looking over some of our recent data, we find that we can expand blood volume very nicely. We can maintain an expanded volume throughout the pregnancy in a malnourished animal; therefore, we have a model now that is malnourished, but has a normal blood volume. However, these fetuses do not grow very well, so it may very well be that increases in blood flow cause the increase in blood volume, rather than the reverse. If that is the case, then we have to find ways to increase blood flow, independent of expanding blood volume, in order to theoretically improve transport. We are not able to do that yet.

MARIA MICHEJDA: Did you try to induce hypertension?

MYRON WINICK: No, but Stan James at our institution is trying to do that very thing, to increase blood flow by inducing hypertension; he is doing that in sheep.

Nutritional Taurine and Central Nervous System Development[a]

JOHN A. STURMAN

Department of Developmental Biochemistry
Institute for Basic Research in Developmental Disabilities
New York State Office of Mental Retardation &
Developmental Disabilities
Staten Island, New York 10314

Taurine, 2-aminoethanesulfonic acid, is a simple amino sulfonic acid that has been known since the early 19th century.[1] At 25°C the pK_a is 1.5 and the pK_b is 8.74, so that at physiological pHs it exists as a zwitterion (FIG. 1). It is one of the most abundant amino acids in living organisms, although it is rarely found in plants.[2] Taurine participates in few biochemical reactions, conjugation with bile acids in the liver being the best documented. It is not found as a constituent of proteins, although it has been identified in some peptides.[3-8] Taurine is chlorinated by neutrophils and this reaction removes the hypochlorous acid generated by the myeloperoxidase system;[9,10] thus taurine may prevent cellular damage in some mammalian tissues by functioning as an antioxidant.[11,12] Although taurine is one of the most abundant amino acids in many mammalian tissues, its presence in brain has attracted particular attention for it is the amino acid present in the greatest concentration in the newborn brain of all mammals (TABLE 1).[13,14] During development, the taurine concentration in brain decreases slowly to the levels present in mature brain, which are still greater than most other amino acids with the notable exception of glutamic acid. It should be noted that although the concentration of taurine decreases postnatally, the total taurine present in brain may increase, as illustrated for rat brain (FIG. 2).[15] Other tissues do not have a similar developmental change, as illustrated for monkey liver (FIG. 3).[16]

One of the confounding aspects of the changes in taurine concentrations during development is the wide range of values found in different species; thus, the concentration of taurine in brain of some species may be as high as 20 μmoles/g at birth and in others may be as low as 4 μmoles/g. Similarly, the concentration of taurine in mature brain may be as high as 9 μmoles/g in some species, and as low as 1 μmole/g in others. If such data are plotted as a proportion of the taurine concentration in mature brain for a given species against the proportion of the time of weaning in that same species (FIG. 4), similar, straight-line graphs are obtained.[16] This association with time of weaning led us to examine the milk as a potential source of taurine. We found that taurine was a major constituent of the free-amino-acid pool in milk of many mammals (TABLE 2).[17] In some species it is the free amino acid present in the greatest concentration, in others it is exceeded only by one amino acid, and in some it is not among the most abundant amino acids (FIG. 5).

A major advance occurred in taurine research with the observation by Hayes and coworkers that the retinal degeneration observed in cats fed synthetic diets

[a] This research has been supported by the New York State Office of Mental Retardation and Developmental Disabilities and by NIH Grants HD-11129 and HD-16634.

196

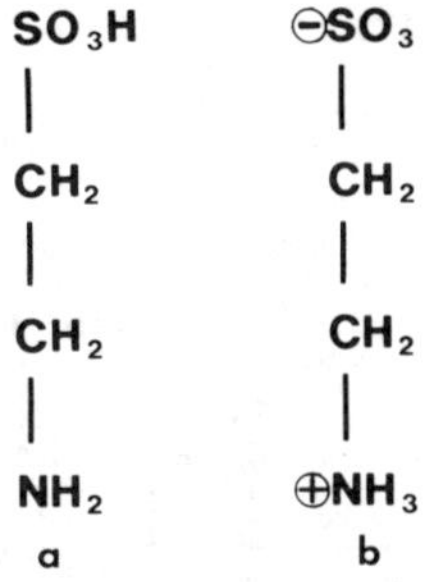

FIGURE 1. Formula of taurine **(a)** and as it exists at physiological pH **(b).**

$$SO_3H$$
$$|$$
$$CH_2$$
$$|$$
$$CH_2$$
$$|$$
$$NH_2$$

a

$$^{\ominus}SO_3$$
$$|$$
$$CH_2$$
$$|$$
$$CH_2$$
$$|$$
$$^{\oplus}NH_3$$

b

TABLE 1. Concentration of Taurine in Adult and Newborn Brain

Species	Adult[a]	Newborn[a]
Mouse	8.6	15.3
Rat	4.4	16.6
Gerbil	6.5	21.2
Guinea pig	1.0	2.0
Rabbit	1.2	5.1
Dog	1.3	6.8
Cat	2.3	9.2
Chick	2.3	8.9
Monkey[b]	2.3	6.9
Man[b]	1.4	3.3[c]

[a] μmol/g wet weight.
[b] Occipital cortex.
[c] Gray matter only, mean of five children one to five years of age.

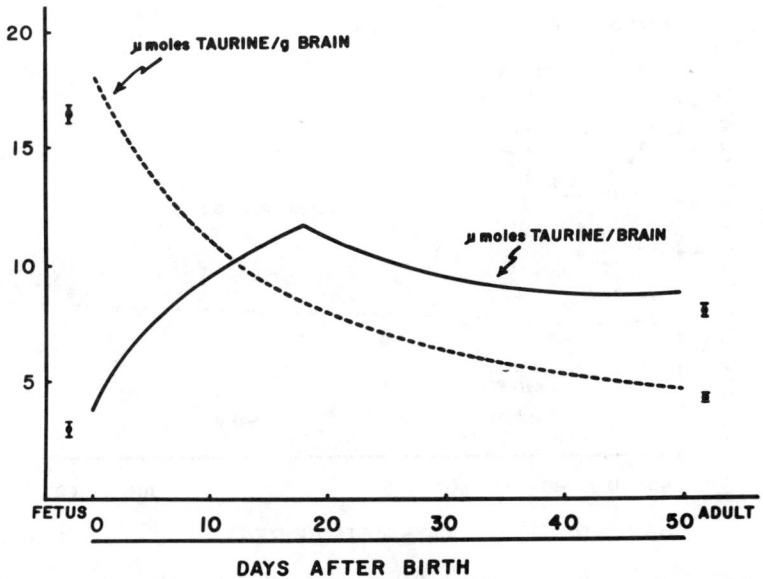

FIGURE 2. Concentration of taurine and total content of taurine in rat brain as a function of age. From Sturman and Hayes[15] with permission.

197

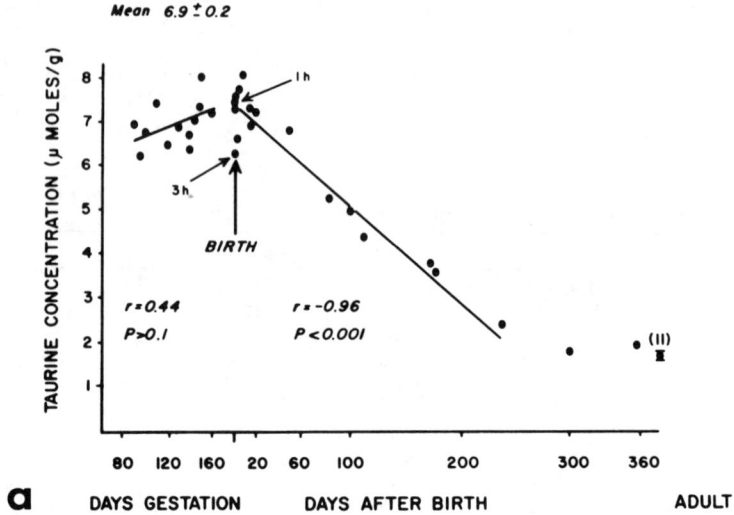

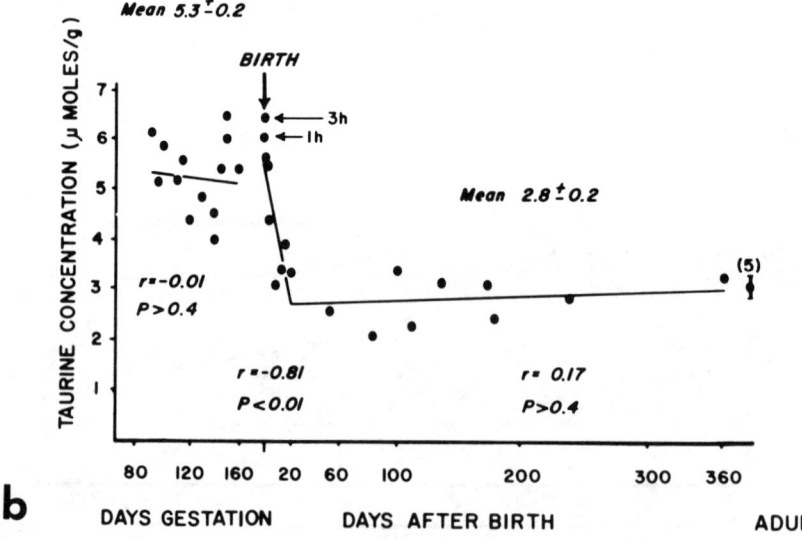

FIGURE 3. Concentration of taurine in Rhesus monkey occipital lobe **(a)** and liver **(b)** as a function of age. The line of best fit was determined by computer using the method of least squares. From Sturman and Gaull[16] with permission.

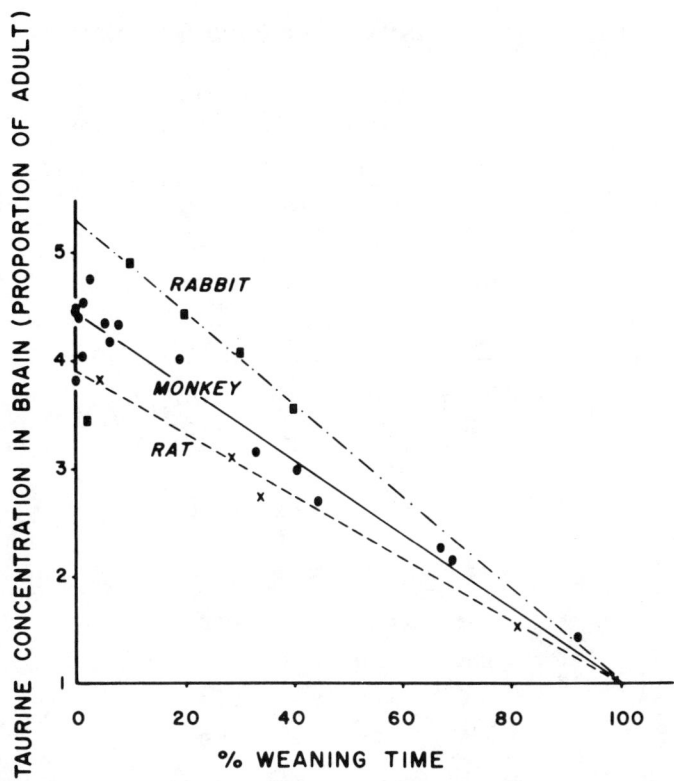

FIGURE 4. Concentrations of taurine in brain of infant monkeys (●); rabbits (■) and rats (×) plotted as proportion of the concentration of taurine in the brain of the appropriate adult as a function of weaning time for each species. The line of best fit for each set of data was determined by computer using the method of least squares. From Sturman and Gaull[16] with permission.

TABLE 2. Taurine in Milk

Species	Less Than Five Days after Births[a]	More Than Five Days after Birth[a]
Gerbil		595
Cat		287
Dog	264	191
Mouse		75
Rhesus monkey	61	56
Baboon		38
Man	41	34
Chimp	71	26
Gorilla		17
Guinea pig		17
Rat	63	15
Java monkey		14
Rabbit		14
Sheep	68	14
Horse		3
Goat		2
Cow	31	1
Pig	56	

[a] μmol/100 ml.

Gerbil	Cat	Dog	Rhesus Monkey
TAU	TAU	TAU	TAU
ETOHNH₂	GLY	ETOHNH₂	GLU
SER	GLUNH₂	GLY	SER
GLY	SER	GLU	GLY

Rat	Baboon	Man	Chimp
ETOHNH₂	GLU	GLU	GLU
TAU	TAU	TAU	TAU
ALA	GLY	ALA	ALA
GLY	SER	GLUNH₂	ASP

Guinea Pig	Rabbit	Horse	Cow
GLY	GLY	GLU	ETOHNH₂
SER	GLU	GLUNH₂	GLU
THR	ALA	SER	GLY
ALA	SER	THR	SER

FIGURE 5. The most abundant free amino acids in milk of various species.

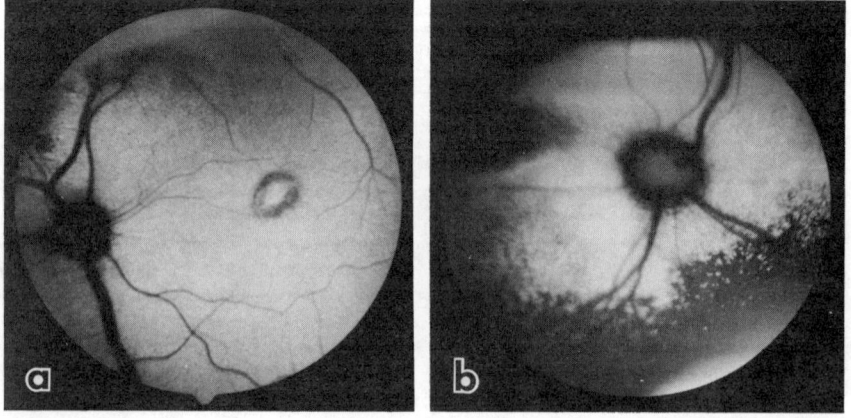

FIGURE 6. Fundus photographs of eyes from taurine-deficient cats **(a)** small lesion early in deficiency **(b)** larger lesion that extends across the retina from the nasal side to the temple side at a later stage of deficiency.

resulted from taurine depletion.[18] This observation has since been confirmed in a number of laboratories, including our own.[19-26] This degeneration, visible with an ophthalmoscope when the retinal taurine concentration has been reduced 50% or more, is initially characterized by a hyperreflective granular zone in the area centralis, and the size of the lesion increases with time into an oval zone that extends across the entire retina (FIG. 6). A second tissue that has been documented to suffer structural damage as a result of taurine depletion is the tapetum lucidum.[27,28] The tapetum consists of layers of cells behind the retina that reflect light back through the retina and maximize retinal sensitivity. In the taurine-depleted cat, the regular lattice array of tapetal rods is severely disorganized, and some of the rods are shorter and thicker than normal, and appear to have "collapsed" into electron-dense globules (FIG. 7). Further investigation demonstrated

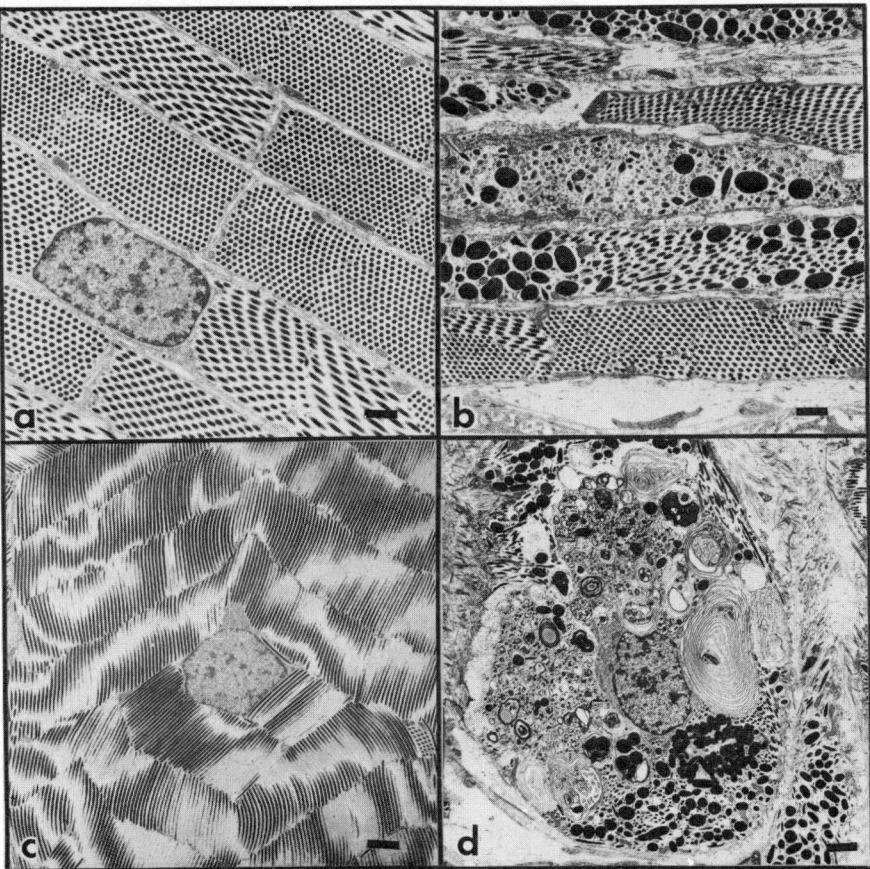

FIGURE 7. Electron micrographs of cross sections of center of tapetum from **(a)** cat fed synthetic diet supplemented with taurine, and **(b)** cat fed synthetic diet alone (scale bars 1 μm). Electron micrographs of tangential sections of center of tapetum from **(c)** cat fed synthetic diet supplemented with taurine, and **(d)** cat fed synthetic diet alone (scale bars 2 μm). The tapetal cell illustrated in **(d)** is an advanced state of degeneration. From Sturman *et al.*[28] with permission.

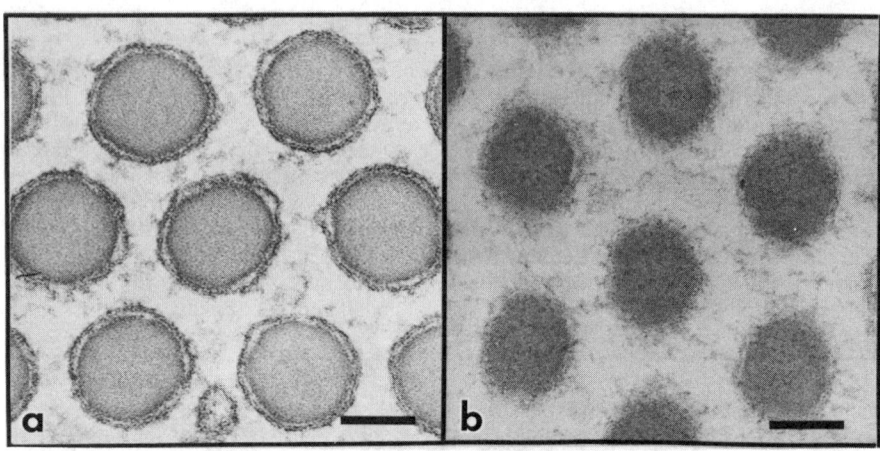

FIGURE 8. High-magnification electron micrograph from ultra-thin (gray interference color) cross sections of center of tapetum from **(a)** cat fed synthetic diet supplemented with taurine, and **(b)** cat fed synthetic diet alone (scale bars 0.1 μm). From Sturman *et al.*[28] with permission.

that the tapetal rods are surrounded by a membrane and that taurine depletion leads first to disruption and disorganization of this membrane followed by the more severe degeneration and disorganization (FIG. 8).

It has been demonstrated that taurine is transported axonally in developing optic axons to a much greater extent than in mature optic axons,[29,30] and it has been proposed as an inhibitory neurotransmitter.[31,32] No other functions for brain taurine or any effects of depletion of brain taurine have been reported before the following studies. Female cats were depleted of taurine by feeding a completely defined synthetic diet for at least six months, at which time they had plasma taurine concentrations less than 1 μmol/100 ml and had ophthalmoscopically visible retinal lesions. Other females were fed the same diet containing 0.05% taurine. There was no difference in food consumption between the two groups. All cats came into estrus normally and were bred with male cats maintained on a taurine-supplemented diet. Pregnancies were confirmed by x-ray four to six weeks postconception. The taurine-depleted females experienced difficulty completing normal pregnancies and suffered fetal resorptions, abortions, stillbirths, and low birth weights of live kittens at term (TABLE 3).[33] Females fed the taurine-supplemented diet, with one exception, encountered no such difficulties.

TABLE 3. Outcome of Pregnancies

Maternal Diet	Pregnancies	To Full Term	Kittens Stillborn	Kittens Liveborn	Survived to Weaning
Taurine-supplemented queens	18	17	0	71	67
Taurine-deficient queens	18	6	6	18	8

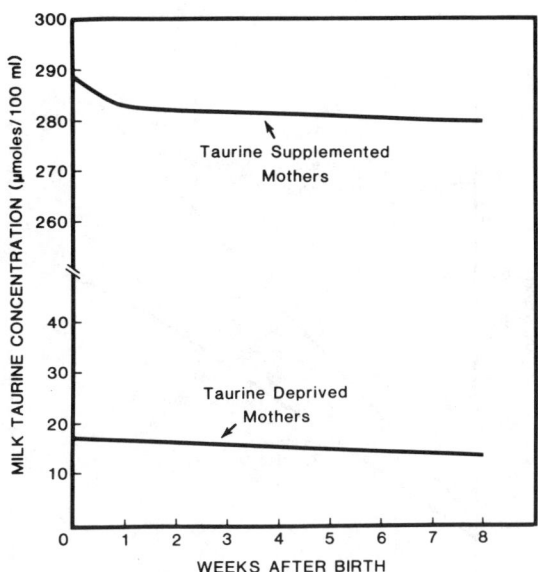

FIGURE 9. Concentration of taurine in milk of taurine-supplemented and taurine-deprived mothers. The curve for the taurine-deprived mothers was derived from the twice-weekly determinations from six queens that carried lactation through for the full eight weeks and the curve for the taurine-supplemented kittens was derived from the twice-weekly determinations from 12 queens that carried lactation through for the full eight weeks. From Sturman *et al.*[33] with permission.

The concentration of taurine in the milk of the taurine-depleted females was markedly lower than that in the milk of the taurine-supplemented females (FIG. 9). The kittens suckling the taurine-depleted mothers grew at a reduced rate (FIG. 10) and exhibited abnormal back leg development and a peculiar gait with the legs splayed (FIG. 11). Examination of such kittens revealed reduced muscle bulk and tone in the rear legs, reduced patellar and achilles tendon jerks, and a very weak grip. X-rays showed obvious kyphosis in the thoracic region (FIG. 12). Histological examination of the cerebellum revealed a persistence of cells in the external granule cell layer (FIG. 13) that contained numerous mitotic figures indicating that cells in this layer are still dividing (FIG. 14). The kittens from taurine-supplemented mothers had apparently normal cerebellums, with no persistence of cells in the external granule cell layer, and no mitotic figures. In normally developing kittens, the width of the cerebellar external granule cell layer increases to a maximum at two weeks after birth at which time cell proliferation ceases and the thickness of the layer decreases as the cells migrate.[34] The morphological changes in the cerebellum are accompanied by significantly smaller taurine concentrations (TABLE 4). More recently we have examined the visual cortex region of the brains of the same kittens and have found an abnormal ontogeny of neurons here also.[35] At eight weeks after birth, there are very few pyramidal and nonpyramidal neurons in the taurine-depleted kittens, and those present have heavily spined dendritic processes with poor arborization (FIG. 15). These changes are also accompanied by significantly smaller concentrations of taurine (TABLE 5).

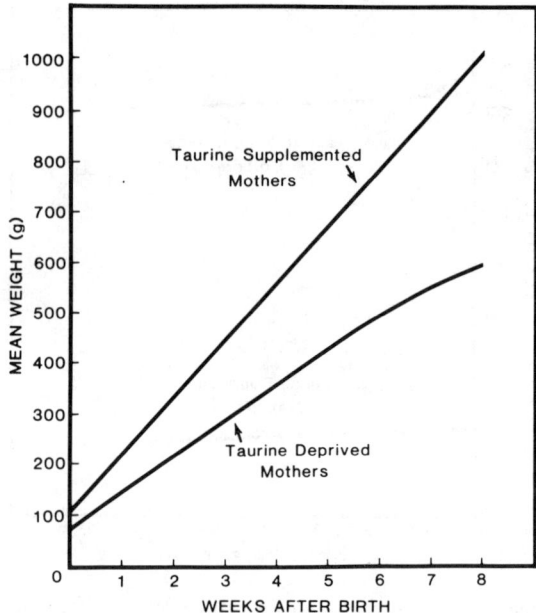

FIGURE 10. Weight of kittens from taurine-supplemented and taurine-deprived queens throughout lactation. The curve for the kittens from taurine-deprived mothers was derived from twice-weekly weights of the eight kittens that survived to the end of lactation, and the curve for the kittens from taurine-supplemented mothers was derived from semiweekly weights of 24 kittens. From Sturman *et al.*[33] with permission.

TABLE 4. Taurine Concentration in Eight-Week-Old Kitten Cerebellum[a]

Source of Kittens	Taurine Concentration (μmoles/g wet weight)
Taurine-supplemented queens (10)	5.0 ± 0.4
Taurine-deprived queens (7)	1.6 ± 0.3

[a] Each value represents the mean $\pm$ SEM from the number of kittens in parentheses.

TABLE 5. Taurine Concentration in Kitten Occipital Cortex

Source of Kittens (eight weeks old)	Taurine Concentration[a] (μmoles/g wet weight)
Taurine-supplemented queens	5.0 ± 0.5
Taurine-deprived queens	1.0 ± 0.3

[a] Each value represents the mean $\pm$ SEM from eight kittens.

FIGURE 11. Photographs of kittens from taurine-deprived mothers illustrating the abnormal back leg development, the excessive abduction, and peculiar gait. The kittens are from different taurine-deprived queens. Both were photographed eight weeks after birth. From Sturman et al.[33] with permission.

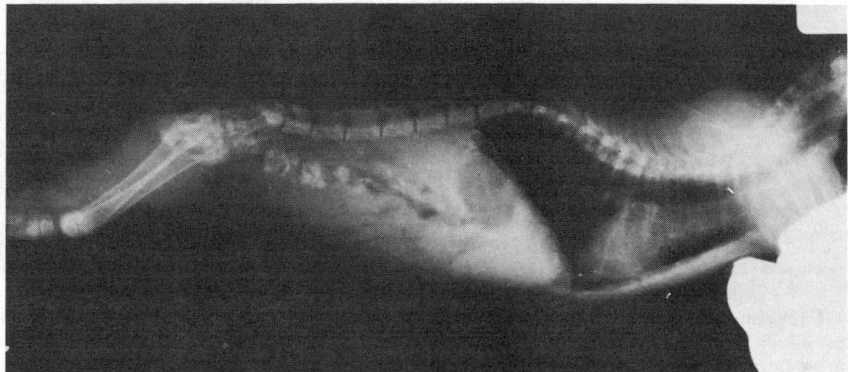

FIGURE 12. X-ray of kitten from taurine-deprived mother showing the grossly apparent thoracic kyphosis. This is from the kitten on the left side of FIGURE 3, taken eight weeks after birth. From Sturman et al.[33] with permission.

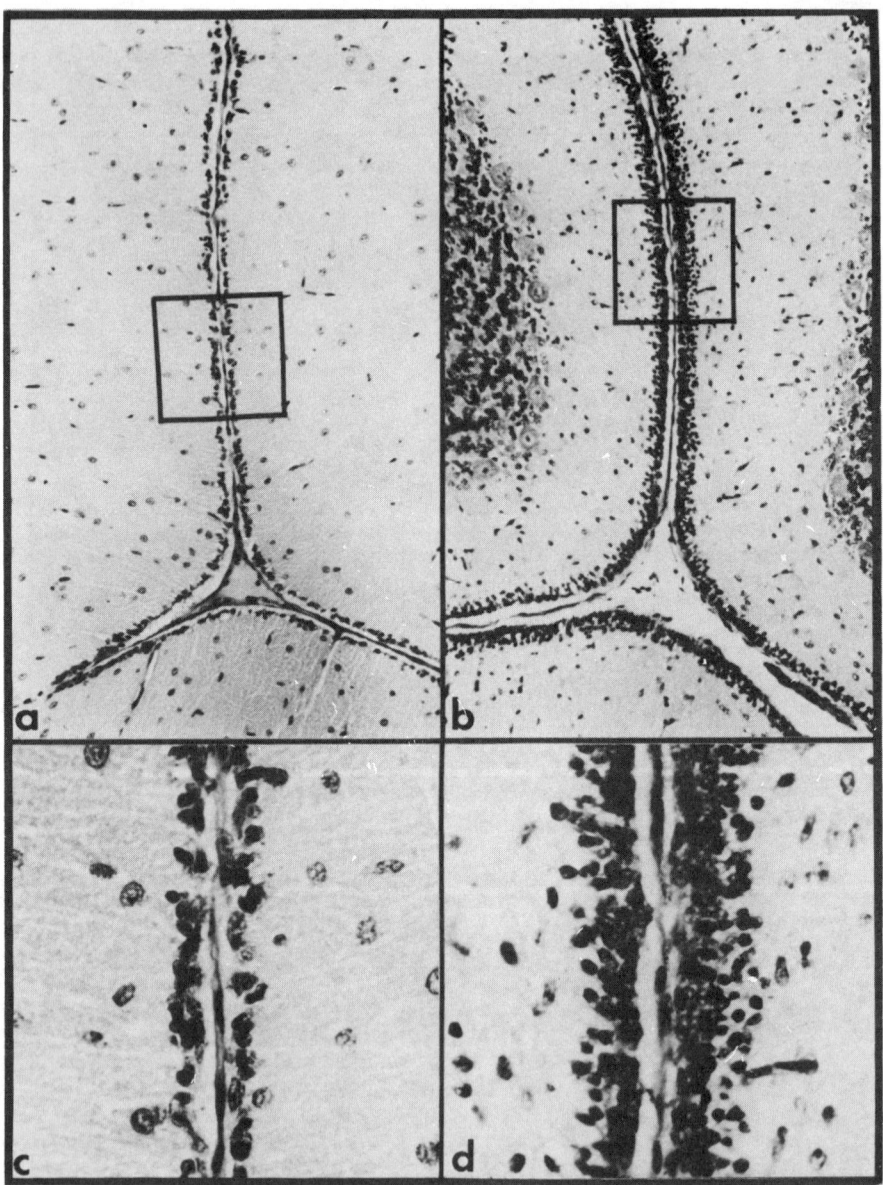

FIGURE 13. Light micrographs of the external granule cell layer of midline saggital sections of cerebellum (lobe VI) of eight-week-old kittens from taurine-supplemented (a,c,) and taurine-depleted (b,d) queens. Note the increased thickness of the external granule cell layer of the taurine-deficient kittens as compared to the controls. a, b, Original magnification ×40; c, d, original magnification ×100; reproduced here at 75% of original size. From Sturman *et al.*[33] with permission.

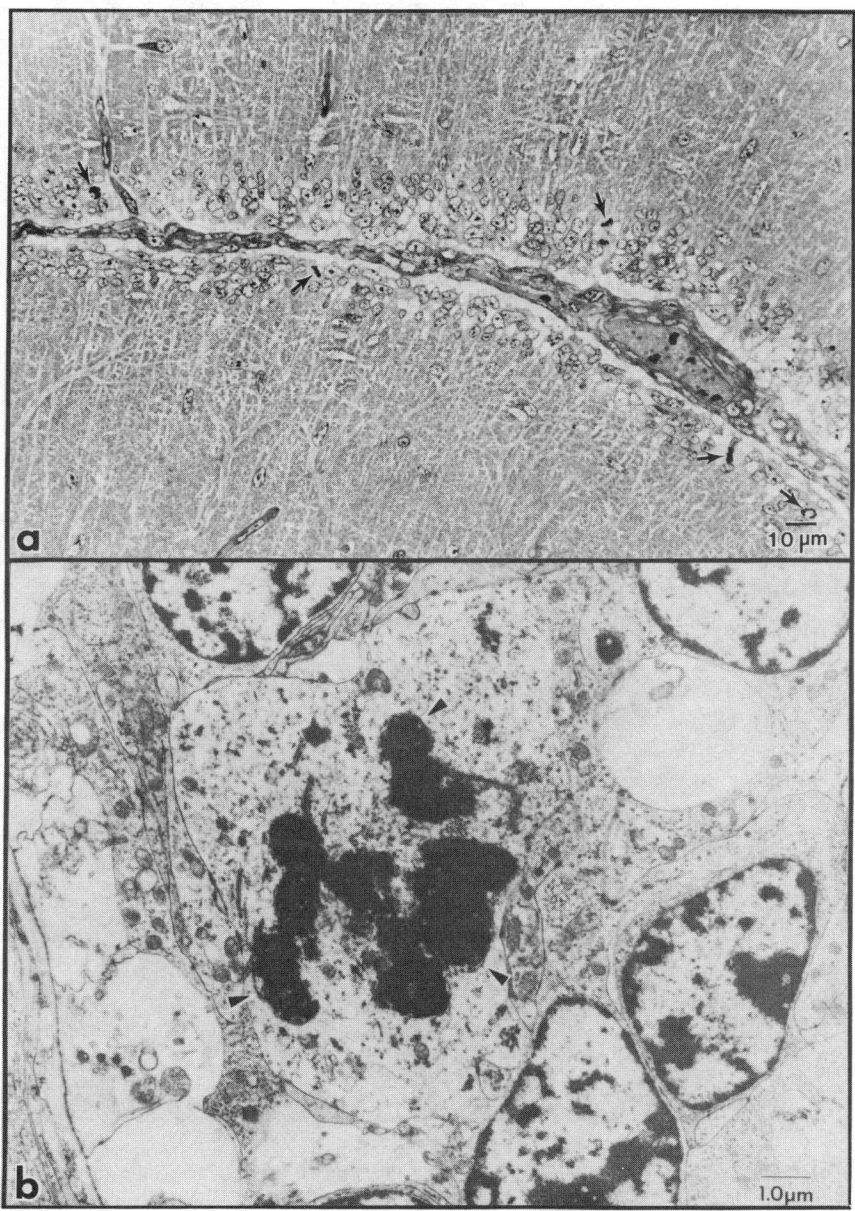

FIGURE 14. Mitotic figures are still found frequently in cells in the external granule cell layer of the eight-week-old kittens from taurine-depleted mothers. (a) Light micrograph of 1-μm section stained with toluidine blue shows several mitotic figures (arrows). (b) Electron micrograph shows a cell in late prophase. Remnants of the nuclear membrane can be seen in close approximation to the condensed chromatin (arrowheads). From Sturman *et al.*[33] with permission.

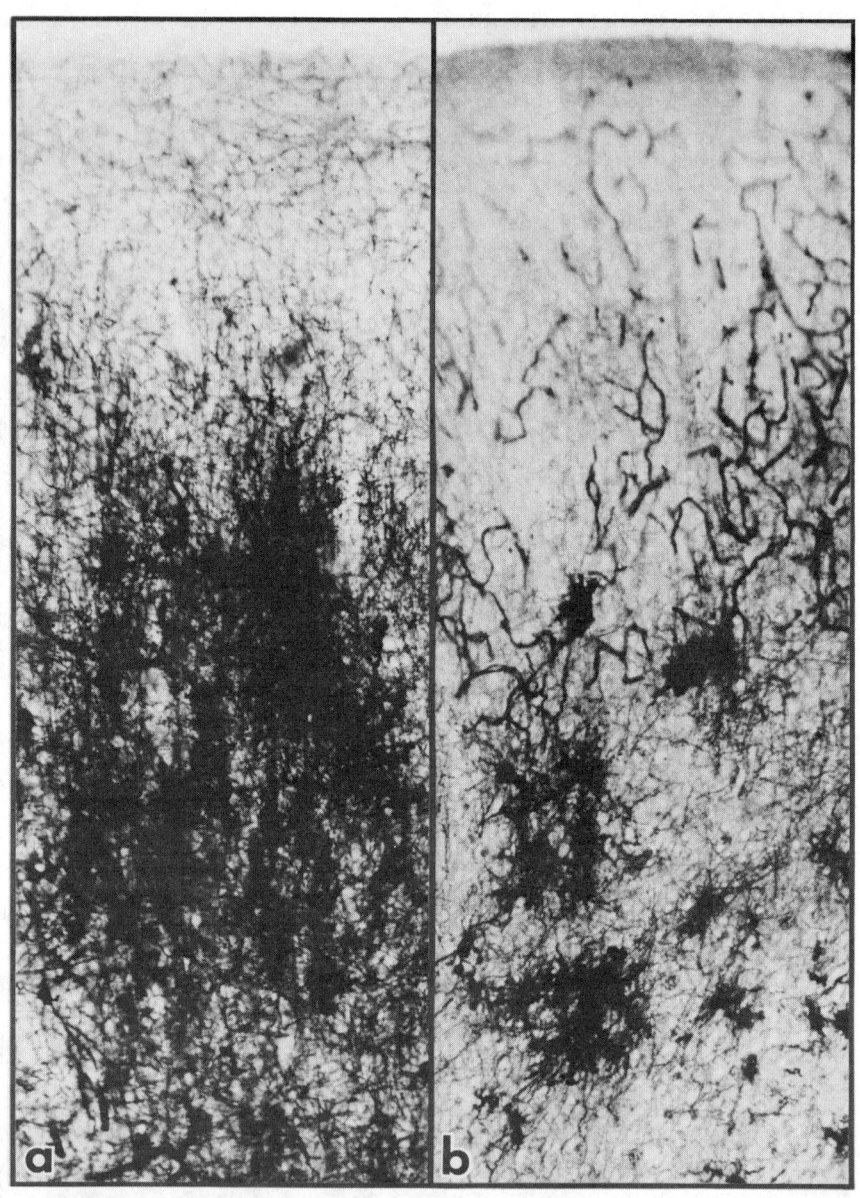

FIGURE 15. 80-μm-thick sections of the silver-impregnated visual cortex of eight-week-old kittens using the rapid Golgi method.

(a) A region from a kitten from a taurine-supplemented mother showing the extensive arborization of axonal and dendritic processes.

(b) A similar region from a kitten from a taurine-depleted mother showing the paucity of neurons. Their dendrites are thick with conspicuous spines and poor arborization. Original magnification, 96×; reproduced here at 90% of original size. Courtesy of Dr. Thomas Palackal.

These studies in the developing kitten indicate that normal concentrations of taurine are necessary for normal central nervous system development, and that subnormal taurine concentrations result in brain abnormalities that include impaired migration and differentiation of neural cells and various neurological abnormalities. We have also observed severe hydrocephalus in an aborted fetus and in a surviving liveborn kitten, and one stillborn kitten with anencephaly (FIG. 16). These severe manifestations may be more frequent since aborted fetuses and stillborn kittens are usually eaten by the mother unless removed immediately, and may account for some of the excessive reproductive wastage.

The results obtained describing the changes in the cat and kitten consequent to dietary taurine depletion take on a greater importance in the light of the findings in nonhuman primates and in man. Human term and preterm infants fed synthetic formulas derived from cow's milk (which contain virtually no taurine) exhibit reduced plasma taurine concentrations and reduced urinary taurine excretion compared with similar infants fed human milk.[36,37] Other recent reports have noted that human infants fed totally using solutions of nutrients administered parenterally (which contain no taurine) also have reduced concentrations of taurine in their plasma and have ophthalmologically and electrophysiologically demonstrable retinal damage;[38–41] thus it appears that taurine deficiency in human infants may be a cause for concern even though the manifestations may not be as dramatic as documented for the cat and kitten. This possibility was supported by a recent report in which Rhesus monkeys were raised from birth on a synthetic human infant formula, either alone or supplemented with taurine.[42] Those fed the taurine-free formula had a selective decrease in plasma taurine concentrations accompanied by degeneration of the retinal cone photoreceptor cells, whereas the monkeys fed the taurine-supplemented formula did not (FIG. 17).

As a result of the recent studies, the U.S.F.D.A. in July 1984 allowed the addition of taurine to synthetic human infant formulas. Taurine is now added to such formulas in Japan, Canada, and Europe, and also is added to parenteral solutions for infants; thus any ill effects from a postnatal deficiency of taurine have been minimized. There is still some cause for concern over infants born to mothers eating vegetarian diets that contain little or no taurine. Such infants may be exposed to a low taurine environment *in utero* as well as a lower taurine concentration in the mother's milk postnatally. A wide range of pediatric prob-

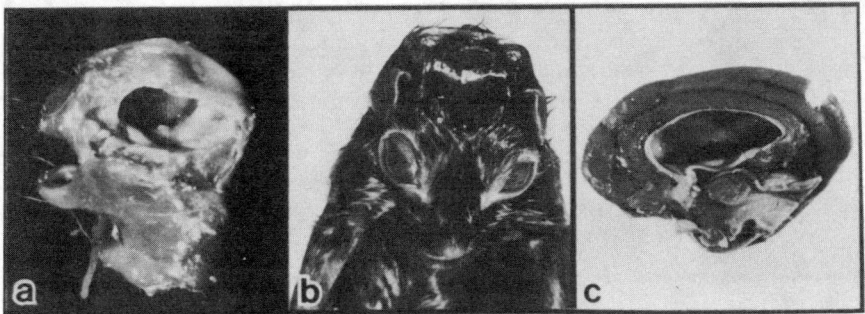

FIGURE 16. Abnormal offspring from taurine-depleted mothers. **(a)** Fetus aborted preterm showing extreme hydrocephalus. **(b)** Full-term stillborn showing anencephaly. **(c)** Brain from a one-year-old kitten showing severe hydrocephalus.

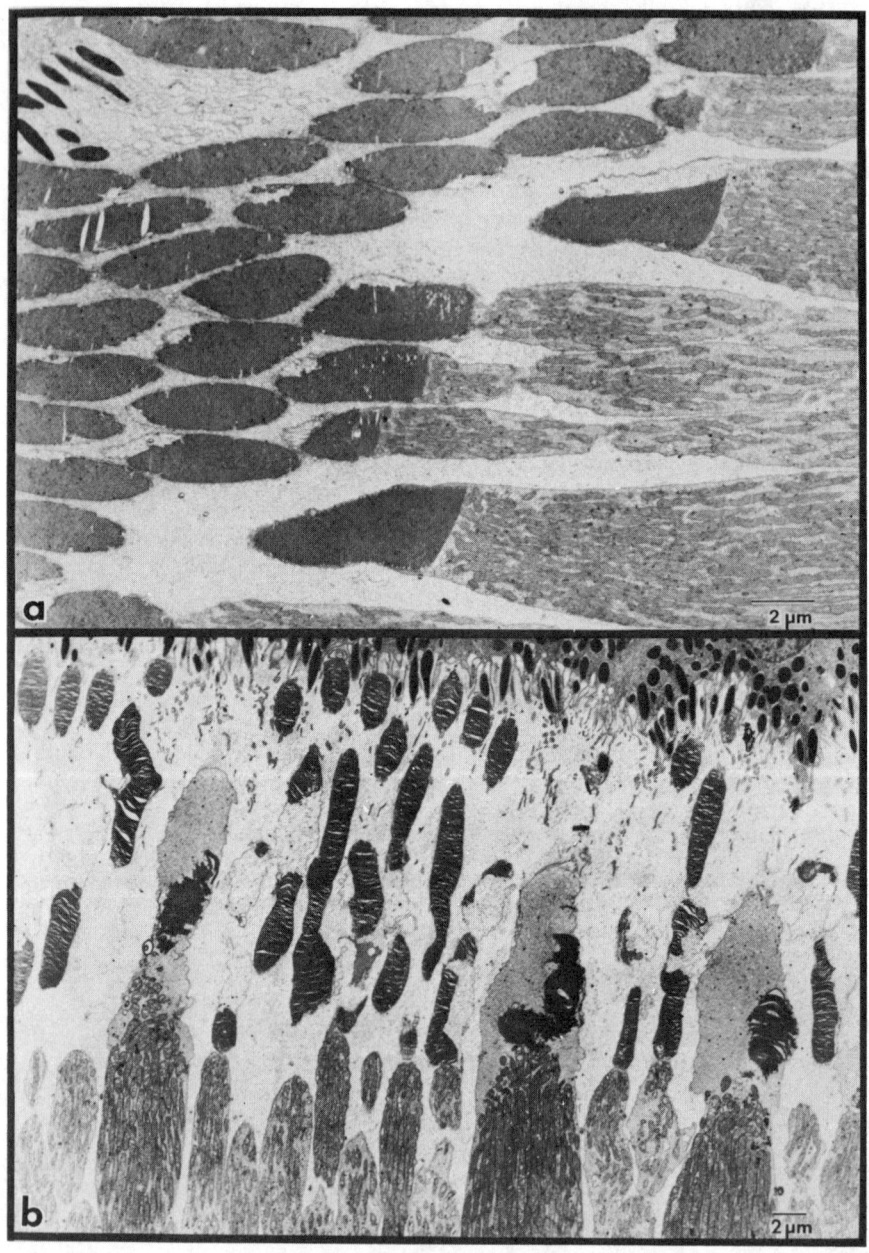

FIGURE 17. Electron micrographs of central retina from infant rhesus monkeys raised on Nutramigen supplemented with taurine (a) or Nutramigen alone (b). From Sturman et al.[42] with permission.

lems have been observed in children from strict vegetarian communities, including death. These problems have usually been attributed to malnutrition and to nutritional deficiencies of currently defined primary nutrients, but a taurine deficiency effect cannot be ruled out.[43,44] It is worth noting that humans suffering from malnutrition in most parts of the world eat little or no animal flesh or fish, which would supply taurine, and are likely to be taurine-depleted. Experimentally, it has been demonstrated that a dietary taurine deficiency is exacerbated by a concurrent protein deficiency.[45,46] It may be that dietary taurine supplements would prove beneficial to humans existing on marginal diets, especially during pregnancy and lactation.

ACKNOWLEDGMENTS

The photography and art work of Mr. Richard Weed and the secretarial assistance of Mrs. Ann Parese are gratefully acknowledged.

REFERENCES

1. TIEDEMANN, F. & L. GMELIN. 1827. Einige neue Bestandtheile der Galle des Ochsen. Ann. Physik. Chem. **9:** 326–337.
2. JACOBSEN, J. G. & L. H. SMITH. 1968. Biochemistry and physiology of taurine and taurine derivatives. Physiol. Rev. **48:** 424–511.
3. REICHELT, K. L. & E. KVAMME. 1973. Histamine-dependent formation of N-acetyl-aspartyl peptides in mouse brain. J. Neurochem. **21:** 849–859.
4. REICHELT, K. L. & P. D. EDMINSON. 1974. Biogenic amine specificity of cortical peptide synthesis in monkey brain. FEBS Lett. **47:** 185–189.
5. FEUER, L. 1977. Theoretical background of the recognition of a new bioactive substance, litoralon, isolated from the parathyroid gland. Further theoretical considerations. Biologia (Budapest) **25:**3–33.
6. LAHDESMAKI, P., K AIKRAKSINEN, M. VARTIAINEN & P. HALONEN. 1980. Characterization of two synaptosomal peptides in calf brain. Acta Chem. Scand. (B) **34:** 343–348.
7. MARNELA, K.-M. & P. LAHDESMAKI. 1983. Mass spectral and hydrolytic determination of amino-acid sequences in synaptosomal peptides from calf brain. Neurochem. Res. **8:** 933–941.
8. MARNELA, K.-M., H. R. MORRIS, M. PANICO, M. TIMONEN & P. LAHDESMAKI. 1985. Glutamyl-taurine is the predominant synaptic taurine peptide. J. Neurochem. **44:** 752–754.
9. ZGLICZYNSKI, J. M., T. STELMASZYNSKA, J. DOMANSKI & W. OSTROWSKI. 1971. Chloramines as intermediates of oxidation reaction of amino acids by myeloperoxidase. Biochim. Biophys. Acta **235:** 419–424.
10. WEISS, S. J., R. KLEIN, A. SLIVKA & M. WEI. 1982. Chlorination of taurine by human neutrophils. Evidence for hypochlorous acid generation. J. Clin. Invest. **70:** 598–607.
11. WRIGHT, C. E., T. LIN, Y. Y. LIN & J. A. STURMAN. 1984. Taurine reacts with HOCl in cultured cells. Fed. Proc. **43:** 616.
12. WRIGHT, C. E., T. T. LIN, Y. Y. LIN, J. A. STURMAN & G. E. GAULL. 1985. Taurine scavenges oxidized chlorine in biological systems. Prog. Clin. Biol. Res. **179:** 137–147.
13. STURMAN, J. A. & K. C. HAYES. 1980. The biology of taurine in nutrition and development. Adv. Nutr. Res. **3:** 231–299.
14. STURMAN, J. A. 1983. Taurine in nutrition research. Prog. Clin. Biol. Res. **125:** 281–295.

15. STURMAN, J. A., D. K. RASSIN & G. E. GAULL. 1977b. Taurine in developing rat brain: Maternal fetal transfer of [35S]taurine and its fate in the neonate. J. Neurochem. **28:** 31–39.
16. STURMAN, J. A. & G. E. GAULL. 1975. Taurine in the brain and liver of the developing human and monkey. J. Neurochem. **25:** 831–835.
17. RASSIN, D. K., J. A. STURMAN & G. E. GAULL. 1978. Taurine and other free amino acids in milk of man and other mammals. Early Hum. Dev. **2:** 1–13.
18. HAYES, K. C., R. E. CAREY & S. Y. SCHMIDT. 1975a. Retinal degeneration associated with taurine deficiency in the cat. Science **188:** 949–951.
19. BERSON, E. L., K. C. HAYES, A. R. RABIN, S. Y. SCHMIDT & G. WATSON. 1976. Retinal degeneration in cats fed casein. II. Supplementation with methionine, cysteine, or taurine. Invest. Ophthalmol. **15:** 52–58.
20. SCHMIDT, S. Y., E. L. BERSON & K. C. HAYES. 1976. Retinal degeneration in cats fed casein. 1. Taurine deficiency. Invest. Ophthalmol. **15:** 45–52.
21. SCHMIDT, S. Y., E. L. BERSON, G. WATSON & C. HUANG. 1977. Retinal degeneration in cats fed casein. III. Taurine deficiency and ERG amplitudes. Invest. Ophthalmol. Vis. Sci. **16:** 673–678.
22. AGUIRRE, G. D. 1978. Retinal degeneration associated with the feeding of dog food to cats. J. Am. Vet. Med. Assoc. **172:** 791–796.
23. ANDERSON, P. A., D. H. BAKER, J. E. CORBIN & L. C. HELPER. 1979. Biochemical lesions associated with taurine deficiency in the cat. J. Anim. Sci. **49:** 1227–1234.
24. BURGER, I. H. & K. C. BARNETT. 1979. Essentiality of taurine for the cat. Kal-Kan Symposium for the Treatment of Dog and Cat Diseases **3:** 64–70.
25. BARNETT, K. C. & I. H. BURGER. 1980. Taurine deficiency retinopathy in the cat. J. Small Anim. Pract. **21:** 521–534.
26. RICKETTS, J. D. 1983. Feline central retinal degeneration in the domestic cat. J. Small Anim. Pract. **24:** 221–227.
27. WEN, G. Y., J. A. STURMAN, H. M. WISNIEWSKI, A. A. LIDSKY, A. C. CORNWELL & K. C. HAYES. 1979. Tapetum disorganization in taurine-depleted cats. Invest. Ophthalmol. Vis. Sci. **18:** 1201–1206.
28. STURMAN, J. A., G. Y. WEN, H. M. WISNIEWSKI & K. C. HAYES. 1981. Histochemical localization of zinc in the feline tapetum: Effect of taurine depletion. Histochemistry **72:** 341–350.
29. POLITIS, M. J. & N. A. INGOGLIA. 1979. Axonal transport of taurine along neonatal and young adult rat optic axons. Brain Res. **166:** 221–231.
30. STURMAN, J. A. 1979. Taurine in the developing rabbit visual system: Changes in concentration and axonal transport including a comparison with axonally transported proteins. J. Neurobiol. **10:** 221–237.
31. MANDEL, P. & H. PASANTES-MORALES. 1978. Taurine in the nervous system. Rev. Neurosci. **3:** 157–193.
32. OJA, S. S. & P. KONTRO. 1978. Neurotransmitter actions of taurine in the central nervous system. *In* Taurine and Neurological Disorders. A. Barbeau & R. J. Huxtable, Eds. Raven Press. New York. pp. 181–200.
33. STURMAN, J. A., R. C. MORETZ, J. H. FRENCH & H. M. WISNIEWSKI. 1985. Taurine deficiency in the developing cat: Persistence of the cerebellar external granule cell layer. J. Neurosci. Res. **13:** 405–416.
34. SMITH, D. E. & I. DOWNS. 1978. Postnatal development of the granule cell in the kitten cerebellum. Am. J. Anat. **151:** 527–537.
35. PALACKAL, T., J. A. STURMAN, R. C. MORETZ & H. M. WISNIEWSKI. 1985. Feline maternal taurine deficiency: Abnormal ontogeny of visual cortex. Trans. Am. Soc. Neurochem. **16:** 186.
36. GAULL, G. E., D. K. RASSIN, N. C. R. RAIHA & K. HEINONEN. 1977. Milk protein quantity and quality in low-birth-weight infants. III. Effects on sulfur amino acids in plasma and urine. J. Pediatr. **90:** 348–355.
37. JARVENPAA, A., D. K. RASSIN, N. C. R. RAIHA & G. E. GAULL. 1982. Milk protein quantity and quality in the term infant. II. Effects on acidic and neutral amino acids. Pediatrics **70:** 221–230.

38. RIGO, J. & J. SENTERRE. 1977. Is taurine essential for the neonates? Biol. Neonate **32:** 73–76.
39. GEGGEL, H. S., M. E. AMENT, J. R. HECKENLIVELY & J. D. KOPPLE. 1982. Evidence that taurine is an essential amino acid in children receiving total parenteral nutrition. Clin. Res. **30:** 486A.
40. GEGGEL, H. S., J. R. HECKENLIVELY, D. A. MARIN, M. E. AMENT & J. D. KOPPLE. 1982. Human retinal dysfunction and taurine deficiency. Doc. Ophthalmol. Proc. Ser. **31:** 199–207.
41. GEGGEL, H. S., M. E. AMENT, J. R. HECKENLIVELY, D. A. MARTIN & J. D. KOPPLE. 1985. Nutritional requirement for taurine in patients receiving long-term parenteral nutrition. N. Engl. J. Med. **312:** 142–146.
42. STURMAN, J. A., G. Y. WEN, H. WISNIEWSKI & M. NEURINGER. 1984. Retinal degeneration in primates raised on a synthetic human infant formula. Int. J. Dev. Neurosci. **2:** 121–129.
43. ZMORA, E., R. GORODISCHER & J. BAR-ZIV. 1979. Multiple nutritional deficiencies in infants from a strict vegetarian community. Am. J. Dis. Child. **133:** 141–144.
44. SHINWELL, E. D. & R. GORODISCHER. 1982. Totally vegetarian diets and infant nutrition. Pediatrics **70:** 582–586.
45. NEURINGER, M., D. DENNY & J. STURMAN. 1979. Reduced plasma taurine concentration and core electroretinogram amplitude in monkeys fed a protein-deficient semipurified diet. J. Nutr. **109:** XXVI.
46. VAN GELDER, N. M. & M. PARENT. 1981. Effect of protein and taurine content of maternal diet on the physical development of neonates. Neurochem. Rev. **6:** 539–549.

DISCUSSION

QUESTION: Were the infant monkeys preterm or full term in your studies of taurine deficiency?

JOHN STURMAN: We have only studied monkeys who are born at term to normally nourished mothers; these infants are separated from their mothers and hand-raised using a human infant formula, either alone or supplemented with taurine. The original study examined monkeys that were killed 26 months after birth, a longer period than human infants would normally be raised solely on a formula. We did not examine the brains of the monkeys, but it should be pointed out that if they did have a persistence of cerebellar external granule-cell layer, as found in kittens, we would not have observed it, since such cells would have degenerated by this time. We are actually in the process of examining the brains of infant monkeys raised for three months and six months on human infant formulas. Preliminary results indicate no obvious histologic abnormalities in the cerebellum, and quite to our surprise, we find marked changes in the visual cortex after postnatal taurine deprivation. These results will have to be completed and evaluated carefully before a firm conclusion can be reached.

QUESTION: Did you look at the taurine concentration in the full-term versus preterm milk?

JOHN STURMAN: In the few samples from the various species that we have obtained, preterm milk and colostrum tend to contain more taurine than does later milk.

PART V. DIETARY AND GENETIC THERAPY OF INBORN ERRORS OF METABOLISM

Introduction

DONALD A. SNIDER

Institute for Basic Research in Developmental Disabilities
New York State Office of Mental Retardation &
Developmental Disabilities
Staten Island, New York 10314

This session includes papers on the broad topic of therapy of inborn errors of metabolism.

Dr. Hugo Moser, Director of the Kennedy Institute for the Handicapped in Baltimore and a leading authority on inherited disorders that cause mental retardation, described at the workshop why there is excitement about genetic replacement and implantation, and also what will need to occur before this becomes a reality.[a]

Among the highlights of the workshop were the presentations and discussion on dietary therapy of phenylketonuria (PKU) and maternal PKU. In his paper, Dr. Edward McCabe of the University of Colorado describes his efforts to produce identical levels of metabolic control of PKU in breast-fed and non-breast-fed babies.

Young women with PKU who benefited from prevention programs (i.e., phenylalanine-free diets) as infants are now in their child-bearing years. Dr. Reuben Matalon of the University of Illinois College of Medicine at Chicago addresses this problem in his paper, which discusses strategies for dietary treatment in the disease. Drs. McCabe and Matalon plus other participants (including Dr. Robert Guthrie, who developed the technique for mass screening of newborns for PKU) vigorously debated the clinical and scientific merits of the rigor and timing of dietary controls.

---

[a] A published version of Dr. Moser's remarks can be found in: ASBURY, A. K., G. M. MCKHANN & K. I. MCDONALD, Eds. 1985. Diseases of the Nervous System. Saunders. Philadelphia, PA.

Issues in the Dietary Management of Phenylketonuria: Breast-Feeding and Trace-Metal Nutriture[a]

EDWARD R. B. McCABE[b] AND LINDA McCABE

B. F. Stolinsky Research Laboratories
Department of Pediatrics
University of Colorado Health Sciences Center
Denver, Colorado 80262

INTRODUCTION

Phenylketonuria (PKU) has been treated by dietary manipulation for more than three decades.[1] Current challenges in the dietary management of the child with PKU include facilitation of initial parental acceptance of the special diet, which will strongly influence the family's initial and long-term adjustment to this disorder, and refinement of the nutritional characteristics of these formulas in order to optimize the patient's physical and mental development. We will discuss the use of breast-feeding in the management of the infant with PKU as a means for facilitating maternal acceptance of the restricted diet. We will also review trace-metal nutriture, and more specifically zinc nutriture, in children managed with these artificial formulas and will discuss the need for further examination of the nutritional characteristics of these special diets.

BREAST-FEEDING THE INFANT WITH PKU

The low phenylalanine or phenylalanine-free formulas used in the management of the individual with PKU require supplementation with phenylalanine in order to provide adequate amounts of this amino acid for normal growth and development. This supplemental phenylalanine is provided as measured amounts of complete, natural proteins. Traditionally, infants with PKU received their required phenylalanine from prescribed volumes of infant formulas or cow's milk. This regimen demanded abrupt weaning of any breast-feeding infant in order to initiate dietary restriction when an elevated blood phenylalanine was confirmed; however, with the renewed interest in breast-feeding in the 1970s, we encountered an increased number of infants who were being breast-fed at the time of diagnosis and whose mothers were reluctant to discontinue nursing. Therefore, we under-

[a] This work was supported by grants from the National Institute of Child Health and Human Development (HDMR 5P30 HD0424) and the General Clinic Research Centers Program at the Division of Research Resources (RR 69), National Institutes of Health, from the Bureau of Community Health Services (MCJ 000252) and from the Milupa Corporation, and by donations to the Friends of Metabolic Research.

[b] Present address: Institute for Molecular Genetics, Room 304 E, Baylor College of Medicine, Texas Medical Center, Houston TX 77030.

took a trial to determine the feasibility of breast-feeding as supplementation to the standard PKU formulas. Our early experiences and recommendations have been summarized previously[2] and a detailed description of our experience is in preparation.[3]

During the period from September, 1977, through June, 1984, we cared for 28 infants who required dietary management after confirmation of positive newborn screens for PKU: 18 were managed by supplemental breast-feeding and 10 were managed in the traditional manner. In the breast-fed group, 16 were weaned at ages ranging from 1.3 to 25.5 months, with a mean of 8.9 ± 7.3 months, and two continued to breast-feed when the study was ended.

Serum phenylalanine concentration was used as a method of assessing metabolic control in these patients. When the monthly mean serum phenylalanine concentrations were compared for breast-fed and control groups, no significant differences were observed (TABLE 1). The most striking difference between these two groups of infants was their respective protein intakes: the breast-fed group showed consistently lower protein intakes than the control group (TABLE 2). However, using growth parameters as crude estimates of nutritional adequacy, no differences were observed between the breast-fed and control groups (TABLE 3).

Early in our experience with breast-milk supplementation of infants with PKU, we developed the clinical impression that these patients seemed to have an exaggeration of the physiological anemia of infancy. At that time, we were in-

TABLE 1. Comparison of Serum Phenylalanine Concentrations by Month for Breast-Fed and Formula-Fed Infants with PKU[a]

	Serum Phenylalanine in mg/dl $\bar{x} \pm$ SD (No.)	
Month	Breast-Fed	Formula-Fed
1	16.4 ± 4.7 (16)	17.9 ± 6.5 (10)
2	7.9 ± 2.4 (17)	7.5 ± 2.3 (10)
3	7.8 ± 3.2 (16)	7.9 ± 3.4 (10)
4	10.2 ± 2.9 (13)	10.4 ± 3.6 (10)
5	9.9 ± 2.5 (12)	9.8 ± 3.6 (10)
6	9.2 ± 2.6 (11)	11.3 ± 4.2 (9)

[a] No significant differences between groups.

TABLE 2. Comparison of Daily Protein Intakes by Month for Breast-Fed and Formula-Fed Infants with PKU[a]

	Protein Intake in g/kg · day $\bar{x} \pm$ SD (No.)	
Month	Breast-Fed	Formula-Fed
1	2.5 ± 0.5 (13)	3.5 ± 0.4 (10)
2	2.4 ± 0.3 (14)	3.6 ± 0.4 (10)
3	2.1 ± 0.4 (13)	3.4 ± 0.7 (9)
4	1.9 ± 0.2 (10)	3.0 ± 0.6 (9)
5	1.9 ± 0.2 (7)	3.0 ± 0.4 (8)
6	1.8 ± 0.1 (5)	2.9 ± 0.2 (6)

[a] All comparisons show significant differences with $p < 0.0001$.

TABLE 3. Comparison of Growth Parameters between Breast-Fed and Formula-Fed Infants with PKU[a]

Age in Months	Length in cm		Weight in kg		Head Circumference in cm	
	Breast-Fed	Formula-Fed	Breast-Fed	Formula-Fed	Breast-Fed	Formula-Fed
Birth	51.5 ± 3.8 (18)	50.2 ± 3.4 (9)	3.36 ± 0.63 (16)	3.02 ± 0.70 (10)	33.8 ± 2.2 (13)	33.5 ± 2.4 (8)
1	53.9 ± 4.4 (14)	52.2 ± 3.8 (7)	3.92 ± 0.88 (14)	3.69 ± 0.93 (7)	36.0 ± 2.3 (14)	35.7 ± 2.4 (7)
2	56.6 ± 3.9 (14)	55.6 ± 4.3 (6)	5.03 ± 0.78 (14)	4.19 ± 0.72 (8)	38.2 ± 1.9 (13)	37.2 ± 2.4 (6)
3	60.7 ± 2.4 (11)	60.2 ± 3.4 (8)	5.93 ± 0.54 (10)	5.43 ± 0.87 (7)	39.6 ± 1.0 (10)	39.6 ± 1.4 (8)
4	63.6 ± 4.2 (13)	60.4 ± 2.8 (6)	6.65 ± 0.98 (13)	5.34 ± 0.63 (6)	41.0 ± 1.8 (12)	39.9 ± 1.9 (6)
5	66.6 ± 2.6 (3)	64.9 ± 3.8 (5)	7.65 ± 0.91 (3)	6.73 ± 1.08 (5)	41.5 ± 0.7 (2)	41.9 ± 1.4 (5)
6	66.8 ± 3.4 (9)	65.1 ± 3.4 (7)	7.86 ± 1.12 (9)	6.45 ± 0.68 (8)	42.6 ± 1.7 (9)	41.7 ± 1.4 (7)

[a] All data reported as x̄ ± SD (No.) and no significant differences noted between any comparisons.

structing the mothers to supplement each low-phenylalanine formula feeding with breast-feeding within 15 minutes, rationalizing that this would provide a complete amino-acid mixture for protein synthesis.[2] We became concerned that some constituent of the casein-hydrolysate-based low-phenylalanine formula (Lofenalac, Mead Johnson Nutritional Division) that we were using exclusively at that time in infancy might decrease the bioavailability of iron from breast milk.[2,4] To test this clinical impression, we focused on the iron status of these infants.

We compared hemoglobin concentration, mean corpuscular volume, serum iron, total iron-building capacity (TIBC), percent iron saturation and serum ferritin between the breast-fed and formula-fed infants (TABLE 4). The mean corpuscular volume showed statistically significant lower mean values in the breast-fed group at three and six months (<0.05 and >0.0001, respectively). Lower means were found for the breast-fed infants, comparing hemoglobin concentration, serum iron, TIBC, and percent saturation with formula-fed infants, but none of these differences reached statistical significance.

Interpretation of these data on the influence of breast-feeding on the iron status of the infant with PKU is made somewhat difficult because of the relatively small numbers of patients studied and by the clinical manipulation of these patients. When a patient was observed to have a hematocrit approaching 30% or a hemoglobin approaching 10 g/dl, he or she was generally started on iron supplementation. In addition, our feeding recommendations changed in response to our concern that the formula might decrease iron bioavailability from breast milk. We began to recommend separation of breast and bottle feedings. This seemed to diminish the problem of anemia in the infant with PKU.

We conclude from this trial of supplemental breast-feeding in the management of PKU that this approach would appear to have no adverse nutritional consequences. We have been unable to identify any clinically significant problems associated with supplemental breast-feeding in this population. And we have been able to diminish somewhat the emotional impact accompanying the diagnosis of PKU for those mothers who have eagerly anticipated the opportunity to breast-feed their babies. It is recognized by all health professionals dealing with chronic illness that parents experience a loss when their infant is diagnosed with a chronic

TABLE 4. Comparison of Indices of Iron Nutriture Obtained at Three and Six Months of Age between Breast-Fed and Formula-Fed Infants with PKU[a]

	Age (mo.)	Breast-Fed	Formula-Fed
Hemoglobin	3	11.4 ± 1.0 (9)	12.2 ± 0.9 (4)
concentration (g/dl)	6	12.6 ± 0.7 (7)	13.0 ± 0.8 (5)
Mean corpuscular	3	82.3 ± 4.3 (8)	87.8 ± 3.3 (4)[b]
volume (fl)	6	76.8 ± 2.4 (7)	84.1 ± 2.9 (5)[c]
Serum iron (μg/dl)	3	66.0 ± 17.5 (6)	79.3 ± 18.2 (3)
	6	54.4 ± 19.6 (5)	72.8 ± 23.1 (5)
Total iron	3	404 ± 47 (5)	499 ± 195 (3)
binding capacity (μg/dl)	6	413 ± 45 (5)	387 ± 21 (5)
Percent iron	3	15.8 ± 4.5 (5)	16.7 ± 2.5 (3)
saturation (%)	6	14.4 ± 3.8 (5)	18.6 ± 6.3 (5)
Ferritin (ng/ml)	3	77.6 ± 28.5 (5)	60.3 ± 47.0 (3)
	6	19.8 ± 6.9 (5)	29.0 ± 21.3 (5)

[a] All data given as $\bar{x}$ ± SD (No.).
[b] $p < 0.05$.
[c] $p < 0.0001$.

disease. This loss need no longer be compounded for the breast-feeding mother by demands that the infant be weaned abruptly. By decreasing the psychological stress and supporting the mother's self-esteem during this traumatic period, it is hoped that better long-term adjustment can be achieved.

TRACE-METAL NUTRITURE IN THE CHILD TREATED FOR PKU

While considering the iron status of infants with PKU managed with a combination of breast milk and the low-phenylalanine formula, we had consulted with Dr. Michael Hambidge, a colleague in the B. F. Stolinsky Laboratories specializing in trace-metal nutriture. These continuing discussions have led to a series of studies on the trace-metal status, and more specifically the zinc status, of children with PKU treated with artificial formulas.

Observations comparing the breast-fed and formula-fed groups were interesting but inconclusive (TABLE 5). The mean plasma zinc values were lower at three and six months in those infants who were solely formula fed. Whereas these differences did not reach statistical significance, the mean plasma zinc for the formula-fed group was clearly in the deficiency range (<62 mg/dl) at six months; however, this work supported other reports that human milk is a good source of highly bioavailable zinc.[5,6]

In collaboration with workers at Emory University, we looked at a larger group of treated PKU patients with respect to their zinc and copper nutriture.[7,8] The study population consisted of 21 infants and children with PKU and one child with hyperphenylalaninemia all treated with Lofenalac as their primary protein source. Zinc and copper intakes did not differ significantly from the RDA; however, the PKU subjects' plasma and hair zinc values were significantly lower ($p <$ 0.05) than those of a group of normal children (TABLE 6). Plasma zinc values were below the lower limit of normal (68 μg/dl) for 52% of the PKU patients and 64% of these patients had hair zinc values below the lower limit of normal (70 μg/dl). The plasma copper concentration was also significantly lower ($p <$ 0.05) for the patients with PKU. A significant positive correlation ($r =$ 0.634, $p <$ 0.001) was observed between plasma zinc and copper, suggesting that a similar mechanism might be affecting the plasma concentrations of both of these trace metals in the treated PKU patients.

In another study at the University of Colorado Health Sciences Center, we examined the zinc status of 50 individuals with PKU between March 1978 and June, 1982 who were treated with Lofenalac or Phenyl-Free (Mead Johnson Nutritional Division).[9] There was a trend to lower plasma zinc for the PKU patients

TABLE 5. Comparison of Plasma Zinc Concentrations[a] at Three and Six Months of Age between Breast- and Formula-Fed Infants with PKU[b]

	Plasma Zinc in μg/dl $\bar{x} \pm$ SD (No.)	
Age	Breast-Fed	Formula-Fed
3 mo.	80.1 ± 16 (7)	73.0 ± 9.0 (3)
6 mo.	73.9 ± 22.0 (4)	60.3 ± 5.5 (3)

[a] Plasma zinc concentrations: Normal = 68–110 μg/dl; borderline = 62–68 μg/dl.
[b] No significant differences noted.

TABLE 6. Comparison of Zinc and Copper Status at Three and Six Months of Age between PKU Patients and Normal Controls[a,b]

Trace Metal	PKU Patients	Controls	p
Zinc—plasma (μg/dl)	66.6 ± 3.3 (21)	84.2 ± 2.9 (26)	<0.05
Zinc—hair (μg/g)	70.2 ± 11.5 (14)	130.7 ± 8.3 (42)	<0.05
Copper—plasma (μg/dl)	87.6 ± 6.6	121.5 ± 3.1	<0.05

[a] Adapted from Acosta et al.[7]
[b] All data given as $\bar{x}$ ± SD (No.).

on diet (76 ± 15 μg/dl, No. = 63) compared with those off diet (88 ± 36 μg/dl, No. = 9), but there was no difference in hair zinc between those on and off diet (107 ± 49 μg/dl, No. = 41 and 111 ± 53 μg/dl, No. = 4, respectively). However, one problem in the interpretation of any study such as this has to do with those patients for whom dietary restriction and the special formula are prescribed and who allege to be following the diet, but who are in fact noncompliant and may be obtaining nutrition from other sources. To look at this, we divided the subjects into those with low serum phenylalanine (low phe, ≤10 mg/dl) and those with high serum phenylalanine (high phe >10 mg/dl) (TABLE 7). The low-phe group, who presumably were more compliant with the special diet, had a significantly lower mean plasma zinc value, the mean (70 μg/dl) falling just above the lower limit of normal (68 μg/dl). There was no difference in hair zinc between these two groups. We next looked at the correlation between serum phenylalanine and plasma zinc in PKU patients on diet. In 63 paired observations a significant positive correlation was observed (r = 0.2615, $p < 0.05$). We conclude that there is a probable relationship between compliance with intake of a low-phenylalanine or phenylalanine-free formula and lower serum zinc.

We also looked at zinc nutriture as part of a study of a new formula introduced to the United States for management of PKU (PKU-2, Milupa Corp.).[10] Ten patients who had been managed on one of the Mead Johnson products were enrolled in the study and data were collected at 0, 4, 8, and 12 months on PKU-2. Two to four individuals had clearly low plasma zinc values at each time interval and one or two individuals were borderline low at 8 and 12 months. Three to five had low hair zinc at each interval.

These studies show that there is an association between decreased plasma zinc and/or hair zinc, as measures of zinc nutriture, and various formulas used in the

TABLE 7. One-way Analysis of Variance Comparing Plasma and Hair Zinc Concentrations between Patients with Low Serum Phenylalanine (≤10 mg/dl) and Patients with High Serum Phenylalanine (>10 mg/dl)

Zinc Concentration	Number	Mean	Standard Deviation	Probability
Plasma zinc (μg/dl)				
Low Phe	19	70	12	<0.05
High Phe	44	78	16	
Hair zinc (μg/g)				
Low Phe	11	101	60	NS
High Phe	30	109	46	

management of the infant or child with PKU. Others have also described negative balances and low blood concentrations of essential trace elements in individuals with PKU treated with synthetic and semisynthetic diets.[11-14] Trace elements reported to be in negative balance in these patients included zinc and copper,[11] and reports of trace elements with decreased blood concentrations in addition to those described above[3,7-10] included zinc[13] and selenium.[12-14] These biochemical deficiencies have been related to deficits in the composition of these special diets, as well as to decreased bioavailability of individual trace elements.[6-8,11,12] The studies that we have presented here, using products with no apparent compositional deficits and documented intakes appropriate to the RDA, tend to implicate bioavailability as a major consideration. One of the products that we have studied, Lofenalac, has been shown to have a markedly detrimental effect on zinc bioavailability.[6,15]

An obvious approach to improving zinc nutriture from these formulas would be for the manufacturers to simply add zinc to their products, without further investigation of the mechanisms underlying altered trace mineral abnormalities; however, such an approach may have other untoward nutritional effects. There appears to be a relationship between zinc and iron absorption and/or utilization;[19,20] other similar, but as yet undefined, relationships undoubtedly exist. Arbitrarily increasing zinc in these products, without regard to the cause of the observed abnormalities, might lead to even more serious consequences.

To date, none of these investigations has shown any obvious clinical signs or symptoms characteristic of trace-metal deficiency; however, one might raise concern about more subtle problems related to these biochemically determined deficiencies. While the outcome of treated children with PKU is markedly improved compared with those who are untreated,[16] the treated individuals still exhibit learning problems and subtle neuropsychological deficits.[17,18] While these residual deficits are generally considered a genetic consequence of the PKU, these children are continuously maintained on artificial formulas as their primary protein source. We are obligated to optimize the nutritional consequences of their special diets in order to assure that even subtle abnormalities in growth and intellectual development are not related iatrogenically to these prescribed formulas.

SUMMARY

We have reviewed our experience with supplemental breast-feeding of the infant with PKU. Our results indicate no harmful nutritional effects of breast-feeding the child with PKU, in comparison with the traditional approach using formula or cow's milk for supplementation. In addition, breast-feeding may provide a source of emotional support for the mother during this difficult period of initial diagnosis and management. It is hoped that this may improve the family's adjustment to this chronic illness. Our work with breast-feeding led us to a consideration of trace-metal nutriture in children treated with these synthetic and semisynthetic formulas. The results of these investigations suggest that there is a biochemically significant decrease in the bioavailability of zinc when these artificial formulas are used. While no clinical trace-metal deficiency has been described in treated PKU patients, we suggest that these nutritional deficits may relate to subtle abnormalities exhibited by these patients.

REFERENCES

1. BICKEL, H., J. GERRARD & E. M. HICKMAN. 1953. Influences of phenylalanine intake on phenylketonurics. Lancet **2:** 812.
2. ERNEST, A. E., E. R. B. McCABE, M. R. NEIFERT & M. E. O'FLYNN. 1979. Guide to Breast Feeding the Infant with PKU. DHHS Publication No. HSA 79-5110. U.S. Government Printing Office. Washington, D.C.
3. McCABE, E. R. B., A. E. ERNEST, M. R. NEIFERT, P. GARRY & L. McCABE. The management of breast-feeding among infants with phenylketonuria. In preparation.
4. PETERS, T., L. APT & J. F. ROSS. 1971. Effect of phosphates upon iron absorption studied in normal human subjects and in an experimental model using dialysis. Gastroenterology **61:** 315–322.
5. HAMBIDGE, K. M., P. A. WALRAVENS, C. E. CASEY, R. M. BROWN & C. BENDER. 1979. Plasma zinc concentrations of breast-fed infants. Pediatrics **94:** 607–608.
6. CASEY, C. E., P. A. WALRAVENS & K. M. HAMBRIDGE. 1981. Availability of zinc: Loading tests with human milk, cow's milk, and infant formulas. Pediatrics **68:** 394–396.
7. ACOSTA, P. B., P. M. FERNHOFF, H. S. WARSHAW, K. M. HAMBIDGE, A. ERNEST, E. R. B. McCABE & L. J. ELSAS. 1981. Zinc and copper status of treated children with phenylketonuria. J. Parent. Ent. Nutr. **5:** 406–409.
8. ACOSTA, P. B., P. M. FERNHOFF, H. S. WARSHAW, L. J. ELSAS, K. M. HAMBIDGE, A. ERNEST & E. R. B. McCABE. 1982. Zinc status and growth of children undergoing treatment for phenylketonuria. J. Inher. Metab. Dis. **5:** 107–110.
9. NORD, A., L. McCABE, A. ERNEST & E. R. B. McCABE. Biochemical and nutritional status of treated and untreated children with phenylketonuria and hyperphenylalaninemia. In preparation.
10. McCABE, E. R. B., A. NORD, A. ERNEST & L. McCABE. 1986. Evaluation of a new formula for the treatment of children with phenylketonuria. Submitted.
11. ALEXANDER, F. W., B. E. CLAYTON & H. T. DELVES. 1974. Mineral and trace-metal balances in children receiving normal and synthetic diets. Q. J. Med. **43:** 89–111.
12. LOMBECK, I., K. KASPEREK, L. E. FEINENDEGEN & H. J. BREMER. 1975. Serum selenium concentrations in patients with maple syrup urine disease and phenylketonuria under dieto-therapy. Clin. Chim. Acta **64:** 57–61.
13. LOMBECK, I., K. KASPEREK, L. E. FEINENDEGEN & H. J. BREMER. 1978. Trace-element disturbances in dietetically treated patients with phenylketonuria and maple-syrup urine disease. Monogr. Hum. Genet. **9:** 114–117.
14. LOMBECK, I., K. KASPEREK, H. D. HARBISCH, K. BECKER, E. SCHUMANN, W. SCHROTER, L. E. FEINENDEGEN & H. J. BREMER. 1978. The selenium state of children. II. Selenium content of serum, whole blood, and hair and the activity of erythrocyte glutathione peroxidase in dietetically treated patients with phenylketonuria and maple-syrup urine disease. Eur. J. Pediatr. **128:** 213–233.
15. CASEY, C. E., A. E. ERNEST, K. M. HAMBIDGE, E. R. B. McCABE & P. A. WALRAVENS. 1980. Zinc absorption and low-phenylalanine diets. Pediatr. Res. **14:** 497.
16. WILLIAMSON, M. L., R. KOCH, C. AZEN & C. CHANG. 1981. Correlates of intelligence test results in treated phenylketonuric children. Pediatrics **68:** 161–167.
17. BRUNNER, R. L., M. K. JORDAN & H. K. BERRY. 1983. Early-treated PKU: Neuropsychologic consequences. J. Pediatr. **102:** 381–385.
18. PENNINGTON, B. F., W. J. VAN DOORNINCK, L. L. McCABE & E. R. B. McCABE. 1985. Neuropsychological deficits in early-treated phenylketonuric children. Am. J. Ment. Def. **89:** 467–474.
19. SOLOMON, N. W. & R. A. JACOB. 1981. Studies on the bioavailability of zinc in humans: Effects of heme and nonheme iron on the absorption of zinc. Am. J. Clin. Nutr. **34:** 475–482.
20. HAMBIDGE, K. M., N. F. KREBS, M. A. JACOBS, D. FAVIER, L. GUYETTE & D. N. IKLE. 1983. Zinc nutritional status during pregnancy: A longitudinal study. Am. J. Clin. Nutr. **37:** 429–442.

Maternal PKU: Strategies for Dietary Treatment and Monitoring Compliance

REUBEN MATALON,[a] KIMBERLEE MICHALS,[b] AND
LINDA GLEASON[c]

[a]*Department of Pediatrics*
[b]*Department of Nutrition and Medical Dietetics*
[c]*Center for Handicapped Children*
University of Illinois at Chicago
Health Sciences Center
Chicago, Illinois 60680

INTRODUCTION

The success in newborn screening and early treatment of phenylketonuria (PKU) has resulted in a population of children and adolescents with PKU with normal intelligence. Most of these individuals discontinued dietary treatment in early childhood and are now reaching child-bearing age. It has been noted that increased blood phenylalanine levels during pregnancy, maternal PKU, have deleterious effects on the outcome of that pregnancy. A retrospective survey by Lenke and Levy showed an increased incidence of mental retardation, microcephaly, congenital heart disease, and low birth weight among offspring of mothers with PKU.[1]

Recently, the National Collaborative Study for PKU and other investigators have indicated that diet discontinuation might have negative effects on some patients with PKU.[2-6]. Therefore, clinics that are treating patients with PKU are facing two challenges: the first, how to return and keep patients on a phenylalanine-restricted diet, and the second, how to prevent the deleterious effects of high phenylalanine level during pregnancy.

The task of returning patients back to diet involves the proper motivation of adolescents and the proper tools for monitoring dietary compliance. Our experience with adolescents returning to diet has met only partial success.[7] The pregnancy of a woman with PKU poses the challenge of finding the level of blood phenylalanine that will prevent deleterious effects, designing an adequate diet that will support the pregnancy, and finding the tools for monitoring dietary compliance and fetal growth.

The purpose of this report is to present the problems associated with returning patients to diet, parameters of measuring dietary compliance, and strategies for the treatment of maternal PKU.

METHODS

Individuals Selected to Return to Diet Study

Thirty-two patients with PKU had discontinued the low-phenylalanine diet at a mean age of six years and two months. After a mean of seven years and three

223

months of unrestricted diet, the patients were counseled to resume a phenylalanine-restricted diet. Resumption of the special diet was done in stages. Milk and milk products were substituted by a phenylalanine-free formula (PKU-2, Milupa Corporation or Phenyl-Free, Mead-Johnson). Gradually meats and other high-protein foods were eliminated and the patients were counseled with the goal to reduce daily phenylalanine intake to 300–600 mg/day. Monthly blood phenylalanine samples were requested in order to monitor dietary compliance. Dietary intake was recorded two days before blood drawing.

Subject with Maternal PKU

The pregnant woman with PKU was 21 years old. She was diagnosed and treated for PKU at the age of three months, and diet was discontinued at seven years of age. Attempts to reinstitute diet therapy for PKU started at the age of 17 through 19 years of age but failed. She came to the clinic at 16 weeks gestation seeking counsel. After informed consent, diet therapy was initiated.

Diet requirements for this patient were based on ideal weight for height and gestational age of 55 kg instead of her actual weight of 68 kg. The patient used a newly available formula, PKU-3 (Milupa Corp.), a phenylalanine-free, low-calorie formula supplemented with tyrosine, vitamins, and minerals. This formula is designed to meet pregnancy needs. The patients ate foods low in protein content to meet her phenylalanine and caloric needs. The diet prescription provided 70 g of protein, 400 mg of phenylalanine, and 5,500 mg of tyrosine per day.

RESULTS

Diet Resumption

Of the 32 patients, 15 failed to resume any diet restrictions after a one- to six-month trial. These individuals were off dietary restriction an average of 10 years and 9 months.

Seventeen patients were able to resume some diet restrictions for one to six years (mean 3½ years). These individuals were off dietary restriction for five years and six months, a considerably shorter period than the group that failed to resume any diet restrictions. The mean phenylalanine blood level for these 17 patients for the past year was 13.6 mg/dl (range 4.0–19.8 mg/dl). The desired treatment range set by this clinic was 2–10 mg/dl. Although blood specimens were requested once a month, nine of the 17 patients sent blood specimens every four months, six patients sent blood monthly, and two patients sent weekly blood specimens. FIGURE 1 shows the fluctuations in weekly blood phenylalanine levels from one of these patients. The data in this figure show considerable fluctuations even in a one-week period. The phenylalanine level can drop from above 10 mg/dl to below 10 mg/dl in one week.

Dietary Treatment of Maternal PKU

The subject started a self-imposed low-protein diet before examination in the clinic. Three days after dietary resumption, her blood phenylalanine dropped to 1 mg/dl. The clinic goal was to maintain blood phenylalanine less than 6 mg/dl.

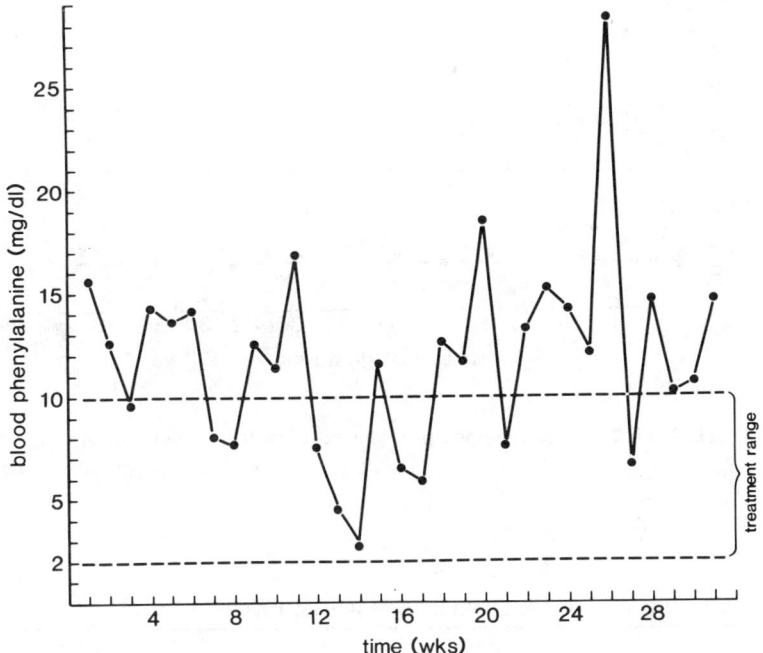

FIGURE 1. Fluctuations in weekly blood phenylalanine on a 13-year-old patient with PKU trying to return to dietary therapy.

Blood phenylalanine levels were monitored twice a week and blood phenylalanine was maintained below 6 mg/dl throughout the pregnancy. On one occasion, blood phenylalanine reached 10 mg/dl when the patient did not follow the diet restrictions. Blood phenylalanine levels throughout the pregnancy are summarized in FIGURE 2.

Blood tyrosine levels were measured on a monthly basis in order to ascertain adequacy of dietary treatment. There is a theoretical possibility that a deficiency of tyrosine may develop during treatment of pregnancy with PKU. In this case, the range of tyrosine was 0.3–0.8 mg/dl. Although these levels are low for normal adults, they were considered adequate for the pregnancy and no tyrosine supplementation was used.

In order to assure compliance and to examine for the excess levels of the neurotoxic metabolites of phenylalanine, phenyllactate, phenylacetate, phenylpyruvate, and phenylethylamine were determined in the urine of the patient on a weekly basis. These metabolites reflected blood phenylalanine levels and although they declined sharply following initiation of the diet, they remained above normal for the whole pregnancy. A sample of these levels is summarized in TABLE 1.

Maternal hemoglobin remained above 12 mg/dl, although iron supplementation was initiated at 30 weeks of gestation because of a drop in ferritin levels. TABLE 2 summarizes these data. Trace metals were obtained at several points during the pregnancy as seen in TABLE 3. Trace metals appeared adequate

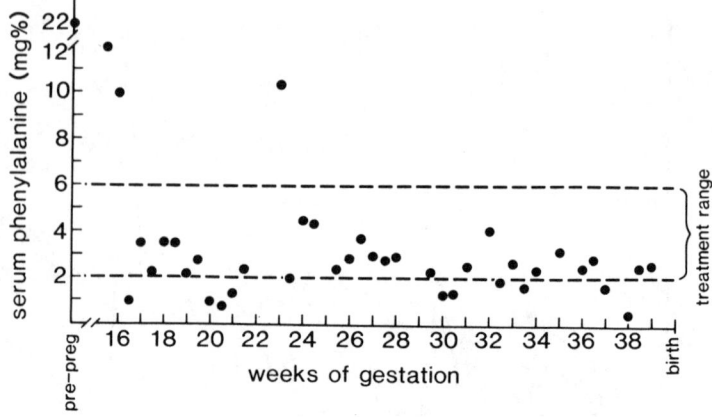

FIGURE 2. Biweekly blood phenylalanine levels in a PKU pregnancy.

TABLE 1. Phenylalanine Metabolites in Maternal PKU

	Blood			Urine		
Weeks	Phe (mg/dl)	Tyr (mg/dl)	PEA[a] (ng/mg creatinine)	PP[b]	PA (mg/gm creatinine)	PL
16	12.0	0.8	—	234.1	251.0	103.0
17	2.1	—	3.3	4.5	0.6	1.2
23	10.4	0.3	6.9	19.3	11.2	57.4
26	2.9	0.8	4.1	2.4	0.5	0.2
28	2.9	0.8	2.5	2.7	3.1	13.5
31	2.4	0.7	2.4	2.9	16.2	4.2
36	2.5	0.3	6.2	2.8	3.1	6.3

[a] PEA = phenylethylamine: normal levels less than 1.5 ng/mg creatinine.
[b] PP = phenylpyruvate; PA = phenylacetate; PL = phenyllactate: All are compounds not present in normal urine.

TABLE 2. Hemoglobin and Ferritin Levels During a Treated Pregnancy[a]

	16–24 Weeks	24–32 Weeks	32–38 Weeks	Normal-Range Pregnancy
Ferritin (μg/ml)	125	50	—	40–130
Hb (μg/dl)	12.5	12.1	13.7	120–16.0
HCT (%)	36.6	35.1	39.5	37–47

[a] Iron supplementation started at 30 weeks.

TABLE 3. Trace-Metal Levels in Blood During a PKU-Treated Pregnancy

Trace Metal	16–24 Weeks	24–32 Weeks	32–38 Weeks
Copper (mcg/dl)	119	—	228
	(165–187)[a]	—	(168–216)[a]
Zinc (mcg/dl)	95	—	105
	(59–77)[a]	—	(47–65)[a]
Selenium (mcg/dl)	13[b]	—	4.1[b]
Manganese (mcg/dl)	1.1	1.2	—
	(1.4 ± 0.9)[a]	(1.4 ± 0.9)[a]	—
Magnesium (mg/dl)	1.9	2.1	—
	(1.2–1.8)[a]	(1.2–1.8)[a]	—

[a] Levels in normal pregnancies.
[b] Selenium—nonpregnant normals 1–21 mcg/dl.

throughout the pregnancy.[8] The formula PKU-3 is not supplemented with selenium and the blood selenium was at the lower range of normal. Vitamins A, D, B_{12}, and folate were examined at 34 weeks' gestation (TABLE 4). These levels were normal for pregnant women.[9] At 20 weeks' gestation, ultrasonography of the head showed biparietal diameter of 47 mm, and at 30 weeks it was 72 mm. Both values are at the lower range of normal.

The baby was born without difficulty after 38 weeks' gestation. The baby was 18 inches long, weighed 5½ pounds, and had a head circumference of 30 cm. The placenta weighed 325 g and had a diameter of 15 cm. Both of these values are small. There was a single umbilical artery. The mother's blood phenylalanine at delivery was 3.4 mg/dl and tyrosine 0.4 mg/dl, the cord blood phenylalanine was 4.2 mg/dl and tyrosine 0.6 mg/dl. The baby had a blood phenylalanine level of 20 mg/dl at 72 hours of age, and a diagnosis of PKU was made on the baby.

On physical exam the baby had a small head with a flat forehead. He had ptosis of the left eye and epicanthal folds as shown in FIGURE 3. He had a strawberry hemangioma of the chest 4 × 4 cm. He also has clinodactyly similar to his mother. The echocardiogram was normal. The dysmorphic features of this baby are similar to those described in maternal PKU.[10]

The mother elected to breast-feed. Colostrum and breast milk were analyzed for phenylalanine, tyrosine, and trace elements (TABLE 5). These amino acids and minerals appeared adequate.[11-12]

TABLE 4. Vitamin Levels During a PKU-Treated Pregnancy

Vitamin	34 Weeks	Pregnant Normal
Vitamin A μg/dl	45.9	≥30
Vitamin D μg/dl	22.0	≥20
Folate ng/ml	19.4	>6
Vitamin B_{12} pg/ml	431	340–656

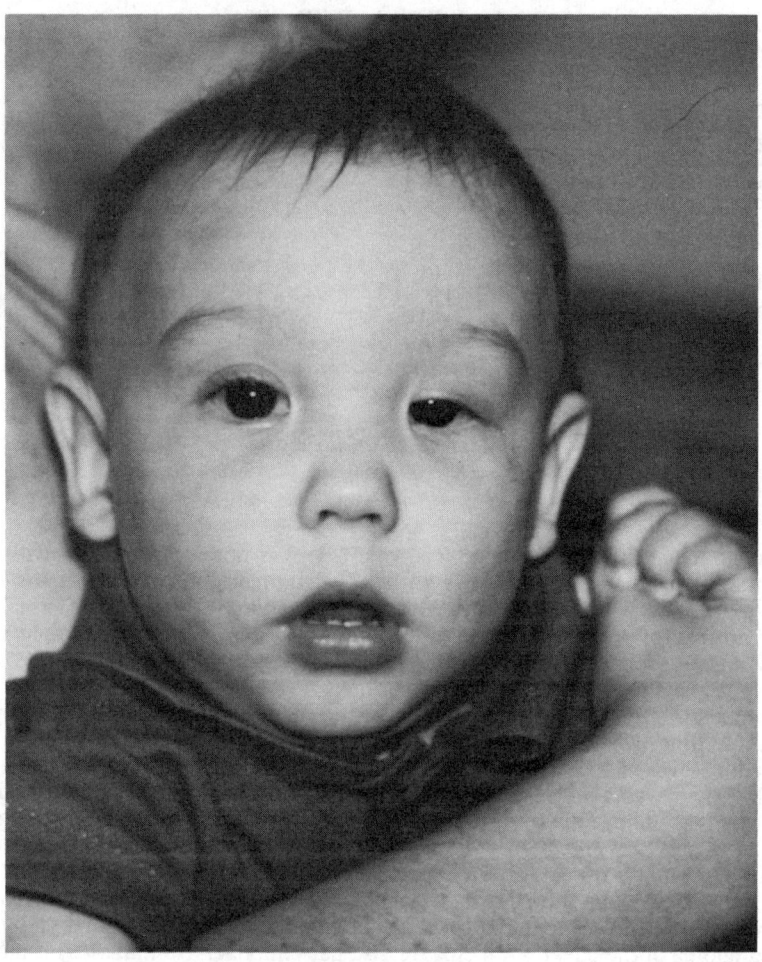

FIGURE 3. The outcome of a late-treated PKU pregnancy. The baby at six months of age showing microcephaly, ptosis of one eye, and facial features of maternal PKU syndrome.

TABLE 5. Amino-Acid and Mineral Content of Maternal PKU Milk and Colostrum[a]

Amino Acid/Mineral	Milk	Colostrum	Reference Value
Phenylalanine (μmol/dl)	0.5	6.0	—
Tyrosine (μmol/dl)	0.5	11.1	—
Taurine (μmol/dl)	36.7	9.3	—
Fe (mcg/ml)	1.40	—	0.28
Zn (mcg/ml)	0.55	2.19	0.82
Cu (mcg/ml)	0.20	0.40	0.193
Mg (mcg/ml)	26.0	28.0	29.6
Se (ng/ml)	9.11	22.31	16.3 (milk)
			41.2 (colostrum)

[a] The trace-metal analyses were performed by M. F. Picciano, University of Illinois at Urbana.

COMMENTS

Treatment of PKU in the United States has been changing toward long-term therapy. This change of policy is not universal and has occurred only recently;[13-14] therefore, it is expected that there will be patients with PKU, especially women reaching the reproductive age, who may seek counsel regarding resumption of diet. The difficulties in the resumption of diet are exemplified in the 32 individuals presented in this paper. Pregnancies are of concern since they may be unplanned and diet treatment may be resumed following conception, as in our case. Pregnancy has proven to be a motivating factor in our patient, which may be the case for other women. Nonetheless, diet resumption before conception is essential if the prevention of the maternal PKU syndrome is to be attained.

The experience with blood phenylalanine determinations on adolescents emphasizes the importance of frequent determinations during pregnancy due to the wide range of fluctuations that can occur in a short period of time. These findings have not been recognized in the past. A suggested protocol for monitoring the pregnancy of a PKU woman is in TABLE 6.

The strategy for treatment of this PKU pregnancy included frequent blood phenylalanine monitoring of twice per week. Since the urinary excretion of metabolites of phenylalanine may lag behind blood phenylalanine levels, examining these metabolites may assure better compliance. Furthermore, of interest is the finding of higher than normal levels of phenyllactate, phenylacetate, and phenylpyruvate in spite of "good" blood phenylalanine levels that were kept less than 4 mg/dl most of the treatment period. The levels of phenylethylamine were also higher than normal levels. These findings may suggest that blood phenylalanine at a given time may not represent fluctuations during an entire day. Since these metabolites are potentially neurotoxic, their follow-up may be of interest in the assessment of the IQ and learning abilities of children born to mothers with PKU.

This experience indicates that the formula used, PKU-3, was adequate to support a pregnancy. The fat-soluble vitamins, folate, B12, and minerals, were adequate. The level of selenium was at the lower range for nonpregnant normals; however, the available selenium data for normal pregnancies is small. Therefore, one cannot reach any conclusions, especially since the baby is now 18 months and has had neither cardiac defects nor cardiomyopathy.

Regarding the adequacy of trace metals and other micronutrients in PKU pregnancy, it is difficult to draw meaningful conclusions. There is a paucity of data regarding these nutrients in pregnancies in general and PKU pregnancies in

TABLE 6. Protocol for Monitoring a PKU Pregnancy

Component Monitored	Biweekly	Weekly	Monthly	Every Two Months
Blood:				
Phenylalanine	x			
Tyrosine			x	
Hb/Hct				x
Ferritin				x
Trace metals				x
Vitamins				x
Urine:				
Amino acids			x	
Organic acids		x		

particular; therefore, further studies are needed to more carefully ascertain their importance and adequacy in PKU pregnancy.

SUMMARY

Blood phenylalanine and the metabolites of phenylalanine can be dramatically lowered in pregnant PKU mothers. Special formulas and a strict protocol should be used to achieve diet control and adequate compliance. Levels of blood phenylalanine of 4 mg/dl and lower can be achieved and are preferable to higher levels. The problem of treatment postconception may lead to limited success, as in our case. In order to achieve optimal results, blood phenylalanine should be controlled at or before conception. Since it is difficult to return patients to diet who have been taken off phenylalanine restriction, diet therapy for PKU should not be discontinued at any age.

REFERENCES

1. LENKE, R. R. & H. L. LEVY. 1980. N. Engl. J. Med. **303:** 1202–1208.
2. WILLIAMSON, M. M., R. KOCH & S. BERLOW. 1979. Pediatrics **83:** 823–824.
3. CABALSKA, B., N. DUSZYNSKA, J. BORZYMOWSKA, K. ZORSKA, A. KUSLACZFOLGA & K. NOZKOQA. 1977. Eur. J. Pediatr. **126:** 253–262.
4. SEASHORE, M. E., E. FRIEDMAN, R. A. NOVELLY & V. CAPOT. 1985. Pediatrics **75:** 226–232.
5. SMITH, I., M. E. LOBASCHER, O. H. WOLFF, H. SCHMIDT, S. GRUBEL-KAISER & H. BICKEL. 1973. Br. Med. J. **2:** 723–726.
6. WAISBREN, S. E., R. R. SCHNELL & H. L. LEVY. 1980. J. Inher. Metab. Dis. **3:** 149–152.
7. MICHALS, K., R. MATALON, M. DOMINIK, V. SCHUETT & E. BROWN. 1985. J. Pediatr. **106:** 933–936.
8. ACOSTA, P. B., L. CASTIGLIONI, F. ROHR, K. MICHALS & E. WENZ. 1984. NICHD-CRMC **83:** 26.
9. Committee on nutrition of the mother and preschool child, food, and nutrition board. 1978. Laboratory indices of nutritional status in pregnancy. Natl. Acad. Sci. USA.
10. LIPSON, A., B. BUEHLER, J. BARTLEY, D. WALSH, J. YU, M. O'HALLORAN & W. WEBSTER. 1984. J. Pediatr. **104:** 216–220.
11. PICCIANO, M. F. 1978. Nutr. Rep. Int. **18:** 5–11.
12. SMITH, A. M., M. F. PICCIANO & J. A. MILNER. 1982. Am. J. Clin. Nutr. **35:** 521–526.
13. SCHUETT, V. E. & E. S. BROWN. 1984. Am. J. Public Health **74:** 5–9.
14. SCHUETT, V. E., E. BROWN & K. MICHALS. 1985. Am. J. Public Health **75:** 39–42.

Dietary and Genetic Therapy of Inborn Errors of Metabolism: A Summary

SELMA E. SNYDERMAN

Department of Pediatrics
New York University Medical Center
New York, New York 10016

It has now been over 50 years since phenylketonuria was first described and over 30 years since the first attempts at therapy, and yet there are still a number of unanswered questions, some of which were highlighted in Dr. McCabe's and Dr. Matalon's presentations. I think the one of prime importance, which has not been touched on at this workshop, and which I will not go into, is the cause of the mental retardation. It can certainly be related to biochemical error, and there are at least a dozen possibilities, but the exact cause is not known; it may be a combination of a number of factors. There are other problems as well that will affect the ultimate outcome. They include the age at which therapy is begun and, as you have heard, the rigidity of the biochemical control.

There are a number of reports that suggest that the treatment of PKU is less than optimal because there are problems even when the child is treated from earliest infancy. Yet many have not been treated early enough: Two months or twenty-one days is not earliest infancy. A great deal can happen in the interim. Perhaps that is one of the reasons for the less-than-perfect treatment results. The rigidity of control can also affect the end result. Shall we keep the phenylalanine level at normal or as close to normal as humanly possible, or are we going to allow it to rise to 10 or 15? The other still unanswered question is, how long should therapy be continued? Is it safe to discontinue at a certain age or should it be continued indefinitely?

Dr. McCabe talked about one aspect of treatment, to allow some degree of breast-feeding when initiating therapy. There are psychological advantages when the mother has strong feelings about breast-feeding. The diagnosis of PKU in an infant who looks well is a devastating experience for parents, and anything that can be done to soften the blow would certainly be advantageous; however, if there is evidence that more rigid control is necessary in the earlier phases of treatment, it may not be in the best interests of the infant to breast-feed. In the study reported today, all of the phenylalanine values in the infants who were breast-fed were 5–6 times above normal. Unless it can be demonstrated that levels of this magnitude are not harmful (and I doubt this), I think it is not justified to promote breast-feeding in the early months of life.

Low blood zinc levels in children who were treated with a number of different dietary products is a provocative finding, especially since the zinc content of these formulas is well within accepted nutritional guidelines. The reason for the low levels is really not known; it is quite possible that the zinc is less bioavailable because of the type of ligand to which zinc is bound. Other possibilities include excess urinary excretion or perhaps formation of some sort of a product between the abnormal metabolite and zinc, which makes it less available for the ordinary biochemical determination. The important issue, however, is not the blood level or the hair level, but the tissue level and its relation to any untoward results.

231

Dr. McCabe suggests that perhaps perceptual motor problems that were observed in a number of treated PKU patients may be the result of this deficiency. I think it is difficult to ascertain this at present because of the other problems that have been mentioned, that is, the age of onset of therapy and the degree of biochemical control. Another factor has to be considered; perceptual motor defects are common in the general population, and perhaps more are found in PKU children because they are watched so carefully and are tested so regularly.

Dr. Matalon has told you about the emergence of a new problem in PKU, one that results from the success of treatment. A new generation of young women who are normal and are no longer spending their lives in institutions for the retarded are leading normal lives and may become pregnant. You already heard about the severe consequences of elevated maternal phenylalanine levels. We expect and hope that the collaborative study that is being organized will be able to answer a number of existing questions. These include the maternal phenylalanine level that may be associated with a poor fetal outcome; how early in pregnancy the treatment must be initiated; how rigidly the phenylalanine level must be controlled; and, is treatment always successful? The exact incidence of these sequelae in untreated women is not known. The best data were obtained in the Lenke-Levy study in which they indicated an incidence of over 90%. This was, however, a retrospective study and has all of the problems associated with a retrospective study, particularly the fact that many of these cases were obtained from the literature, and it is possible that more abnormal than normal results were published. There is certainly no doubt that this is a real and increasing problem although the exact incidence is not known.

Dr. Moser has given us a comprehensive view of the various modalities that can be used to treat metabolic disorders. There are a number of diseases that respond to measures that reduce the accumulation of the toxic metabolite. These include dietary intervention similar to that which is successful in phenylketonuria, the removal of toxic material by chelation therapy as in Wilson's disease, and stimulation of alternate pathways of excretion, such as the use of sodium benzoate in the inherited defects of the urea cycle. The administration of large doses of a cofactor of an enzyme has been of value in a limited number of disorders. The mode of action of these vitamins that act as cofactors is not known. There seems to be some information that they stimulate the small amount of residual enzyme activity that is present.

The most direct form of therapy would be the replacement of the enzyme that is inactive, and this can be done either by administering purified enzymes or by transplanting tissue that contains the desired enzyme. Tissue transplantation of liver, kidney, and bone marrow has been successful in a number of diseases. Transplantation of amniotic membrane and fibroblast cultures have also been investigated. Organ transplant is associated with a number of risks, and the mortality rate in bone marrow transplantation has been unacceptably high to date. The use of purified enzymes has not been very successful because of the difficulties in producing a pure enzyme and the problems inherent in the tissue uptake of the enzyme.

Gene replacement is a possible future mode of therapy and might be feasible in those disorders where there is no reason to control the exact level of the enzyme activity; however, before it can be applied to human disease, much more information about gene expression, methods of targeting the gene to the specific tissue, and any untoward effects is necessary.

There has been great progress in understanding and treating metabolic disorders since phenylketonuria was first recognized. The recent advances in molecular genetics should lead the way to very exciting developments in the near future.

DISCUSSION

ROBERT GUTHRIE: A number of us have felt for a long time that elevated blood phenylalanine levels in the adult with PKU can produce significant effects on behavior. We have carried out unpublished studies of effects of phenylalanine loading on university students and also on parents of children with PKU that showed, with some exceptions, interference with CNS performance at transient levels above 30 mg/dl of blood phenylalanine. Dr. Louis Elsas has unpublished data that show similar effects, and one hears of other examples at meetings.

EDWARD McCABE: When one looks at that slide comparing serum phenylalanine concentrations in breast-fed and nonbreast-fed patients, one would prefer to see mean concentrations of 6 or 7 mg/dl. The reality was that the means were somewhat higher. Obviously, the mean serum phenylalanine concentrations were high in the first month of life, reflecting the pretreatment concentrations and the values obtained before the patients were brought into metabolic control. Very few studies have reported mean serum phenylalanine concentrations for treated PKU patients. The mean values that we observed are similar to those reported by the PKU Collaborative Study. As you will recall, that study had two treatment groups—one targeted for lower blood levels and another targeted for higher blood levels—and it was very difficult to keep the patients in their targeted ranges. We also tried very diligently to maintain our patients' serum phenylalanine levels in what was at that time our desired therapeutic range of 5–10 mg/dl, and if you had asked me what the data would show before we calculated the means, I would have guessed that the values would have been lower than what we found. It would be worthwhile for other investigators to make similar calculations, but the important point is that, in our hands, the breast-fed and non-breast-fed patients achieved identical levels of metabolic control.

REUBEN MATALON: The blood phenylalanine values presented in the tables of Dr. McCabe are too high for our clinic and some other clinics in the country. In our experience, children with PKU whose blood phenylalanine levels remain high, above 10–15 mg/dl, suffer from difficulties in school, short attention span, and hyperactivity; therefore, we have been trying hard to keep blood phenylalanine at or below 5 mg/dl.

Another important point is the frequency of blood monitoring. Spot checks of once a month, or less frequently than that, may not be representative of blood phenylalanine levels during that period of time. Therefore, in order to keep blood phenylalanine at the levels we think are optimal for treatment of PKU, we examine these levels twice weekly in the early months of life and once per week thereafter. All PKU patients who have good compliance then are switched to biweekly tests. Furthermore, the organic acids of phenylalanine are quantitated as another parameter of compliance.

Certainly we have come a long way in the detection and treatment of PKU, but some "fine tuning" is still required.

EDWARD McCABE: There are several points to make regarding Dr. Matalon's comments concerning the phenylalanine levels presented in our tables. While here in this room with Drs. Matalon and Snyderman, I may be in the minority in terms of our preferred therapeutic range for blood phenylalanine, but there remains considerable debate nationally concerning the target phenylalanine level, and I would suggest that the clinics of Drs. Matalon and Snyderman are aiming lower than most. In addition there are no data that I am aware of to support the need for maintenance of lower blood phenylalanine levels. Dr. Matalon should test his clinical impression statistically to see if the lower blood phenylalanine level does improve school performance and attention span and diminish hyperactivity, or even to see if he is achieving his target phenylalanine level. As I noted previously, the mean levels in our patients were higher than I would have expected before doing the calculations, which shows the problem with clinical impression that is not tested objectively.

Concerning the frequency of blood testing, we follow a similar schedule to Dr. Matalon. Our results represent the monthly means for each group calculated from the means for each subject taken from their individual repeated tested levels during that month. The number for the breast-fed group declined with time as subjects were weaned and the number for the formula-fed group decreased to nine in the sixth month because one of these subjects had not yet reached six months of age at the time the data were calculated.

SELMA SNYDERMAN: There has not been a great deal of experience in relating plasma zinc levels in children to the state of zinc nutrition. The determination of tissue levels of, for example, the white blood cells, might be preferable to that of blood; some functional test might be even better. I believe that zinc deficiency has been overdiagnosed in the general public and unnecessary supplementation has been used widely. I would not like to see supplementation used in the treatment of PKU children without a good deal more data demonstrating that it is necessary.

I would like to make a comment about the phenylalanine level of the child under treatment. There is absolutely no reason to believe that the plasma phenylalanine should be maintained at a higher level than that of the normal child. This thought originated when techniques were used to monitor that could not differentiate normal from deficient blood phenylalanine levels, and hence some patients did become phenylalanine deficient. If sensitive enough measures are used, it is possible to maintain plasma phenylalanine levels in the normal range without fear of phenylalanine deficiency.

JOSEPH FRENCH: Dr. McCabe, do you have any metalloenzyme data for this cohort? Superoxidismutase activity (in terms of its thermal stability) would be an easy probe.

EDWARD MCCABE: No, we do not have any metalloenzyme data. The enzyme that some have proposed for assessment of zinc nutriture is alkaline phosphatase, but this has not proven as useful as once hoped. The methods for determining zinc in blood elements are still being perfected and were not used in the studies reported here. We used the accepted methods and standards in evaluating zinc and copper nutriture in these patients.

We agree with Dr. Snyderman that it would be possible to overinterpret the zinc data. We have noted that there were no obvious clinical signs or symptoms of zinc deficiency and have cautioned the manufacturers not to begin adding zinc to their products arbitrarily. We would suggest that these products, and this patient population, may be showing us something about trace-metal bioavailability and possibly about subtle effects of mild deficiency states. If we are going to control the nutritional status of our patients through these special formulas, then we must be assured that we have optimized the nutritional parameters of their diets.

We also agree with Dr. Snyderman that one of the reasons why physicians treating PKU patients have aimed for blood phenylalanine levels higher than the normal physiological range has to do with the laboratory measurement of phenylalanine concentration; however, the difficulties are not only with paper chromatography for phenylalanine quantitation. Neither of the two current methods in common use, that is, the bacterial inhibition assay nor the fluorometric method, is very accurate in the physiological range. In order to obtain target levels in the normal physiological range safely, one must use a good column chromatographic system, which is more labor-intensive and expensive, and may not be available to these patients in all centers. Before implementing a recommendation for treatment aimed at lower levels, which would substantially increase the cost of monitoring these patients nationally, it would seem appropriate to objectively test the clinical impression that these lower levels are truly beneficial.

ROBERT GUTHRIE: At a recent meeting of the Northeast Maternal PKU group, someone, I think it was Dr. Harvey Levy, described an attempt to get a young woman with PKU onto the low-phenylalanine diet to lower her phenylalanine blood levels, so that she could become pregnant without risk to the fetus. She had not been on the diet since childhood and found it very difficult to resume it, but she reported that she felt so much better on the diet that she would stay on it after her baby was delivered.

SELMA SNYDERMAN: We have had that same experience, Dr. Guthrie. I have had two young ladies come to me who wanted to try going back on diet because they were engaged, and they wanted to be sure that they could stand it before they faced the problem of having children. Both were successful in bringing down their levels, and both reported feeling much better, more alive and more awake.

The other interesting experience we have had with adult PKU has been with two mildly retarded young women who had been taken off treatment at about six years of age by two other centers. Both became psychotic 15–20 years after diet termination and both had abnormal electroencephalograms. Although I was doubtful about being able to accomplish anything, I decided to try dietary therapy. The behavior of both improved markedly and the electroencephalogram normalized; therefore, there can be an effect of therapy later in life. It will not change the IQ, but it can affect other parameters.

QUESTION: What are other effects?

SELMA SNYDERMAN: Hyperactivity can be markedly reduced.

JOSEPH FRENCH: Did the alleviation of psychoses correlate in any way with either the CSF or urinary organic acids and monoamine contents?

SELMA SNYDERMAN: We did not have the chance to study CSF. We followed blood phenylalanine level and phenylpyruvic acid in the urine.

JOSEPH FRENCH: What happened to the levels of phenylpyruvic, phenylacetic, and phenyllactic acids?

SELMA SNYDERMAN: The urinary metabolites were reduced at the time of the first measurement, approximately one week after diet was instituted.

SUSAN SKLOWER: Is anyone monitoring organic acids of phenylalanine as a means of controlling their patients?

REUBEN MATALON: I do not have my own data on the excretion of the organic acids of phenylalanine, but Helen Barry from Cincinnati has some data on this relationship. According to those data, phenylpyruvate, phenyllactate, and phenylacetate continue to be excreted even though blood phenylalanine has dropped sharply. This lag in the excretion of these compounds may take from a few days to a week. This is one of the reasons why we are now monitoring these aromatic acids as a measure of good treatment.

SELMA SNYDERMAN: I think we should use it more often.

REUBEN MATALON: The determination of phenylethylamine and the aromatic acids of phenylalanine as a measure of successful treatment is a novelty that we have just begun studying systematically in our clinic. The levels of these compounds are related to blood phenylalanine and other neuropsychological measures. We hope that these studies will help us better adjust the treatment of PKU on an individual basis.

QUESTION: What is the role of phenylalanine ammonia lyase in the treatment of PKU?

SELMA SNYDERMAN: It can lower the blood phenylalanine level, but will not bring it into the therapeutic range.

SUSAN SKLOWER: I would like to end the session with a quote. In 1939, Dr. Jervis, the founding director of the Institute for Basic Research, wrote that "the study of phenylketonuria may throw light on the whole problem of mental deficiency." I think you see that even today. Many questions relating to the biochemical basis of developmental disability remain unanswered, but the study of PKU has led to important concepts in this field.

Introduction

DONALD A. SNIDER

Institute for Basic Research in Developmental Disabilities
New York State Office of Mental Retardation &
Developmental Disabilities
Staten Island, New York 10314

Two papers on topics unrelated to the other sessions are included because of their special significance to the field of mental retardation: anticonvulsant drug therapy and aging. The former is important because up to 50% of institutionalized mentally retarded persons have epilepsy. The latter is important because the life expectancy of mentally retarded persons has increased steadily, and they now constitute a significant proportion of the long-term-care population.

Dr. Robert DeLorenzo, then of the Yale Medical School and now at the Medical College of Virginia, describes his research into developing more effective anticonvulsants, particularly those that treat epileptic seizures effectively with less sedation. This is an issue of tremendous importance since existing anticonvulsants tend to diminish cognition and other aspects of functioning of mentally retarded persons. A variety of disciplinary perspectives are reflected in the discussion following his paper.

Dr. Henryk Wisniewski, Director of the Institute for Basic Research, stresses that in spite of the massive Alzheimer-type changes, the major clinical sign of Alzheimer's disease, progressive dementia, is not commonly observed in most mature and elderly persons with Down's syndrome. He also described the difficulty of early and proper diagnosis of dementia in the low-functioning client.

Dr. Matthew Janicki of the New York State Office of Mental Retardation and Developmental Disabilities offers some summary observations on the presentations by Drs. Robert DeLorenzo and Henryk Wisniewski. Dr. Janicki's remarks focus primarily on the aging of the developmentally disabled population and underscore the importance of research at this end of the age continuum.

A Molecular Approach to the
Development of Anticonvulsants

R. J. DeLORENZO, A. C. BOWLING, AND W. C. TAFT

Departments of Neurology and Pharmacology
Medical College of Virginia
Richmond, Virginia 23298

Seizure disorders are a major clinical complication of mental and developmental retardation. Control of seizure disorders in these individuals becomes a major factor in regulating their quality of life and providing the opportunity for reaching full potential. Although seizures can be partially controlled with currently available medication, these anticonvulsants frequently produce significant sedative side effects that lead to a general impairment of the quality of life. Thus, in developing a treatment protocol that allows routine integration into society for these individuals, it is important to design a therapeutic regimen that separates effective seizure control and the sedative properties that are frequently associated with it.

MOLECULAR APPROACH TO ANTICONVULSANT DRUG DEVELOPMENT

From a historical point of view, anticonvulsants were developed empirically, with no understanding of their molecular effects; however, in the past two decades, the mechanism of action of anticonvulsant compounds has been the object of intense investigation. Recent advances have led to the identification of some of the biochemical and physiological effects of these compounds, as well as a better understanding of the generation and spread of seizure activity. Importantly, this new progress has made possible a molecular approach to the design of better drugs and improved therapeutic protocols. This presentation describes our efforts to identify the cellular effects of anticonvulsant compounds and develop a test system that could be used for screening new derivatives for anticonvulsant activity.

We have employed benzodiazepines (BZs) as a tool for investigating the biochemical and physiological effects of anticonvulsant compounds. BZs are valuable compounds for use as prototypes because there are hundreds of derivatives available for study, and a wealth of literature exists that describes the clinical effects of these compounds. Diazepam (DZ) (Valium) and chlordiazepoxide (Librium) are commonly used BZs that produce both sedative and anticonvulsant actions. In addition, BZs are effective inhibitors of seizure activity in two important models of epilepsy: pentylenetetrazol (PTZ)–induced seizures and maximal electric shock (MES)–induced seizures. MES-induced seizure activity is the principal clinical model for status epilepticus and generalized tonic-clonic seizures; thus, the study of BZ effects on the cellular level may provide important insights into the regulation of neuronal excitability and the treatment of seizure disorders.

BZs produce their clinical effects by binding to specific membrane binding sites. We have been using the BZs to characterize these "anticonvulsant receptors" in brain membrane preparations. Since anticonvulsant therapy is complicated by the multiple clinical effects of anticonvulsant compounds, it is important to determine the molecular basis of the various neuronal effects in order to understand and improve their therapeutic effectiveness. The identification of these anticonvulsant binding sites and correlation with the biochemical and physiological effects of these compounds provide the molecular basis for developing a test system for anticonvulsant compounds.

BENZODIAZEPINE RECEPTORS IN BRAIN

Nanomolar-Affinity Benzodiazepine Receptors

Early studies designed to elucidate the site of action of benzodiazepines led to the identification of specific high-affinity benzodiazepine receptors in brain membrane.[1,2] These nanomolar-affinity BZ receptors exhibit saturable binding in the nanomolar concentration range and have an apparent dissociation constant (K_d) for diazepam of 3 nM.[1,2] Nanomolar-affinity BZ receptors have been linked on the molecular level with the γ-aminobutyric acid (GABA) receptor and the chloride ion channel. BZ binding at the nanomolar-affinity BZ receptor correlates well with some of the pharmacological effects of the BZs, including muscle relaxation, relief of anxiety, and inhibition of PTZ-induced seizure activity.[3] The theoretical mechanism of anticonvulsant action was that BZs, binding to the nanomolar-affinity BZ receptor, potentiate the effects of GABA, resulting in increased Cl⁻ influx and subsequent membrane stabilization. This model system has been supported by more recent studies using a variety of benzodiazepine derivatives, GABAergic ligands and anion channel modulators.

However, certain clinical effects of the BZs and other anticonvulsants do not correlate with BZ binding at the nanomolar-affinity BZ receptor. Specifically, there is no correlation between nanomolar-affinity BZ binding and BZ inhibition of MES-induced seizures, performance in conditioned avoidance testing, and BZ sedative effects. In addition, several types of anticonvulsant compounds do not bind to nanomolar-affinity BZ receptors, including phenytoin, carbamazepine, and barbiturates.[4] These observations suggested that other anticonvulsant mechanisms exist that are not mediated by binding to nanomolar-affinity BZ binding sites.

Low-Affinity Benzodiazepine Binding Sites

We have recently reported the identification and characterization of stereospecific BZ receptors that have micromolar binding affinity for BZs and are pharmacologically distinct from nanomolar-affinity BZ receptors.[5–7] [³H]DZ binding to low-affinity BZ receptors saturates in the micromolar concentration range and has a K_d for diazepam of 45 μM[5] (FIG. 1). Several pharmacological characteristics distinguish nanomolar-affinity and low-affinity BZ binding sites. First, the apparent K_d values for binding differ significantly. Second, at the nanomolar-affinity site, DZ is 10,000 times more potent than Ro5-4864, whereas at the low-affinity site they are essentially equipotent. Third, concentrations of GABAergic ligands that modulate nanomolar-affinity BZ binding and chloride ion fluxes have no effect on low-affinity BZ binding.[5] Thus, BZ binding in the nanomolar and micromolar concentration ranges represents binding to multiple classes of BZ receptors.

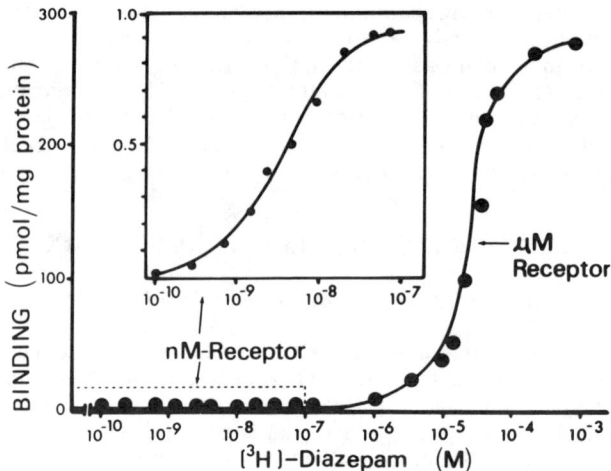

FIGURE 1. Saturation curve of [³H]DZ binding in rat brain membrane demonstrating the presence of two classes of BZ binding sites. The data are taken from [³H]DZ binding assays and show that saturation occurs in the nanomolar range and the micromolar concentration range, corresponding to the binding of the nanomolar-affinity central BZ receptor and low-affinity BZ binding sites. Inset: expanded view of binding in the nanomolar concentration range.

Evidence is accumulating to suggest that binding to low-affinity BZ binding sites may represent the molecular mechanism for BZ effects against MES-induced seizures and sedation. Like diazepam, phenytoin is an extremely effective compound against MES-induced seizures and is a primary drug for treating epilepsy, including status epilepticus and tonic-clonic seizures. In therapeutically significant concentrations, phenytoin inhibits binding of [³H]DZ to low-affinity BZ receptors, indicating that the two drugs may have the same binding sites.[5] In other investigations, we have characterized a micromolar-affinity [³H]phenytoin binding site in brain membrane, and its binding can be blocked by micromolar concentrations of DZ.[8] Thus, on the basis of membrane binding data and clinical efficacy, diazepam and phenytoin have related pharmacological activities. These results suggest that general "anticonvulsant receptors" may exist in brain membrane that bind both benzodiazepine and hydantoin compounds and that may account for some of the pharmacological effects of these drugs on seizure activity and sedation. Identification of the molecular components and the physiological effects of these "anticonvulsant binding sites" provides a molecular approach to investigating the pathophysiology of epilepsy.

Recent evidence from our laboratory suggests that a second type of low-affinity BZ receptor exists in rat brain membrane. A saturation curve of binding over a wide concentration range demonstrates saturation in the high nanomolar and low micromolar concentration ranges. A Scatchard plot of these binding data is biphasic, consistent with the existence of two major classes of binding sites in this concentration range. The Scatchard plot reveals an apparent K_d for diazepam for the high-nanomolar-affinity site of 180 nM, compared to 45 μM for the micromolar-affinity site. Although this appears to be a novel class of BZ binding sites based on DZ affinity, further characterization is necessary to confirm the pharmacological distinction. The affinity of this binding site for BZs suggests that it may be involved in the ability of BZs to limit sustained, high-frequency repetitive firing (SRF) in cultured neurons.[9]

NEURONAL EFFECTS OF LOW-AFFINITY BENZODIAZEPINE BINDING

In order to improve the therapeutic effectiveness of anticonvulsant compounds, it is necessary to identify their precise molecular actions on neuronal membrane and their effects on neuronal function. The physiological changes induced by anticonvulsant binding at low-affinity BZ binding sites may be a foundation for understanding the molecular basis of sedation and MES seizure inhibition. Thus, we have examined the biochemical and physiological effects of BZs in order to determine a possible correlation with BZ binding to lower-affinity BZ receptors.

Micromolar BZ Inhibition of Voltage-Sensitive Ca^{2+} Influx

It has been observed that micromolar concentrations of BZs inhibit depolarization-dependent Ca^{2+} uptake in synaptosomal preparations,[10–12] suggesting that this inhibitory effect on Ca^{2+} uptake may be mediated by low-affinity BZ binding sites (FIG. 2). In addition, BZs inhibit Ca^{2+}-dependent neurotransmitter release in synaptosomes, suggesting that they regulate functionally significant Ca^{2+} channels. To investigate the effects of BZs on synaptosomal C^{2+} uptake, ^{45}Ca flux was monitored under both depolarized and nondepolarized conditions.[7] BZs block the ^{45}Ca uptake component of Ca^{2+} flux and show no effect on Ca^{2+} efflux under these conditions. BZs inhibit the rapid phase of depolarization-sensitive ^{45}Ca uptake and have no significant effect on control synaptosomes. Synaptosomal depolarization can be induced either by presence of high K^+ levels (60–70 mM) or by veratridine (50 μM), and both result in Ca^{2+} accumulation, which is BZ-sensitive. Under these depolarizing conditions, synaptosomal Ca^{2+} uptake proceeds through voltage-sensitive Ca^{2+} and Na^+ channels. The Na^+ channel blocker, tetrodotoxin (TTX), effectively blocks veratridine-induced ^{45}Ca, but TTX does not inhibit high K^+-induced ^{45}Ca uptake, nor does it affect BZ inhibition of high K^+-induced ^{45}Ca uptake. Thus, BZ-sensitive ^{45}Ca uptake in depolarized synaptosomes is not mediated by the TTX-sensitive Na^+ channel and must be associated with the voltage-sensitive Ca^{2+} channel. In support of this conclusion, we observed that BZ-sensitive ^{45}Ca uptake is also sensitive to the Ca^{2+} channel blockers, Mn^{2+} and Co^{2+}. In summary, based on biochemical evidence, these investigations demonstrate that the nature of BZ inhibition of Ca^{2+} accumulation in synaptosomes is blockage of Ca^{2+} uptake induced by depolarization and mediated by voltage-sensitive Ca^{2+} channels.

Several lines of evidence suggest that BZ inhibition of voltage-sensitive Ca^{2+} channels is mediated by low-affinity BZ binding sites.[7] First, BZ inhibition of Ca^{2+} uptake is stereospecific, as shown by the potent inhibition by B10(+) and the lack of effect of B10(−). Second, the BZ potency series for inhibition of micromolar BZ binding and Ca^{2+} uptake closely parallel each other, including Ro5-4864, which binds to the low-affinity BZ receptor but not the high-affinity central BZ receptor. Third, irreversible BZ binding to intact synaptosomes (by photoaffinity

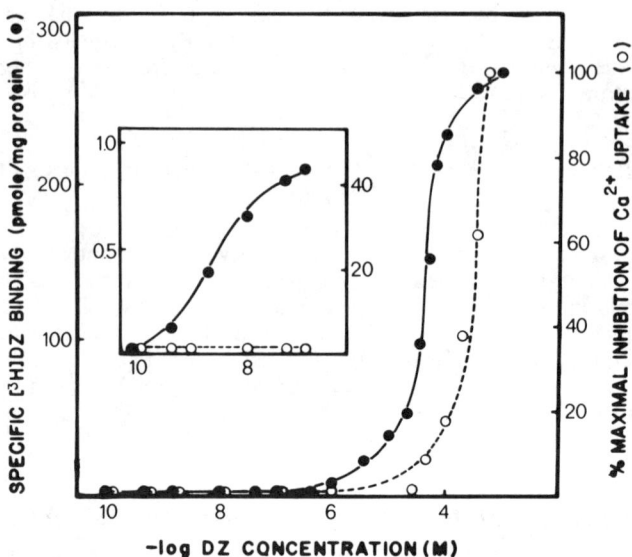

FIGURE 2. Comparison of the saturation binding curve for [³H]DZ to rat brain membrane and DZ inhibition of depolarization-dependent Ca^{2+} uptake. [³H]DZ binding (●) was performed by filtration assay with 300 μg brain membrane protein. Data represent specific binding and are the means of 10 determinations. DZ inhibition of ⁴⁵Ca uptake (○) was determined by quantitation of high K^+-stimulated synaptosomal ⁴⁵Ca uptake in the presence or absence of DZ at varying concentrations. Data represent the means of 12 determinations. Inset: expansion of [³H]DZ binding curve and DZ inhibition of Ca^{2+} uptake at low DZ concentrations.

labeling) results in irreversible inhibition of both Ca^{2+} uptake and low-affinity binding. Thus, membrane-bound BZ, in the absence of free drug, produces inhibition of Ca^{2+} uptake. These results, and the strong pharmacological similarities observed, indicate that micromolar-affinity BZ receptors may mediate BZ inhibition of voltage-sensitive Ca^{2+} channels in synaptosomes.

Electrophysiological Effects of Anticonvulsants

To determine the electrophysiological effects of BZs on Ca^{2+} conductances, we studied the actions of these compounds on identified nociceptive neurons (N cells) in the leech *Macrobdella decora*.[13] These neurons possess Mn^{2+}- and Co^{2+}-sensitive regenerative divalent cation potentials that have the same properties as Ca^{2+} channels described in other preparations from vertebrate and invertebrate phyla.[14,15] We observe that, in micromolar concentrations, BZs reversibly inhibit voltage-gated Ca^{2+} conductance in a dose-dependent manner, indicating that the BZs act as Ca^{2+} channel blockers (FIG. 3). Like Mn^{2+}, BZs inhibit the Ca^{2+} potentials at concentrations that do not significantly affect the resting membrane potential or V_{max} of the Na^+-dependent action potential. These findings suggest

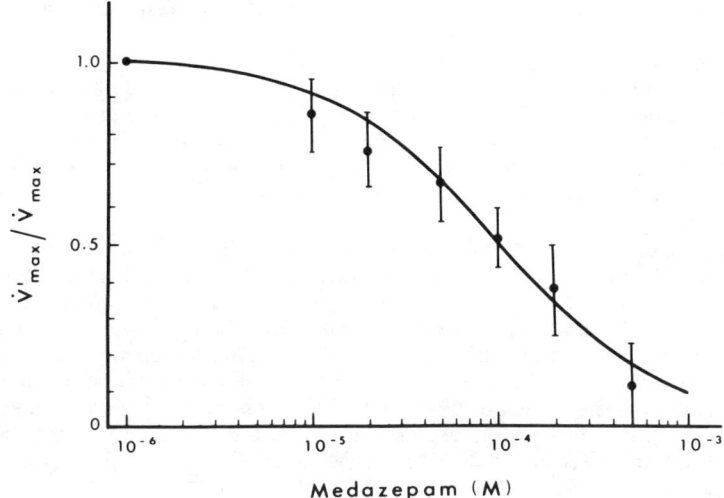

FIGURE 3. Dose-response curve for the effect of medazepam on Ca conductance in the medial N cell from the leech *Macrobdella decora*. The plot shows the ratio of the maximum rate of depolarization (V_{max}) of the action potential in the absence (V_{max}) and presence (V'_{max}) of the drug. The data points represent the mean of 10–15 determinations.

that BZs are not acting nonspecifically or like local anesthetics, but may selectively affect the Ca^{2+} channel in the leech. Micromolar-affinity BZ binding sites were demonstrated in leech neuronal preparations and had essentially the same pharmacological properties as the micromolar-affinity BZ receptor characterized in rat brain. The apparent K_d for BZ binding in leech neuronal membrane correlates well with the apparent K_i for BZ inhibition of the N-cell divalent cation conductance.[13] These findings are consistent with the hypothesis that micromolar-affinity BZ binding sites regulate voltage-sensitive Ca^{2+} channels and that these sites may play a role in regulating neuronal excitability.

The data from these electrophysiological studies do not preclude that BZs may have effects on other ion conductances. It will be important to investigate the effects of BZs in other model systems to more accurately determine their role as neuroactive agents. We are particularly intrigued by the possibility that the high-nanomolar-affinity BZ binding site (K_d 100–200 nM) we have demonstrated in brain membrane may not be involved with inhibition of conductance but may be associated with BZ limitation of Na^+ channel rectification described by Mac-Donald in cultured neurons.[9] His group reports that DZ (250–700 nM) limits sustained repetitive firing (SRF) in cultured neurons and suggests that this may be due to DZ-induced accumulation of Na^+ channels in the inactivated state. The existence of multiple low-affinity BZ binding sites in the high-nanomolar and low-micromolar concentration range suggests that distinguishable sites for BZ modulation of Na^+ channel rectification and Ca^{2+} currents may exist. The investigation of the molecular nature of low-affinity BZ binding sites may provide an important tool for the identification of the molecular components of these neuronal currents.

Benzodiazepine Inhibition of CaM-dependent Kinase Activity

Early studies by DeLorenzo and coworkers[16] demonstrated that CaM-dependent phosphorylation in nerve terminal preparations is inhibited by benzodiazepines (BZs) and other anticonvulsant compounds. A major component of nerve terminal phosphorylation was later shown to be autophosphorylation of an endogenous CaM-dependent protein kinase. Our group has recently purified CaM-dependent kinase from rat brain[17] and examined the effects of BZs on activity of the purified kinase.[18]

Benzodiazepines inhibit autophosphorylation of the two major subunits (rho and sigma) of a highly purified CaM-dependent kinase from rat brain. BZs inhibit the phosphorylation of several major substrates by CaM-dependent kinase including tubulin, MAP 2, and neurofilament protein. The effects of BZs on CaM kinase activity occur at micromolar BZ concentrations and are stereospecific. A variety of BZ derivatives are effective inhibitors of CaM kinase activity including (in order of potency) medazepam, diazepam, B9(+), clonazepam, B9(−), and Ro5-4963. BZ inhibition of CaM kinase appears to be a direct effect on the enzyme itself and not an effect on Ca^{2+}, Mg^{2+}, CaM, or ATP.

SUMMARY

Anticonvulsants are neuronal stabilizing compounds that exhibit multiple clinical effects, including anticonvulsant, anxiolytic, sedative, and muscle-relaxant properties. This complex therapeutic picture complicates the treatment of seizure disorders in individuals with mental and developmental disorders, and frequently impairs the routine integration into society for these individuals. In order to improve the therapeutic effectiveness of these compounds, it is necessary to identify their precise molecular actions on the neuronal membrane and their effects on neuronal function. We have identified two major classes of low-affinity BZ binding sites that seem to function as generalized anticonvulsant receptors and that may mediate the anticonvulsant and sedative effects produced by these compounds. The identification of these binding sites and their anticonvulsant binding profile may clarify the complex picture of anticonvulsant mechanisms and elucidate the site(s) at which anticonvulsants produce their inhibition of MES-induced seizures and sedative effects. We will continue to examine the physiological changes induced by anticonvulsant binding at these BZ binding sites that may be a foundation for understanding the molecular basis of sedation and seizure inhibition. Specifically, we will investigate the specific membrane components associated with the inhibition of Ca^{2+} channels, Na^+ channel rectification, and CaM kinase II. If these goals can be achieved, then model systems could be developed to screen potential anticonvulsant or sedative compounds in the search for more effective therapeutic drugs.

REFERENCES

1. BRAESTRUP, C. & R. F. SQUIRES. 1977. Proc. Natl. Acad. Sci. USA 74: 3805.
2. MOHLER, H. & T. OKADA. 1977. Life Sci. 20: 2101.
3. BRAESTRUP, C. & R. F. SQUIRES. 1978. Eur. J. Pharmacol. 48: 263.
4. GREENBLATT, D. J., E. WOO, M. D. ALLEN, P. J. ORSULAK & R. I. SHADER. 1978. J. Am. Med. Assoc. 240: 1872.

5. BOWLING, A. C. & R. J. DeLORENZO. 1982. *Science* **216:** 1247.
6. DeLORENZO, R. J. & W. C. TAFT. 1985. Workshop Neurotrans. Epilepsy. In press.
7. TAFT, W. C. & R. J. DeLORENZO. 1984. Proc. Natl. Acad. Sci. USA **81:** 3118.
8. BOWLING, A. C. & R. J. DeLORENZO. 1982. Neurology **32:** A224.
9. MacDONALD, R. L. & M. J. McLEAN. 1985. Workshop Neurotrans. Epilepsy. In press.
10. DeLORENZO, R. J. 1981. Cell Calcium **2:** 365.
11. LESLIE, S. W., M. B. FRIEDMAN & R. R. COLEMAN. 1980. Biochem. Pharmacol. **29:** 2439.
12. TAFT, W. C. & R. J. DeLORENZO. 1983. Trans. Am. Soc. Neurochem. **14:** 195.
13. JOHANSEN, J., W. C. TAFT, J. YANG, A. L. KLEINHAUS & R. J. DeLORENZO. 1985. Proc. Natl. Acad. Sci. USA **82:** 3935.
14. KLEINHAUS, A. L. & J. W. PRICHARD. 1977. J. Physiol. **246:** 351.
15. HAGIWARA, S. & L. BYERLY. 1981. Ann. Rev. Neurosci. **4:** 69.
16. DeLORENZO, R. J., S. BURDETTE & J. HOLDERNESS. 1981. Science **213:** 546.
17. GOLDENRING, J. R., B. GONZALES, J. S. McGUIRE & R. J. DeLORENZO. 1983. J. Biol. Chem. **258:** 12632.
18. TAFT, W. C., J. R. GOLDENRING, T. M. BUCKHOLZ & R. J. DeLORENZO. 1985. Pharmacologist **27:** 185.

DISCUSSION

MAZHAR MALIK: Do these drugs bind calcium, and if so, what regulates the calcium levels in the cell?

ROBERT DeLORENZO: That is a very interesting question because it has been noted that Dilantin and other drugs can chelate calcium. The concentrations at which these effects of the benzodiazepines are occurring are reached at a dose where calcium binding is not the major effect. You cannot overcome the ability of these compounds to block calcium uptake by increasing calcium concentrations in the media. For example, if you take a look at calcium uptake with a drug that binds up all the calcium in the solution, there is no calcium left to get into the cell, which would look like you were blocking calcium uptake. We have controlled for this effect as have others who have reproduced this work.

JOSEPH FRENCH: Are any of these effects due to calcium-activity-induced changes on inositol polyphosphates?

ROBERT DeLORENZO: Many of you know about inositol and the C-kinase system. Calcium plays a major role in regulating that system. We have no evidence at the moment that they are affected by these drugs; however, they may very well be. I have tried to indicate that these are effects of the benzodiazepines that are not explainable by binding to the nanomolar receptor. Most investigators would now agree that you cannot explain all of the clinical effects of these drugs by nanomolar binding. Investigators are getting interested in this subject, and in looking at other receptors. If you are interested in the lower-affinity binding sites, I will be glad to help you. I hope it is not unrealistic to say that in the next 5 to 10 years, by studying this system, we will develop compounds to treat epilepsy with less sedation.

QUESTION: Do these drugs bind calmodulin?

ROBERT DeLORENZO: Some drugs that have been shown to have anticonvulsant activities bind to calmodulin. These drugs in very high concentrations will bind and inactivate calmodulin to some extent, but the effects I am showing here result at concentrations below these levels. We found that these drugs (diazepam, for example) will block calmodulin kinase not through an inhibition of calmodulin, but by directly binding to the enzyme. I did not have the time to talk about that, but we do now know that the enzyme itself is actually modulated by these drugs.

JOHN STURMAN: I would like to ask about the nature of receptors in the developing nervous system. Judged by the binding characteristics of some amino-acid receptors, at least, they are not the same in the developing animal as in the mature animal. For example, in mature chick retina, those receptors that bind L-glutamate are specific for this compound, which is not displaced by D-glutamate or a range of other compounds. In the young chick, however, L-glutamate that binds to receptors is displaced by D-glutamate and a range of other compounds. Are there such differential specificities or properties during development of the receptors that you are studying, and if so, could this be used to design different strategies with regard to the use of anticonvulsants for children from those used for adults?

ROBERT DeLORENZO: That is a very important question. If you look at specific models, for example, in the nanomolar benzodiazepine receptor system, it is known that there is not just one receptor, but multiple receptors: receptors that bind GABA benzodiazepines, convulsants, and other compounds. In this system, the chloride channel is surrounded by a group of proteins that are regulated in different ways. What may happen to these receptors during development is not known.

HENRY WISNIEWSKI: We are finding that the mentally retarded person with epilepsy does not respond in the same way to anticonvulsants as the epileptic who is not mentally retarded. I see it from the clinical side of our institute's activities. Our anesthesiologists have found that many times a dose that is normal for the age and weight group produces a very different response in the mentally retarded person.

Discrepancy between Alzheimer-Type Neuropathology and Dementia in Persons with Down's Syndrome[a]

HENRYK M. WISNIEWSKI AND AUSMA RABE

Institute for Basic Research in Developmental Disabilities
New York State Office of Mental Retardation &
Developmental Disabilities
Staten Island, New York 10314

Alzheimer's disease (AD) is a progressive dementing disorder, with distinct neuropathology, that affects about 10% of people over 65 years of age. Down's syndrome (DS), a chromosomal abnormality, is a major known cause of mental retardation; the disorder occurs in one out of every thousand live births.[1] It is commonly thought that persons with DS succumb to AD not only at a much earlier age, but also in much larger numbers than people without DS. A critical examination of this view is the essence of this presentation.

The diagnosis of a definite AD is based on the presence of progressive dementia and the histopathologic demonstration in the brain of senile, or neuritic, plaques and neurofibrillary tangles. The presence of only one of these hallmarks of AD, either dementia or neuropathology, is not sufficient for a diagnosis of definite AD.[2-4] Since people, as they age, develop some plaques and tangles in the cerebral cortex and other brain regions without developing dementia,[5-8] the neuropathological criteria for AD, shown in TABLE 1, take cognizance of this age-associated development of plaques and tangles.

There is unanimous agreement in the neuropathological literature that almost every person with DS who is over the age of 35 develops plaques and tangles.[1,9,10] Even in the 30–40-year age group, the densities of the lesions meet, or even exceed, the criteria recommended for a positive diagnosis of AD in the general population, and the numbers of these two lesions increase with age (TABLE 1).[11,12] The lesion densities in DS brains are in the range of those found in the brains of people with AD.[13-15] Moreover, lesions found in both populations are identical at the light microscopic as well as the electron-microscopic level.[16-18] The other less-frequently studied neuropathological changes associated with AD (granulovacuolar degeneration, Hirano bodies, and cell loss) are also commonly found in mature and older DS brains.[14,18-23]

In spite of the massive Alzheimer-type neuropathological changes, the major clinical sign of AD, progressive dementia, is not commonly observed in most mature and elderly persons with DS.[1,9-12,21] On the contrary, according to physicians, psychologists, and people who care for DS persons on a daily basis, only relatively few suffer from dementia. Available literature suggests that between 15–30% (perhaps even a higher percent) of DS persons develop dementia.[11,12,21,24-29] Although these figures point to a higher rate of AD than in the general population, they nevertheless are far below the near-100% presence of AD neuropathology in persons with DS over 35.

[a] Our work on Alzheimer's disease is supported in part by NIH Grant 5 P01 AG04220.

247

TABLE 1. Comparison of the Neuropathological Guidelines for a Diagnosis of Alzheimer's Disease in the General Population with the Number of Cortical Plaques and Tangles in Down's Brains

Disorder	Plaques/mm^2	Tangles/mm^2
Alzheimer's Disease[a]		
<50 yrs	>2–5	>2–5
50–65 yrs	>8	>8
66–75 yrs	>10	present[b]
Down's Syndrome[c]		
31–40 yrs	7.0	2.2
41–50 yrs	11.8	4.5
51–60 yrs	16.4	13.0
61–70 yrs	17.5	19.1

   [a] From a workshop on the diagnosis of AD, see Wisniewski and Merz.[8]
   [b] No consensus was reached about the number of tangles in this age group since after the age of 70 their number appears to decline and in some cases they may not be found at all.
   [c] Calculated from the data reported by Wisniewski et al.[12]

What are the reasons for this discrepancy? In search for an answer to this question, we will examine whether the discrepancy could be explained by (1) a failure to detect dementia in persons with mental retardation, (2) the relationship of the two major lesions, plaques and tangles, to dementia, and (3) neurotransmitter pathology.

ALZHEIMER DEMENTIA IN PERSONS WITH DOWN'S SYNDROME

Although it is common knowledge that many more people with DS have AD (i.e., have the neuropathology as well as dementia) than the elderly in the general population, we know neither the prevalence nor the incidence of AD among the DS population. As our review of the literature will show, estimates vary widely. This is not surprising, since there are no established criteria or tests for the diagnosis of dementia in mentally retarded persons.

Several reports show that the behavioral competence of older DS persons declines as compared to that of younger DS persons. None of these studies, however, demonstrate the presence of dementia with certainty. One study[30] suggested that DS persons deteriorate prematurely, since after the age of 31 they had lower IQ than would be expected on the basis of probability theory. A second study[31] showed that after 30 years of age institutionalized DS persons were less interested in the external world but showed an increase in self-oriented behaviors. A third study[24] compared an older DS group (over 35) with a young DS group (20–25 years) of the same IQ on a psychological questionnaire, neurological checklist, and EEG; the older DS group showed subtle differences, but no dementia. A fourth study[25] also compared young (below 35) and older (above 35) institutionalized DS persons on various psychological and neurological measures, and again the older DS persons showed a higher incidence of impaired function. Since none of these four studies included age-matched non-DS retarded, institutionalized control groups, they failed to prove that the measured age-associated functional differences are characteristic of persons with DS rather than of the whole institutionalized mentally retarded population.

Another study,[26] using an interview and clinical assessment with institutionalized persons over 45 years of age showed that 25% (2/8) of the DS persons and only 6% (9/147) of the non-DS persons had dementia. This study did not assess younger persons.

Another group of studies, comparing not only young and old DS persons but also young and old mentally retarded people without DS, likewise concluded that the older DS persons were behaviorally less competent than the other groups. The mental age (MA) of older DS persons (40–50 years of age) appeared to decline, while other retarded persons in the same age group showed a slight rise in MA, and the younger persons (20–40 years) in either group showed no change.[32] Thase et al.[27] compared 165 DS and 163 other institutionalized retarded persons (divided into four age groups, matched for IQ) on a structured neuropsychiatric interview. They found that DS persons were significantly more impaired than non-DS persons on several measures (orientation, digit span, visual short-term memory, and others); while the older DS persons were the most impaired, the non-DS retarded groups showed age-related improvement in function. However, the nature of the differences is difficult to interpret since the zero scores of subjects who could not perform a given task have also been included in the group data.

A well-controlled study examined short-term visual memory using a matching-to-sample procedure in old, intermediate, and young DS persons and young and old matched non-DS retarded persons.[28] The old DS persons performed worse than any of the other four groups; they were deficient already at the shortest-retention intervals (0, 5, 10 sec). The old non-DS retarded subjects performed as well as the young non-DS persons.

In a longitudinal study extending over an eight-year period, Dalton and Crapper-McLachlan[29] demonstrated that only 24% (12/49) of the older DS persons showed a decline of visual short-term memory; none of the older age- and IQ-matched non-DS subjects showed such a decline. The authors interpreted this visual retention deficit in the older DS persons as an early sign of dementia of the Alzheimer type. So far they have reported full clinical and neuropathological confirmation of AD for only six of the 12 persons who suffered the visual retention loss.[33]

Although the procedures of Dalton and his associates[28,29] are experimentally sound, one may have some reservations about accepting their view that visual retention deficit is an early and certain indicator of Alzheimer dementia. Even their young DS persons had a visual retention deficit at the longer retention intervals of 15–60 sec.[28] Other data indicate that DS persons have a different memory deficit for visual compared to auditory stimuli.[34] It is conceivable that factors other than dementia may have aggravated the visual memory deficit in some of the older DS persons. We need to know whether the visual retention deficit, observed by Dalton and associates, represents a generalized memory deficit characteristic of AD, or whether it is specific to the visual modality and is characteristic of DS persons.

Miniszek[35] is among the few who have been explicitly concerned with the difficulties of an early and proper diagnosis of dementia in low-functioning DS and other mentally retarded persons. He examined the scores of old DS and non-DS institutionalized persons on Part I of the AAMD Adaptive Behavior Scale (ABS),[36] and found that the old DS persons rated lower than their age-matched peers without DS. Moreover, Miniszek suggested that the scale differentiated between regressed and nonregressed persons, and concluded that the ABS may be a sufficiently sensitive instrument for diagnosing dementia in low-functioning mentally retarded persons.

There are several reports on the neuropathology of older DS persons that also make a retrospective determination of whether or not the persons suffered from dementia. Among them, Ropper and Williams[11] examined brains of individuals over 30 years of age and found that all of the brains had tangles and plaques in the cerebral cortex and most of them in the hippocampus as well. Retrospective diagnosis of dementia was made in 15% (3/20) of the cases. Wisniewski et al.[12] examined brains of DS persons of different ages. Forty-four of 51 brains, all from individuals below 30 years of age, did not have any Alzheimer-type neuropathology. In contrast, the brains of all persons 32 years and older had tangles and plaques, and 26.5% (13/49) of these were retrospectively diagnosed as having been demented. A problem with these neuropathological studies is that their retrospective diagnosis may be inaccurate.

This review should convince anyone that we do not yet know the prevalence of Alzheimer-type dementia among DS persons. We need criteria for diagnosing Alzheimer dementia in DS, as well as for distinguishing the malignant (progressive) dementia from benign cognitive changes with age.[37] Perhaps the criteria proposed for diagnosing AD in the general population[38] could be adapted for persons with DS. Cognitive function and its decline with age and AD should be assessed in persons with DS at all levels of retardation. Existing psychological tests should be adapted for this purpose and, when necessary, new tests should be developed. Activities of daily living and patterns of behavior should be monitored for changes with adaptive behavior scales,[36] and reflexes and other neurological signs should provide useful additional indices of pathology.[1,25] Electrophysiological recording procedures should supplement the behavioral and neurological tests. Only longitudinal studies, in which properly selected subjects are adequately and repeatedly tested, will enable us to determine the incidence of dementia, as well as its onset, course, and severity in persons with DS. Although it is possible that even the highest current estimate underestimates the proportion of elderly persons with DS who become demented later in life, it is difficult to imagine that proper diagnostic studies would raise the estimate to 100% and the dissociation between Alzheimer neuropathology and dementia would be eliminated.

RELATIONSHIP OF PLAQUES AND TANGLES TO DEMENTIA

The idea of a dissociation between dementia and plaques and tangles in persons with DS is predicated on the definition of AD as a joint presence of dementia and a sufficient number of plaques and tangles. In seeking a better understanding of this discrepancy, we will summarize the main facts about the relationship between plaques and tangles and dementia in the general nonretarded population first and then proceed with a similar analysis of the DS data.

Although it has been repeatedly demonstrated that normal, nondemented persons over 50 have some plaques and tangles,[5,7,39,40] a quantitative age-associated relationship does not appear to have been established.[40] Contrary to some speculation,[6] there is no evidence that the presence of a few plaques and/or tangles is associated with a mild age-related decline in intellectual function.[7]

It has, however, been clearly demonstrated that brains of persons with AD-type dementia have many more plaques and tangles than brains from age-matched nondemented persons.[13,15,40-42] (This is illustrated in TABLE 2.) Such data are consistent with the idea, recently advocated by Roth,[43] that there is a threshold number of plaques and tangles that must be exceeded before the appearance of

TABLE 2. Plaques and Tangles (mean number/mm^2) in Brains of Nondemented and Demented People from the General Population and with Down's Syndrome[a]

Study	No.	Mean Age	Cerebral Cortex		Hippocampus	
			Plaques	Tangles	Plaques	Tangles
General Population						
Blessed et al.[13]						
Nondemented	34	79	2.6	—	—	—
Demented	26	79	15.6	—	—	—
Wilcock & Esiri[15]						
Nondemented	12	79	6.7	0.2	1.4	2.2
Demented	12	79	28.9	10.9	6.4	15.4
Down's Syndrome						
Ropper & Williams[11]						
Nondemented	11	53	—	—	13.3	21.1
Demented	1	49	—	—	16.5	4.2
Wisniewski et al.[12]						
Nondemented	26	53	10.4	7.0	11.4	12.6
Demented	13	53	23.8	21.2	27.2	20.6

[a] The values presented here were calculated by us from the published data.

dementia; this idea is also inherent in the proposed age-adjusted criteria values of plaques and tangles for AD (TABLE 1).

It has been widely held that a quantitative relationship exists between the number of plaques and tangles and the degree of dementia.[2–4,6] A closer scrutiny of the basis for such a belief clearly demonstrates two facts: (1) there are relatively few studies with both morphometric counts of plaques and tangles and quantitative measurements of degrees of dementia and/or cognitive function, and (2) the degree of quantitative relationships between plaques and/or tangles and dementia, although often highly significant statistically indicating that the relationship is not a chance result, is at best modest in magnitude. To illustrate the latter point, we will review the main results from two prospective correlational studies.[13,15]

In their well-known and widely cited study of 26 demented and 34 nondemented persons, Blessed et al.[13] correlated individual cerebral cortical plaque (but not tangle) counts with a dementia score based upon the patient's ability to deal with daily tasks. The demented persons had a significantly higher plaque count than the nondemented persons (see TABLE 2). In addition, the correlation (r) between individual plaque counts and dementia scores was 0.77, which is highly significant statistically. The high correlation was in part due to the large number of cases without dementia who had very low plaque counts; when these cases were removed from the calculations, the correlation between plaques and dementia dropped to 0.64, still a highly significant correlation. We calculated the coefficients of determination, r^2, which indicate what percent in the variability of the functional measures could be explained by plaque counts alone. The coefficient of determination for the plaques and dementia scores for all cases was $0.77^2 = 0.59$; for the demented persons only, $0.64^2 = 0.41$. Thus, the variations in the plaque density could only partially explain the variations in the degree of dementia (59% or 41%, respectively). This demonstrates that the underlying pathological process was imperfectly accessed by plaque counts.

An examination of the quantitative relationship between tangles and dementia does not appear to have been reported in a full-length paper until more than a decade later. Wilcock and Esiri[15] calculated the correlations between plaques and tangles in different areas of the cerebral cortex and hippocampus and a four-point dementia score of 14 nondemented and 35 demented cases. As expected, the demented brains had many (and significantly) more plaques and tangles than brains from age-matched, nondemented persons (see TABLE 2). The correlations between tangles and the degrees of dementia were more substantial than those for plaques; the highest correlations were between temporal cortical ($r = 0.59$), para-hippocampal ($r = 0.49$), and frontal cortical ($r = 0.49$) tangle counts and dementia scores. The respective coefficients of determination (r^2) are 0.35, 0.27, and 0.24, indicating that regional tangle scores could, at best, account for only 35% of the variability in the dementia scores. There also were several significant correlations between plaques and the dementia scores, but these correlations were considerably lower than those reported by Blessed et al.[13] Whether this discrepancy was due to differences in the procedure or the cases studied is not known.

The Wilcock and Esiri[15] study also provides a rarely published measure, correlations between plaques and tangles in several different brain regions. The best relationship between the two lesions existed in the hippocampus, the highest correlation being 0.56 ($r^2 = 0.31$). The results of this study not only confirm that factors other than plaques and tangles must be involved in dementia, but also suggest that the relative importance of plaques and tangles for dementia may differ, and that the brain regions in which the lesions occur may also be an important factor.

Although the idea that the topography of the two classic lesions may indeed be an important variable in the relationship of plaques and tangles to dementia is not a new one, the critical regions still remain to be delineated. In AD, senile plaques are distributed in large numbers throughout the neocortex, and they are also found in the limbic system, substantia innominata, hypothalamus, periaqueductal region, and the pontine tegmentum.[2-4] The distribution of neurofibrillary tangles follows essentially the same pattern, but the densities of the two lesions are not perfectly correlated. The highest concentration of tangles, but not of plaques, appears in the hippocampus, parahippocampal areas, and the amygdala, and some investigators advocate the view that the limbic system and temporal cortex are the critical areas for AD.[4-47] In contrast to such a view, other recent reports draw attention to the variable distribution of lesions in AD,[48] as well as demonstrating involvement of still other brain regions, that is, the striatum and diencephalon,[49] and the dorsal raphe nucleus.[50]

To recapitulate the main points about the relationship of plaques and tangles to dementia in the general nonretarded population: Plaques and tangles develop in the brains of normal nondemented elderly persons, but the brains of age-matched persons with Alzheimer dementia have many more of them (TABLE 2). There may be threshold values of plaques and tangles for dementia; however, these values are likely to vary for different brain regions since the brain is a heterogeneous organ and lesions in different regions contribute differentially to a given functional deficit. It may therefore never be possible to determine how many plaques and tangles are necessary before they can impair cognitive function. The degree of dementia (usually assessed near the time of death and thus at the end-stage of the disease) correlates with the number of plaques and tangles, but the correlations are imperfect, suggesting that other factors may be involved and that plaques and tangles may have only a contributory, but not a causal, relationship to dementia.

The relative importance of plaques and tangles in dementia may differ, but their respective roles in this regard have not been established.

We will now consider the relationship of plaques and tangles to dementia in persons with DS. As already stated, plaques and tangles appear at a much younger age and in higher densities in the brains of virtually all persons with DS; moreover, the number of plaques and tangles increases with age (TABLE 1). In spite of the high plaque and tangle counts, the majority of persons with DS do not have dementia. This dissociation between the two classic AD lesions and dementia suggests that, unlike in the general population, there is no relationship between these variables in persons with DS. A new analysis of recently published data by Wisniewski et al.[12] shows, however, that this is not true. As in the nonretarded general population, the brains of nondemented persons with DS had significantly lower plaque and tangle counts in the frontal cortex and the hippocampus than of DS persons with Alzheimer-type dementia (TABLE 2). (The data from the only other relevant morphometric study,[11] also shown in TABLE 2, are not wholly consistent with this finding, but the demented group consists of a single case.) A stepwise discriminant analysis was performed to determine the strength of the relationship between dementia (present/absent) and four measures of neuropathological changes (i.e., plaque and tangle counts in the frontal cortex and hippocampus). Using all four of these variables, 92% (24/26) of the nondemented and 69% (9/13) of the demented DS persons could be classified properly. Adding age as a fifth variable improved the proper classification to 96% (25/26) for the nondemented and 85% (11/13) for the demented. The plaque and tangle counts in the cortex and hippocampus together accounted for 54% of the variance in the dependent variable (dementia present or absent); adding age as a fifth variable raised this figure to 62%. In both analyses, frontal plaques discriminated by far the best accounting for 40% of the variance; the contributions of the other four variables ranged from 2–9%.

The Wisniewski et al.[12] results suggest that the relationship between plaques and tangles and dementia in persons with DS is parallel to that found in the general population (TABLE 2), except that the lesions in persons with DS appear much earlier and are much more numerous. These DS data are clearly consistent with the idea of a threshold density of plaques and tangles for dementia, but the threshold appears to be much higher in DS than in the general population. Whether there also is a significant correlation between tangles and plaques and the degree of dementia in persons with DS, as there is in the general population, is not known since no studies have been done of such a relationship.

Regarding the topography of plaques and tangles in the brains of persons with DS, there appear to be no consistent differences from that found in AD in the general population. Although Ropper and Williams[11] have suggested that plaque and tangle distributions in the hippocampus of DS brains differ from those found in the hippocampus of non-Down's persons with AD, their findings are contradicted by those of other investigators.[20,23] Moreover, the counts reported by Ropper and Williams were extremely variable, suggesting that their reported intrahippocampal differences could have been due to chance. It is not known whether there are any differences in lesion topography between nondemented and demented DS cases. In our opinion, there is no convincing evidence at this time to invoke topographical differences as a possible explanation for the differences in the lesion/dementia relationship in mature DS persons and the general population.

The newly discovered difference in the plaque and tangle counts between demented and nondemented persons with DS and the concept that the relation-ship between plaques and tangles and dementia in DS may be parallel to that found in the general population have an important implication. In future studies of Alzheimer neuropathology, we must compare data of age-matched demented and nondemented DS brains just as we do in the case of nonretarded elderly with and without dementia.

PATHOLOGY OF NEUROTRANSMITTER SYSTEMS

A severe acetylcholine deficiency in the brains of AD patients has been clearly demonstrated by many different groups of investigators;[2-4,42,45,51-53] however, the basic relationship between cholinergic deficiency and AD is not yet understood. There is a loss of neurons in the substantia innominata (the source of cholinergic projection fibers to the cerebral-cortex) and a reduction in cholinergic markers in the cerebral cortex.[54-59] Several investigators have reported significant correla-tions between cortical choline acetyltransferase (CAT) levels and plaque and/or tangle counts, and CAT levels and the degrees of dementia. Perry et al.,[41] using 13 nondemented controls and 18 demented cases with AD, found that cortical CAT activity and cortical plaque counts were highly correlated ($r = -0.82$), while the correlation between CAT and dementia scores (based on mental test perfor-mance) was -0.81. Wilcock et al.[42] examined 12 nondemented and 31 demented brains and also found that cortical, particularly temporal, CAT levels correlated well ($r = -0.67$) with the degree of dementia, and temporal cortical tangle counts correlated ($r = 0.58$) with CAT activity. However, in contrast to Perry et al.,[41] they found no significant correlations between CAT levels and cortical plaque counts. Mountjoy et al.[60] compared 25 nondemented with 25 demented brains and found significant correlations between cortical (frontal and temporal) CAT activ-ity and plaque as well as tangle counts.

Involvement of other neurotransmitter systems can also be expected, even if only as a consequence of the cholinergic deficiencies.[2-4,53] Loss of noradrenergic neurons in AD is suggested by loss of cells in the locus ceruleus,[61,62] and reduced noradrenaline levels in the brain.[63-65] It is not clear whether dopamine levels are reduced in AD: one group has found normal levels,[65,66] while another has reported a decrease.[67] GABA levels may be slightly reduced in AD.[58,60]

Information on neurotransmitter abnormalities in brains of persons with Down's syndrome is still sparse, and it has not yet contributed toward under-standing the relationship between the two classic lesions and dementia in persons with DS. Brains from elderly persons with DS appear to have the same kind of cholinergic deficiencies as observed in AD.[65,68] Similarly, the number of neurons in the nucleus basalis of DS brains is apparently reduced, but the reduction may be slight.[69-70] Noradrenaline, dopamine, and related enzyme levels may also be reduced in elderly DS brains[64,65,71,72] Yates et al.[65] stress that while noradrenaline is reduced in both DS and AD, dopamine may be reduced only in DS and not in AD.

One should keep in mind that all of the above neurochemical findings with DS brains have been based on a very small number of cases (a total of about a dozen elderly brains). In addition, control data on neurotransmitter abnormalities in younger DS brains without AD-type neuropathology are inadequate (a total of about half-a-dozen brains have been studied). Future neurochemical studies should include brains of young DS persons, and should also examine differences in quantified regional neurotransmitter activity between age-matched demented and nondemented brains of older persons with DS.

CONCLUDING REMARKS

Very little is known about the process of dementia in persons with Down's syndrome. Consequently the true prevalence and incidence of Alzheimer's disease in this population are still unknown. Urgently needed are not only criteria for assessment of Alzheimer-type dementia in retarded persons with DS, but also appropriate tests for detecting early changes in cognitive functions and other aspects of adaptive behavior. Prospective long-range studies on the development, course, and severity of dementia can address these needs. The well-known and puzzling discrepancy between the presence of massive Alzheimer neuropathology and absence of dementia in a large portion of mature and elderly persons with DS may be viewed from a different perspective. A new analysis demonstrated that demented persons with DS have significantly higher counts of plaques and tangles than nondemented persons with DS, a relationship that appears to be parallel to that found in elderly demented and nondemented persons from the general population. Although the lesion counts in persons with DS are higher and occur much earlier than in the general population, the fact that the demented DS persons have higher lesion counts than the nondemented persons suggests that the neuropathology and dementia in DS are not dissociated. Persons with DS appear to have a higher threshold for dementia as measured by plaque and tangle counts. Speculations about why this might be so is a story for the future. In the meantime, we would like to urge that new studies on the neuropathology and neurochemistry of Alzheimer's disease in persons with Down's syndrome include comparisons among at least three different groups of persons with DS: (1) demented mature and elderly persons with plaques and tangles, (2) nondemented age-matched persons with plaques and tangles, and (3) young adult persons without AD-type pathology to provide baseline controls for such characteristics of DS as reduced neuron counts,[73-75] and possibly lowered levels of some neurotransmitters.[65]

Properly focused studies of age-associated functional changes and AD in the population with DS will not only enhance our understanding of AD in general, but will help care providers plan for and manage some of the problems of aging in this mentally retarded population. These would be significant accomplishments. The fact that the numbers of aged persons with DS and other developmental disabilities are steadily increasing due to advances in medicine and habilitative technology adds a note of urgency to our recommendations.

ACKNOWLEDGMENTS

Dr. Gene Fisch suggested and carried out the discriminant analyses. Dr. Wayne Silverman made valuable editorial comments. Dr. Krystyna Wisniewski graciously granted our request to reanalyze the results of her study, and Dr. G. Y. Wen was most helpful in providing the raw data. We thank them all.

REFERENCES

1. WISNIEWSKI, K. E. & H. M. WISNIEWSKI. 1983. Age-associated changes and dementia in Down's syndrome. *In* Alzheimer's Disease: The Standard Reference. R. Reisberg, Ed. The Free Press. New York. pp. 319–326.
2. TERRY, R. D. & P. DAVIES. 1980. Dementia of the Alzheimer type. Annu. Rev. Neurosci. **3:** 77–95.
3. TERRY, R. D. & R. KATZMAN. 1983. Senile dementia of the Alzheimer type: Defining a disease. *In* The Neurology of Aging. R. Katzman & R. D. Terry, Eds. Davis. Philadelphia. pp. 51–84.
4. REISBERG, B. 1983. An overview of current concepts of Alzheimer's disease, senile dementia, and age-associated cognitive decline. *In* Alzheimer's Disease: The Standard Reference. B. Reisberg, Ed. The Free Press. New York. pp. 3–20.
5. TOMLINSON, B. E., G. BLESSED & M. ROTH. 1968. Observations on the brains of nondemented old people. J. Neurol. Sci. **7:** 331–356.
6. ULRICH, J. 1985. Alzheimer changes in nondemented patients younger than sixty-five: Possible early stages of Alzheimer disease and senile dementia of the Alzheimer type. Ann. Neurol. **17:** 273–277.
7. CREASEY, H. & S. I. RAPOPORT. 1985. The aging human brain. Ann. Neurol. **17:** 2–10.
8. WISNIEWSKI, H. M. & G. S. MERZ. 1985. Neuropathology of the aging brain and dementia of the Alzheimer type. *In* Aging 2000: Our Health Care Destiny. Volume I: Biomedical Issues. C. M. Gaitz & T. Samorajski, Eds. Springer. New York. pp. 231–243.
9. WHALLEY, L. J. 1982. The dementia of Down's syndrome of Alzheimer's disease and its relevance to aetiological studies of Alzheimer's disease. Ann. N.Y. Acad. Sci. **396:** 39–53.
10. EPSTEIN, C. J. 1983. Down's syndrome and Alzheimer's disease: Implications and approaches. *In* R. Katzman, Ed. Biological Aspects of Alzheimer's Disease. Banbury Report No. 15. Cold Spring Harbor Laboratory. pp. 169–182.
11. ROPPER, A. H. & R. S. WILLIAMS. 1980. Relationship between plaques, tangles, and dementia in Down's syndrome. Neurology **30:** 639–644.
12. WISNIEWSKI, K. E., H. M. WISNIEWSKI & G. Y. WEN. 1985. Occurrence of neuropathological changes and dementia of Alzheimer's disease in Down's syndrome. Ann. Neurol. **17:** 278–282.
13. BLESSED, G., B. E. TOMLINSON & M. ROTH. 1968. The association between quantitative measures of dementia and of senile changes in the cerebral grey matter of elderly subjects. Br. J. Psychiatry **114:** 797–817.
14. BALL, M. J. 1977. Neuronal loss, neurofibrillary tangles, and granulovacuolar degeneration in the hippocampus with aging and dementia. Acta Neuropathol. (Berlin) **37:** 111–118.
15. WILCOCK, G. K. & M. M. ESIRI. 1982. Plaques, tangles, and dementia: A quantitative study. J. Neurol. Sci. **56:** 343–356.
16. SCHOCHET, S. S., P. W. LAMPERT & W. F. McCORMICK. 1973. Neurofibrillary tangles in patients with Down's syndrome: A light and electron microscopic study. Acta Neuropathol. (Berlin) **23:** 342–346.
17. ELLIS, W. G., J. R. McCULLOCH & C. L. CORLEY. 1974. Presenile dementia in Down's syndrome: Ultrastructural identity with Alzheimer's disease. Neurology **24:** 101–106.
18. BURGER, P. C. & F. S. VOGEL. 1973. The development of the pathologic changes of Alzheimer's disease and senile dementia in patients with Down's syndrome. Am. J. Pathol. **73:** 457–468.
19. SOLITAIRE, G. B. & J. B. LAMARCHE. 1966. Alzheimer's disease and senile dementia as seen in mongoloids: Neuropathological observations. Am. J. Ment. Def. **70:** 840–848.
20. OLSON, M. I. & C. M. SHAW. 1969. Presenile dementia and Alzheimer's disease in mongolism. Brain **92:** 147–156.
21. MALAMUD, N. 1972. Neuropathology of organic brain syndromes associated with aging. *In* Aging and the Brain. C. M. Gaitz, Ed. Plenum Press. New York. pp. 63–87.

22. BALL, M. J. & K. NUTTALL. 1980. Neurofibrillary tangles, granulovacuolar degeneration, and neuron loss in Down syndrome: Quantitative comparison with Alzheimer dementia. Ann. Neurol. **7:** 462–265.
23. BALL, M. J. & K. NUTTALL. 1981. Topography of neurofibrillary tangles and granulovacuoles in hippocampi of patients with Down's syndrome: Quantitative comparison with normal aging and Alzheimer's disease. Neuropath. Appl. Neurobiol. **7:** 23–20.
24. OWENS, D., J. D. DAWSON & S. LOSIN. 1971. Alzheimer's disease in Down's syndrome. Am. J. Ment. Def. **75:** 606–612.
25. WISNIEWSKI, K., J. HOWE, D. G. WILLIAMS & H. M. WISNIEWSKI. 1978. Precocious aging and dementia in patients with Down's syndrome. Biol. Psychiatry **13:** 619–627.
26. REID, A. H. & P. G. AUNGLE. 1974. Dementia in aging mental defectives: A clinical psychiatric study. J. Ment. Def. Res. **18:** 15–23.24.
27. THASE, M. E., R. TIGNER, D. SMELTZER & L. LISS. 1984. Age-related neuropsychological deficits in Down syndrome. Biol. Psychiatry **19:** 571–585.
28. DALTON, A. J., D. R. CRAPPER & G. R. SCHLOTTERER. 1974. Alzheimer's disease in Down's syndrome: Visual retention deficits. Cortex **10:** 366–377.
29. DALTON, A. J., & D. R. CRAPPER-MCLACHLAN. 1984. Incidence of memory deterioration in aging persons with Down's syndrome. In Perspectives and Progress in Mental Retardation, Vol. II: Biomedical Aspects. J. M. Berg, Ed. University Park Press. Baltimore. pp. 55–62.
30. NAKAMURA, H. 1961. Nature of institutionalized adult mongoloid intelligence. Am. J. Ment. Def. **66:** 456–458.
31. FRANCIS, S. H. 1970. Behavior of low-grade institutionalized mongoloids: Changes with age. Am. J. Ment. Def. **75:** 92–101.
32. DEMAINE, G. C. & A. B. SILVERSTEIN. 1978. MA changes in institutionalized Down syndrome persons: A semi-longitudinal approach. Am. J. Ment. Def. **82:** 429–432.
33. WISNIEWSKI, K. E., A. J. DALTON, D. R. MCLACHLAN, G. Y. WEN & H. M. WISNIEWSKI. 1985. Alzheimer disease in Down syndrome: Prospective clinico-pathological studies. Neurology **35:** 957–961.
34. MCDADE, H. L. & S. ADLER. 1980. Down syndrome and short-term memory impairment: A storage or retrieval deficit? Am. J. Ment. Def. **84:**561–567.
35. MINISZEK, N. A. 1983. Development of Alzheimer disease in Down syndrome individuals. Am. J. Ment. Def. **87:** 377–385.
36. FOGELMAN, C. J., ED. 1975. Revision: Nihira, K., R. Foster, M. Shellhaas & H. Leland. AAMR Adaptive Behavior Scale. American Association on Mental Deficiency. Washington, D.C.
37. MILLER, E. 1981. The nature of cognitive deficit in senile dementia. In Clinical Aspects of Alzheimer's Disease and Senile Dementia. N. E. Miller & G. D. Cohen, Eds. Raven Press. New York. pp. 103–120.
38. MCKANN, G., D. DRACHMAN, M. FOLSTEIN, R. KATZMAN, D. PRICE & E. M. STADLAN. 1984. Clinical diagnosis of Alzheimer's disease: Report of the NINCDS-ADRDA work group under the auspices of the department of health and human services task force on Alzheimer's disease. Neurology **34:** 939–944.
39. WISNIEWSKI, H. M. & G. S. MERZ. 1984. Aging, Alzheimer's disease, and developmental disabilities. In Aging and Developmental Disabilities. M. P. Janicki & H. M. Wisniewski, Eds. Brookes. Baltimore. pp. 177–184.
40. TOMLINSON, B. E. & G. HENDERSON. 1976. Some quantitative cerebral findings in normal and demented people. In Neurobiology of Aging. R. D. Terry & S. Gershon, Eds. Raven Press. New York. pp. 183–204.
41. PERRY, E. K., B. E. TOMLINSON, G. BLESSED, G. BERGMAN, P. H. GIBSON & R. H. PERRY. 1978. Correlations of cholinergic abnormalities with senile plaques and mental test scores in senile dementia. Br. Med. J. **2:** 1457–1459.
42. WILCOCK, G. K., M. M. ESIRI, D. M. BOWEN & C. C. T. SMITH. 1982. Alzheimer's disease: Correlation of cortical choline acetyltransferase activity with the severity of dementia and histological abnormalities. J. Neurol. Sci. **57:** 407–417.
43. ROTH, M. 1985. Correlation of neuropathological and clinical findings in dementia.

XIII International Congress of Gerontology. July 12–17. New York. Book of Abstracts: 19.
44. BALL, M. J., M. FISMAN, V. HACHINSKI, W. BLUME, A. FOX, V. A. KRAL, A. J. KIRSHEN, H. FOX & H. MERSKEY. 1985. A new definition of Alzheimer disease: A hippocampal dementia. Lancet i: 14–16.
45. WILCOCK, G. K. 1983. The temporal lobe in dementia of Alzheimer's type. Gerontology 29: 320–324.
46. KEMPER, T. L. 1983. Organization of the neuropathology of the amygdala in Alzheimer's disease. In R. Katzman, Ed. Biological Aspects of Alzheimer's Disease. Banbury Report No. 15. Cold Spring Harbor Laboratory. Cold Spring Harbor, NY. pp. 31–36.
47. BURGER, P. C. 1983. The limbic system in Alzheimer's disease. In Biological Aspects of Alzheimer's Disease. R. Katzman, Ed. Banbury Report No. 15. Cold Spring Harbor Laboratory. Cold Spring Harbor, NY. pp. 37–44.
48. ULRICH, J. & H. B. STAEHELIN. 1984. The variable topography of Alzheimer-type changes in senile dementia and normal old age. Gerontology 30: 210–214.
49. RUDELLI, R. D., M. W. AMBLER & H. M. WISNIEWSKI. 1984. Morphology and distribution of Alzheimer neuritic (senile) and amyloid plaques in striatum and diencephalon. Acta Neuropathol. (Berlin) 64: 273–281.
50. YAMAMOTO, T. & A. HIRANO. 1985. Nucleus raphe dorsalis in Alzheimer's disease: Neurofibrillary tangles and loss of large neurons. Ann. Neurol. 17: 573–577.
51. COYLE, J. T., D. L. PRICE & M. R. DELONG. 1983. Alzheimer's disease: A disorder of cholinergic innervation. Science 219: 1184–1190.
52. PRICE, D. L., P. J. WHITEHOUSE, R. G. STRUBLE, A. W. CLARK, J. T. COYLE, M. R. DELONG & J. C. HEDREEN. 1982. Basal forebrain cholinergic systems in Alzheimer's disease and related dementias. Neurosci. Comments 1: 84–92.
53. REISBERG, B., ED. 1983. Alzheimer's Disease: The Standard Reference. Section IV. The Neurochemistry of Alzheimer's Disease. The Free Press. New York. pp. 81–138.
54. WILCOCK, G. K., M. M. ESIRI, D. M. BOWEN & C. C. T. SMITH. 1983. The nucleus basalis in Alzheimer's disease: Cell counts and cortical biochemistry. Neuropathol. Appl. Neurobiol. 9: 175–179.
55. ROGERS, J. D., D. BROGAN & S. S. MIRRA. 1985. The nucleus basalis of Meynert in neurological disease: A quantitative morphological study. Ann. Neurol. 17: 163–170.
56. ARENDT, T., V. BIGL, A. TENNSTEDT & A. ARENDT. 1985. Neuronal loss in different parts of the nucleus basalis is related to neuritic plaque formation in cortical target areas in Alzheimer's disease. Neuroscience 14: 1–14.
57. BOWEN, D. M., J. S. BENTON, J. A. SPILLANE, C. C. T. SMITH & S. J. ALLEN. 1982. Choline acetyltransferase activity and histopathology of frontal neocortex from biopsies of demented patients. J. Neurol. Sci. 57: 191–202.
58. ROSSOR, M. N., N. J. GARRETT, A. L. JOHNSON, C. Q. MOUNTJOY, M. ROTH & L. L. IVERSEN. 1982. A postmortem study of the cholinergic and GABA systems in senile dementia. Brain 105: 313–330.
59. NAGAI, T., P. L. MCGREER, J. H. PENG, E. G. MCGREER & C. E. DOLMAN. 1983. Choline acetyltransferase immunohistochemistry in brains of Alzheimer's disease patients and controls. Neurosci. Lett. 36: 195–199.
60. MOUNTJOY, C. Q., M. N. ROSSOR, L. L. IVERSEN & M. ROTH. 1984. Correlation of cortical cholinergic and GABA deficits with quantitative neuropathological findings in senile dementia. Brain 107: 507–518.
61. BONDAREFF, W., C. Q. MOUNTJOY & M. ROTH. 1981. Selective loss of neurons of origin of adrenergic projection to cerebral cortex (nucleus locus coeruleus) in senile dementia. Lancet i: 783–784.
62. TOMLINSON, B. E., D. IRVING & G. BLESSED. 1981. Cell loss in locus coeruleus in senile dementia of Alzheimer type. J. Neurol. Sci. 49: 419–428.
63. MANN, D. M. A., J. LINCOLN, P. O. YATES, J. E. STAMP & S. TOPER. 1980. Changes in the monoamine-containing neurones of the human CNS in senile dementia. Br. J. Psychiatry 136: 533–541.

64. YATES, C. M., I. M. RITCHIE, J. SIMPSON, A. F. J. MALONEY & A. GORDON. 1981. Noradrenaline in Alzheimer-type dementia and Down syndrome. Lancet ii: 39–40.
65. YATES, C. M., J. SIMPSON, A. GORDON, A. F. J. MALONEY, Y. ALLISON, I. M. RITCHIE & A. URQUART. 1983. Catecholamines and cholinergic enzymes in presenile and senile Alzheimer-type dementia and Down's syndrome. Brain Res. 280: 119–126.
66. YATES, C. M., Y. ALLISON, J. SIMPSON, A. F. J. MALONEY & A. GORDON. 1979. Dopamine in Alzheimer's disease and senile dementia. Lancet ii: 851–852.
67. WINBLAD, B., R. ADOLFSSON, A. CARLSSON & C. G. GOTTFRIES. 1982. Biogenic amines in brains of patients with Alzheimer's disease. In Alzheimer's Disease: A Report of Progress in Research. S. Corkin, K. L. Davis, J. H. Growdon, E. Usdin & R. J. Wurtman, Eds. Aging, Vol. 19: 25–33. Raven Press. New York.
68. YATES, C. M., J. SIMPSON, A. F. J. MALONEY, A. GORDON & A. H. REID. 1980. Alzheimer-like cholinergic deficiency in Down syndrome. Lancet ii: 979.
69. KIRKPATRICK, J. B. & P. HICKS. 1984. Nucleus basalis in Down's syndrome. J. Neuropath. Exp. Neurol. 43: 307.
70. PRICE, D. L., L. C. CORK, R. G. STRUBLE, P. J. WHITEHOUSE, C. A. KITT & L. C. WALKER. 1985. The functional organization of the basal forebrain cholinergic system in primates and the role of this system in Alzheimer's disease. Ann. N.Y. Acad. Sci. 444: 287–295.
71. MANN, D. M. A., J. LINCOLN, P. O. YATES & C. M. BRENNAN. 1980. Monamine metabolism in Down syndrome. Lancet ii: 1366–1367.
72. MANN, D. M. A., P. O. YATES & J. HAWKES. 1982. Plaques and tangles and transmitter deficiencies in dementia. J. Neurol. Neurosurg. Psychiatry 45: 563–564.
73. SYLVESTER, P. E. 1983. The hippocampus in Down's syndrome. J. Ment. Def. Res. 27: 227–236.
74. Ross, M. H., A. M. Galaburda & T. L. Kemper. Down's syndrome: Is there a decreased population of neurons? Neurology 34: 909–916.
75. Wisniewski, K., M. Laure-Kamionowska & H. M. Wisniewski. 1984. Evidence of arrest of neurogenesis and synaptogenesis in brains of patients with Down's syndrome. N. Engl. J. Med. 311: 1187–1188.

DISCUSSION

JOSEPH FRENCH: What is the status of the prion protein in the pathogenesis of this lesion? Is it possible that a gene product can give rise to life?

HENRY WISNIEWSKI: The term *prion* was developed to denote the infectious component of the unconventional slow infections. This term can only contribute to the current discourse on the nature of these agents if, as the term implies, they are composed only of protein. This is because there are already words—virus and virino—that cover situations in which there is a protein plus nucleic acid. The concept that these agents could be "protein only" seems even less likely today than when the term prion was first coined. Recent evidence that the gene, messenger RNA, and an unmodified version of the protein are found in normal cells strongly suggests that there must be another macromolecule (most probably nucleic acid) that participates in modifying the protein and establishes the heritable characteristics of the various slow-infection agents. Clearly, there is a protein associated with these infectious agents; the issue is to find the other macromolecule(s) associated with it.

Antibody raised to this protein does not cross-react with amyloid plaque material or PHF of Alzheimer's disease; thus, the protein(s) associated with the unconventional slow-infection agents do not appear to be related to the pathogenesis of these lesions in Alzheimer's disease.

Dr. Richard Carp has reviewed the status of the facts and hypotheses related to prions in a recent article in the *Journal of General Virology;* those of you who are interested in the matter are directed to that article (1985, **66:** 1357–1368).

MAZHAR MALIK: Could an animal model be used to learn about the cause of Alzheimer's disease?

HENRY WISNIEWSKI: Yes, if one wants to look into infectious causes of Alzheimer's disease, one should use a strain of mice that are susceptible to amyloid formation. We are using two such strains.

EDMUND JENKINS: What are your plans to investigate the genetics of Alzheimer's disease?

HENRY WISNIEWSKI: We will be concentrating on the 21st chromosome because it is associated with accelerated occurrence of symptoms of Alzheimer's disease. Down's syndrome also shows the phenotypic expression of accelerated aging processes. Dr. Ted Brown's chapter in the new book *Aging and Developmental Disabilities* edited by M. P. Janicki and H. M. Wisniewski (Paul H. Brooks Publishing Co., Baltimore, 1985) describes the phenotypic expression of all of the syndromes of accelerated aging and suggests some possible lines of inquiry.

Some Comments on Aging and a Need for Research

MATTHEW P. JANICKI

New York State Office of Mental Retardation &
Developmental Disabilities
Albany, New York 12229

Most of the research on persons who are mentally retarded has focused on the early childhood segment of the lifespan; I am pleased that Dr. Henry Wisniewski was able to offer a contribution that put the latter segment of lifespan development into perspective. Little attention has been paid to date to the concerns related to aging in the field of mental retardation. Indeed, because of how the field evolved—primarily from a concern over conditions at birth and childhood disorders—it has not effectively confronted the fact that mentally retarded people do grow old.

The "graying" phenomenon among this nation's population, in general, is a growing concern. The proportion of elderly persons in this country represents an ever-growing population. That segment of this nation's population age 60 and older has doubled in number since 1900 and is expected to triple by 2030.[1] In 50 years we are going to have three times as many persons in this age group as we did in 1900. This is a critical concern for a number of reasons, most obviously because of the social and health-care implications. From our perspective this growth raises another concern: there is a significant number of persons among this population who are mentally retarded.

This marked growth in the population of elderly persons is one of the issues that we must address. As researchers we must acknowledge this fact and begin to develop a research agenda that will address a range of issues related to aging among persons with mental retardation. There virtually has been no significant research effort with populations that are mentally retarded, or for that matter, otherwise chronically impaired.[2] There are numerous studies that examine life-span development, but these have generally been concerned with early development.

Aging, as a phenomenon among retarded persons, is of interest for a number of reasons. One reason for interest is related to what we are beginning to learn about premature aging.[3] We know that, on the average, individuals who have Down's syndrome had a life expectancy of about 11 years in the 1930s; now that life expectancy has risen to about 50 years—still significantly below the mid-70s, the life expectancy of the average American. This greater longevity among persons with Down's syndrome has exposed another facet of this syndrome: premature aging. Another reason for interest in the aging of the mentally retarded is, as the service aspect of the field of mental retardation has matured, mentally retarded persons have begun to experience better overall health and greater longevity.[4] However, these factors also pose an interesting dilemma that is related to still another reason for our increased interest in this area. As mentally retarded persons age, they begin experience conditions not previously observed in this population. Indeed, the observation of premature aging among persons with Down's syndrome as well as the association of Down's syndrome and Alzheimer's disease are good examples of this.[5]

We are just beginning to learn about some of the demographic aspects of this population. It has been estimated that elderly mentally retarded persons represent a population of at minimum 200,000 persons in the United States.[6] That is not a large number when compared to the number of elderly persons in the general population; however, these individuals represent a significant demand for community care services as well as on the nation's long-term care resources.

We have also observed that, in relation to the overall number of known older mentally retarded persons, there is a disproportionate number of older mentally retarded persons in the nation's long-term care institutions. Some studies[7,8] have shown this number to be in excess of 60% (in contrast, only about 5% of the general older population is institutionalized). This may be a factor of the strategies employed in deinstitutionalization (i.e., younger persons have been discharged first); however, it is still a concern. Given the continued deinstitutionalization efforts currently underway in the United States, it is quite conceivable that many of this nation's public institutions for mentally retarded persons will eventually evolve into geriatric care facilities; however, being aged does not by necessity mean that a mentally retarded person should be in an institution. In this vein, data that I have seen seem to indicate that there is not a high rate of new admissions among older persons to such facilities.

In terms of comparative cohort studies, we have observed certain patterns of behavioral skill maintenance and decline among older mentally retarded persons. For example, although premature aging has been observed among certain syndromes associated with mental retardation, there may not be substantial evidence to believe that such a phenomenon is prevalent among all mentally retarded persons. Large-scale population studies[9] among mentally retarded persons in New York, for example, have shown that most behavioral skill levels are maintained among aging mentally retarded persons well into their 70s—equivalent to that of nondisabled age peers. When early decline is observed, it is usually in areas related to gross motor and overall independent living skills. Certain differences were observed in the patterns that appear to be a function of the level of intellectual impairment. Persons with profound intellectual deficits, those with organically derived retardation, showed less decline in all areas over time. Persons with mild or moderate intellectual deficits, those with environmentally linked retardation, showed patterns more like that of the non–mentally disabled population (that is, more marked decline over time).

We have also begun to observe changes in service demands that are associated with legislated changes in our social environment. Public Law 94-142 (Education of All Handicapped Children Act) mandates a free and appropriate education for all handicapped children in this country. What will be the long-term effects of this legislation? Will the enhanced environment of learning and the enrichment of health care, as well as other beneficial conditions, produce a healthier and longer-lived mentally retarded population—one that will be an ever-greater portion of the elderly population of 2030?

The results of these preliminary investigations have begun to suggest a research agenda in the area of aging and mental retardation. Such an agenda can be dichotomized into those efforts addressing basic research questions and those efforts addressing social or applied research areas. In the basic research arena, we need to better understand the mechanisms that set off biological aging, particularly the premature aging phenomenon observed among persons with Down's syndrome. We also need to know whether there are differential aging patterns among other genetically linked syndromes. Another area of inquiry is the relationship between Down's syndrome and Alzheimer's disease.

We need to look at the long-term effects of early intervention. Earlier we heard some excellent presentations on the secondary effects during pregnancy among women treated for phenylketonuria (PKU). This is one latent, albeit adverse, effect of early intervention: will there be others as these individuals live into their 40s, 50s, and into old age? Are there other similar, yet still unknown, phenomena among older adults treated for other conditions while they were infants or children?

In the area of social or applied research, there is a multitude of aspects that require study. For example, gerontologists are looking at self-perceptions of age and aging among older individuals in the general population. We do not yet know how the cognitive limitations of most retarded persons affect their self-perceptions of lifespan development and, in particular, aging. Other psychological phenomena warrant investigation; for example, how do cognitive abilities wane? How are death and dying perceived?

Studies, such as by Bell and Zubek,[10] have noted differences in cognition, memory, and fine motor skills among study populations of institutionalized late-middle-aged retarded persons. Were these results a function of the individuals' residential environment, their early life history, or their age? Studies are needed that examine the long-range effects of institutionalization. Are there differential effects on the developmental process, over a lifespan, among persons who have been institutionalized as opposed to those who have lived in the community all their lives?

More significantly, longitudinal studies are needed that examine the aging process and its accompanying decline patterns among particular groups of individuals. Further, there is a need to examine the relationship of mortality and morbidity patterns to certain syndromes linked to mental retardation.

Aging among the general population has many unanswered questions. Aging among mentally retarded persons may even have more. Nevertheless, there are areas that require attention. What I have tried to do is offer some comments on this area as well as offer some suggestions in terms of a research agenda.

REFERENCES

1. SIEGEL, J. S. & C. M. TAEUBER. 1982. The 1980 census and the elderly: New data available to planners and practitioners. Gerontologist 22: 144–150.
2. SELTZER, M. M. 1985. Research in social aspects of aging and developmental disabilities. *In* Aging and Developmental Disabilities: Issues and Approaches. M. P. Janicki & H. M. Wisniewski, Eds. Paul H. Brookes. Baltimore. pp. 161–173.
3. WISNIEWSKI, K., J. HOWE, D. G. WILLIAMS & H. M. WISNIEWSKI. 1978. Precocious aging and dementia in patients with Down's syndrome. Biol. Psychiatry 13: 619–627.
4. LUBIN, R. A. & M. KIELY. 1985. Epidemiology of aging in developmental disabilities. *In* Aging and Developmental Disabilities: Issues and Approaches. M. P. Janicki & H. M. Wisniewski, Eds. Paul H. Brookes. Baltimore. pp. 95–114.
5. MINISZEK, N. A. 1983. Development of Alzheimer disease in Down syndrome individuals. Am. J. Ment. Def. 87: 377–385.
6. JACOBSON, J. W., M. S. SUTTON & M. P. JANICKI. 1985. Demography and characteristics of aging and aged mentally retarded persons. *In* Aging and Developmental Disabilities: Issues and Approaches. M. P. Janicki & H. M. Wisniewski, Eds. Paul H. Brookes. Baltimore. pp. 115–142.
7. JANICKI, M. P. & A. E. MACEACHRON. 1984. Residential, health, and social service needs of elderly developmentally disabled persons. Gerontologist 24: 128–137.

8. SUTTON, M. S. 1983. Treatment issues and the elderly institutionalized developmentally disabled individual. Paper presented at the annual meeting of the American Psychological Association. Anaheim, CA.
9. JANICKI, M. P. & J. W. JACOBSON. 1986. Generational trends in sensory, physical, and behavioral abilities among older mentally retarded persons. Am. J. Ment. Def. **90:** 490–500.
10. BELL, A. & J. P. ZUBEK. 1960. The effect of age on the intellectual performance of mental defectives. J. Gerontol. **15:** 285–295.

Introduction

DONALD A. SNIDER

*Institute for Basic Research in Developmental Disabilities
New York State Office of Mental Retardation &
Developmental Disabilities
Staten Island, New York 10314*

The subject of the early detection of mental retardation and developmental dis-abilities is approached in this session from psychological, clinical, morphological, and biochemical perspectives. Unfortunately, the program had to be modified because Dr. Alfred Baumeister was in an automobile accident and unable to attend the meeting. Dr. Marcel Kinsbourne not only introduced the session and commented on Dr. Vietze's paper, but also graciously agreed to expand some of his comments into the brief paper included in these proceedings.

Dr. Peter Vietze, Director of the Mental Retardation Research Centers Pro-gram, approaches the diagnostic problem from an information-processing per-spective and reviews the history of traditional infant tests. Dr. Vietze suggests some directions for developing new behavioral techniques to identify infants at risk for mental retardation.

Dr. Marcel Kinsbourne of the Eunice Kennedy Shriver Center in Waltham, Massachusetts comments on Dr. Vietze's paper and offers perspectives on what psychometric tests measure—outcomes or learning potential.

Dr. Krystyna Wisniewski of the Institute for Basic Research describes her use of the skin biopsy in the IBR's Diagnostic and Research Clinic and compares her findings using the electron microscope with those employing biochemical assays.

Dr. Edwin Kolodny, Director of the Shriver Center, observes that the great challenge to us is to diagnose mental retardation at the point where cognitive functioning is beginning to decline and noncognitive development is continuing to advance. He reviews clinical, biochemical, and ultrastructural approaches to di-agnosis.

Information-Processing Approaches to
Early Identification of Mental Retardation

PETER M. VIETZE

Mental Retardation Research Centers Program
Center for Research on Mothers and Children
National Institute of Child Health and Human Development
Bethesda, Maryland 20205

DEBORAH L. COATES

Psychology Department and
Center for the Study of Youth Development
Catholic University
Washington, D. C.

The purpose of this paper is to suggest some directions for developing new behavioral techniques to identify infants at risk for mental retardation. Initially, the history of traditional infant tests will be reviewed and reasons for their lack of success in predicting intellectual performance in later childhood will be suggested. Next, a number of basic psychological procedures, which may be labeled "information-processing procedures," will be described. These have promise for predicting intellectual performance where the traditional tests have failed and therefore might be used for early identification of mental retardation. Finally, the steps that might be taken to further develop these information-processing methods into a test battery to be used in early identification of mental retardation will be briefly outlined. This final section assumes that early screening is valuable even though we might not have completely effective treatments for "curing" mental retardation detected using these information-processing approaches.

TRADITIONAL INFANT TESTS

Although interest in mental testing appeared in the late 19th century, the first intelligence test as it is known today was developed in France by Binet and Simon to determine which children would be most likely to benefit from or succeed in school.[1] This initial effort to sort children according to their performance on a test led ultimately to the development of many intelligence tests. In some of his writings, Binet took issue with German and American test developers as focusing excessively on sensory processes and advocated inclusion of memory, attention, and comprehension items as important functions to be included in mental tests.[2] This advice went unheeded and led to the loss of momentum of the American testing movement around the turn of the century.

The original project by Binet and Simon (with some assistance from the young Jean Piaget) was later transformed into the Stanford-Binet Intelligence Test by Terman and his associates.[3] This test survives today as one of the best tests of childhood and adult intelligence. Within a short period of time after the initial development of the Binet test, a pediatrician at Yale University, Arnold Gesell, began developing a test to be used clinically to describe children. Other individuals also engaged in similar efforts, but in his earliest work, Gesell undertook to collect comprehensive norms by testing 50 children at each of ten age levels, 0, 4, 6, 9, 12, 18, 24, 36, 48, and 60 months.[4] At the time this was considered to be an extraordinary standardization effort.

Gesell designed his test to be a "developmental schedule," not an infant intelligence test, and included four areas of development: motor development test, language development, adaptive behavior, and personal–social behavior. Later, Gesell tested another group of 107 children longitudinally, although not all the children were tested at every age. At any one age, between 26 and 49 children were examined on the items in the Gesell test.[5] Gesell's test included items such as watching the infant visually track an object, having the child stack blocks, hiding objects and having the child find them, among others.

Other people who developed infant tests soon after Gesell either borrowed many of his items or devised similar items. However, what was common to all infant tests—and even true for most intelligence tests for older children—was the fact that scoring was on a pass-fail basis. The child being examined was tested to see whether he or she could pass the items. The score basically consisted of the number of items passed. There was no attempt to see if the child could learn to perform a task; thus, there was really no direct test of the child's learning ability, only an assumption that was based on the age norms established for a large number of children. The child was considered to be normal or not normal developmentally depending upon the number of items he or she could pass. Little other information was provided by most developmental or intelligence tests except for those that, like the Gesell Schedules, had subscores. These were useful in establishing whether a child was consistent across domains represented by these subscores.

Gesell's test was only one of a variety of tests that could be used to test children under six. Other tests were developed to test infants. Some of these were intended to be downward extensions of intelligence tests. Several tests were designed to test infants exclusively and some of these were expected to represent infant intelligence while others were not. Among the infant tests available in English were the Griffiths scale,[6] the Cattell Infant Intelligence Scale,[7] and the Bayley Scales of Infant Development.[8]

The most widely used infant test, the Bayley Scales of Infant Development, (and its forerunner, the California First-Year Mental Test) was designed to produce a mental- and a motor-scale score. Because of the way these tests were constructed, however, they yielded scores that had considerable test-retest reliability;[8,9] thus, psychometrically they were quite sophisticated. Many studies have been conducted to examine the stability of infant tests and intelligence scales over time. Intelligence scales were found to be very stable after the preschool age. They also were found to predict school achievement, job status, and many other outcomes important to a technocratic world. Infant tests, whether they purported to measure infant intelligence or not, were consistently found to be lacking in their ability to predict later intelligence.

There are a number of reasons for the low predictive validity of infant developmental scales to later IQ. One important factor is that infant behavior is variable. This means that at any one age, unless an infant is tested repeatedly within a few days, any test score may not be representative of that infant's true performance. The most dramatic example of how multiple testings can improve predictability is a study by Lester.[10] In this study, Lester had newborns tested three times on the Newborn Behavioral Assessment Scales[11] and calculated a score for each scale based on the average of all three tests. He found that these composite scores predicted later infant intelligence much better than any one of the individual scores alone; however, it is rare for an investigator to provide multiple assessments on his or her subjects even if the interest is in predicting later intelligence. Another major problem with the predictive validity of infant tests is that although there are invariant sequences for many abilities and skills, the phasing of these sequences in terms of age is not reliable before the age of two. Some infants may have developed certain abilities before 12 months of age while others may not; thus, it may be difficult for the infant test to accurately measure the infant's ability. This, of course, will affect how the infant test predicts later intelligence.

The heterogeneity of a group of infants in which an attempt is made to predict later intelligence affects the strength of the prediction. If children with very low scores are included, the prediction from infancy to later childhood becomes quite high; thus, if a group of infants tested at 12 months included low-birth-weight or premature infants who test low, it is likely that the correlation between infant test scores and test scores at later ages will be higher than if the group included no low-functioning infants.[12] It is likely that these low-testing infants have some sort of neurological impairment and fail to recover relative to their higher-functioning peers. The fact that there might be discontinuity in the distribution of abilities assessed by infant tests could account for the differential predictability of these infant tests. If there were tests that measured underlying psychological processes and for which there was continuity within a sampling distribution, it is possible that infant test scores might predict later developmental test scores better than do traditional infant tests.

INFORMATION-PROCESSING TECHNIQUES

In contrast to the traditional infant tests that provide merely summary scores on a pass/fail basis for a mixture of test items, information-processing paradigms provide measures of the process underlying the test. Such measures indicate how a subject has learned a particular skill or how long it takes to solve a problem. For example, the number of trials it takes for an infant to learn a particular association is far more instructive than knowing only whether the association was learned or not. Since information-processing tests might measure underlying psychological dimensions whereas traditional infant tests do not, it is more likely that the information-processing test will predict outcome of later IQ or developmental tests. This is especially true since IQ tests at later ages are better indices of learning ability since cognitive skills are better established and items represent these skills more adequately. Having early measures of information processing might therefore predict IQ later better than traditional infant tests.

Information-processing measures are promising since they might be sensitive to identifying infants early who may not be learning adequately. The infant who is not able to perform well on information-processing measures may perform adequately on the less-sensitive traditional infant test. It is also possible that some physically handicapped infants, such as infants with cerebral palsy or neural-tube defects (e.g. spina bifida), might have intact processing abilities as determined by information-processing tests.[13] An information-processing test might identify infants who are at risk for mental retardation and also select infants who are believed to be handicapped who may function normally.

Another major advantage of information-processing measures used as infant tests is that the former may suggest specific behavioral intervention. If an infant cannot classify objects, one sort of teaching or intervention strategy might be used, whereas if an infant can classify objects but is not able to remember objects, another sort of teaching strategy might be suggested. This is generally not possible with the traditional infant test since the individual items are rarely represented in the scores and only limited information is provided regarding how the infant performed. Of course the way in which the information-processing measure is constructed will facilitate its usefulness in selecting intervention techniques.

Information-processing techniques are those in which the infant must analyze some aspect of the situation presented and make a specific behavioral response. In some cases, this analysis may be almost automatic, but in most cases a specific judgment and selection will be made. Basic research with infants over the last twenty-five years has led to the development of several behavioral techniques designed for infants that might be labeled information-processing techniques. Although some of these may not be commonly recognized as information-processing paradigms due to the automatic nature of the processing, it is appropriate to include them here since they may prove useful in efforts to develop an information-processing test battery. Each of these techniques will be briefly described, and how they might be used in an assessment battery will be explained.

CONDITIONING PARADIGMS

The first two paradigms are basic forms of learning. Both of these procedures rely on characterizing and manipulating stimuli and responses available to the infants. It is possible to test infants very early in life using these two paradigms, so they show promise for use as early identification tests. They also have extensive literature available from studies with animals and older human subjects. Finally, being learning paradigms, they each lend themselves well to corrective intervention.

Classical Conditioning

This is a basic form of learning discovered by the Russian physiologist Ivan Pavlov. It depends on the fact that certain stimuli produce certain responses reflexively. Pavlov studied this phenomenon in dogs using the salivary reflex. He discovered that previously neutral stimuli could be paired with stimuli that elicit specific responses until the formerly neutral stimulus gained the power to elicit a similar response. This process has been studied in many species and in humans of all ages. In the late 1930s Dorothy Marquis conducted a series of experiments with human infants in an attempt to demonstrate classical conditioning in newborn infants.[14] Her results were equivocal, and even today it is not widely acknowledged that newborns can be conditioned in the classical sense.[15] Recent evidence from one laboratory indicates that under some conditions newborns can be conditioned.[16] Nevertheless, infants beyond the newborn period show evidence of classical conditioning and this paradigm has been used to explore infant responsiveness to a variety of stimulus dimensions. In adapting this paradigm to be used as an information-processing test, the number of trials to conditioning might be used as the dependent measure for a number of different modalities—those that are involved in the various reflexes—auditory, visual, tactile, and vestibular.

Although it would be useful to have normative data for various reflexive responses to different unconditioned stimuli, it would be impossible to use this paradigm as part of an information-processing test for identifying infants at risk for developmental disabilities and mental retardation without such norms.

Instrumental Conditioning

This form of learning was derived from the Law of Effect and developed in the research on operant conditioning by B. F. Skinner. The basic paradigm is now quite familiar and consists of a response followed by a "reinforcing stimulus," which increases or decreases the probability of the response occurring again. Although much of Skinner's research was conducted with pigeons, a number of infant-behavior researchers have demonstrated operant or instrumental conditioning. These studies take a variety of forms and use a variety of reinforcing stimuli; operant responses that have been studied include sucking, head movement, arm movement, leg movement, eye movement, and vocalizations. Reinforcing stimuli that have been researched include auditory stimuli, gustatory stimuli, visual stimuli, and tactile stimuli.[18] Although there is still a great deal of basic information to be learned from infants using this paradigm, enough is known presently so that it could be included in an information-processing battery to identify infants with developmental disabilities and mental retardation. The major dependent variable that could be used would be length of time to reach a certain criterion of conditioning. This paradigm has been used with infants ranging in age from newborns to 12-month-olds.

Unfortunately, there appear to be no studies extant that show the relationship between either form of conditioning just discussed and later IQ. This could be due to the fact that investigators who have studied this problem have found no relationships between infant learning and later performance on infant tests. It is also possible that such relationships have never been studied. There are convincing arguments that infants who perform well on tasks of early learning such as are presented in either classical- or operant-conditioning paradigms should perform well on tests of developmental status or IQ later. McCall has presented convincing data from several longitudinal studies that instrumental learning tasks are related to later IQ and that such tasks are representative of one of the early precursors of preschool IQ.[19] Despite more extensive supportive data, it is clear that these two paradigms might contribute to an assessment battery designed to identify infants at risk for mental retardation and developmental disabilities that relies on information processing.

ATTENTIONAL PARADIGMS

The next three paradigms that will be considered rely on the fact that the infant is an active organism who attends to stimuli in the environment presumably in an effort to gain or process information. Each of these paradigms involves presenting a stimulus or stimuli to the infant and monitoring the infant's attentional response to the stimulus presented. In some cases, comparison of performance on successive trials must be made in order to obtain the measure of information processing desired. These paradigms have also been studied in different-aged infants so there is an accumulated knowledge base indicating performance at various ages using different stimuli.

Visual Fixation

This paradigm is the simplest of the attentional paradigms. It consists merely of presenting a stimulus to the infant and measuring the length of time it takes for the infant to first notice the stimulus and then the length of time the infant spends attending to the stimulus. This technique has been used to differentiate low-birth-weight infants from their normal-birth-weight peers.[20] Basically, the shorter the latency and the longer the length of the first look at the stimulus, the more intact information processing is assumed; however, the way in which these responses vary in relation to different stimulus dimensions has also been studied. Varying the physical complexity of the stimuli on some recognized dimension may affect the levels of attention observed. There has also been different experience at different infant ages so that the younger infants look longer than do the older infants. This may be used as a guide in making judgments of relative cognitive maturity among the infants of interest—those at risk for mental retardation and developmental disabilities. It has been found, for example, that infants with Down's syndrome tend to look longer at visual stimuli presented to them.[21–23] It is not clear what the reason for this longer attentional response is, although it might be that infants with Down's syndrome take longer to process information. The visual-fixation paradigm is the simplest to administer of the attentional paradigms.

Preference for Novelty

The preference-for-novelty paradigm is one of the best researched and best developed of all the techniques being reviewed for applicability to early assessment. It was first devised by Robert Fantz in his early work on infant perception.[24] Fantz had previously worked with chickens and had developed this technique to study chick perception. More than 25 years ago, Fantz began experimenting with human infants to test whether infants demonstrate the pattern selectivity he observed in chickens. His discoveries helped to encourage other researchers to study a large variety of phenomena in human infants. It is especially fitting that one of his associates, Joseph Fagan, has worked to develop an information-processing test based on his and Fantz's pioneering work. The preference-for-novelty paradigm is based on the fact that infants spend more time attending to a novel visual stimulus than to a familiar one; thus, if an infant is exposed to a stimulus for a period of time and then that stimulus is paired with a different one, the infant will attend longer to the "novel" stimulus. This can only occur if the infant discriminates between the two stimuli. The score derived from this test is the percent of time the infant spends attending to the novel stimulus. By varying the stimuli presented to the infants, we can learn about the infant's ability to discriminate. Furthermore, if the time between exposure to the familiar stimulus and test are varied, information about infant memory can be studied. Fagan, Fantz, and their colleagues have studied all of these phenomena. It is based on these many studies that Fagan has come to develop his infant memory test. Of all the information-processing measures that will be discussed, this one alone has been developed to the point where it is virtually ready to be used as an infant-assessment instrument. Fagan[25] has reported excellent predictive validity between his infant memory test and later subsequent IQ measures and has developed norms for infants between three and nine months of age. Fagan believes that this test alone may be sufficiently useful as an infant test of information processing. The authors do not share this view since a variety of skills and their assessment would appear to cover a broader spectrum of abilities. In addition to the scientific development of this paradigm, Fagan has also begun to develop a portable prototype that could be used in a clinical setting and will be testing this unit in field sites shortly. Among the predictive validity studies Fagan has conducted, he has shown that his test is more specific and more sensitive in predicting mental retardation and normality than are the Bayley Scales of Infant Development.

Habituation Paradigm

Long ago it was discovered that many organisms will show response decrement to a particular stimulus after they have been repeatedly exposed to it. This phenomenon has been demonstrated in human infants as well. Researchers have observed habituation to both visual and auditory stimuli and have used this paradigm to study infant response to various stimulus dimensions including form, color, shape, facial expressions, and phonemes. Habituation has been researched with infants varying in age from newborn to over a year. A number of different versions of the paradigm have been studied since the criterion for response decrement is arbitrary.[26] The basic features of this technique consist of repeated exposure of a stimulus to the infant and monitoring of the infant's attentional response. When the level of attention reaches the criterion being used—often 50% less attention than observed initially—then a test stimulus is presented. If attention is shown to recover to the test, then it is inferred that the infant habituated to the first stimulus and that the infant detected the difference between the two stimuli. Leslie Cohen, who has pioneered advances in infant habituation has most recently begun using this paradigm to study how infants categorize stimuli.[27] This is based on the discovery that different stimuli belonging to the same class of stimuli will also lead to habituation; thus, a series of animals presented randomly will elicit shorter and shorter looks until the habituation criterion is reached. By varying the characteristics of the novel stimulus presented after habituation is achieved, inferences about what the infants habituated to can be made. Using this paradigm, Cohen has hypothesized that there is a developmental progression in information processing from processing features of stimuli to processing patterns as a whole to processing categories; thus, the same pictures may be found to be processed differently by different infants at different ages.[28] The infant-habituation paradigm represents a potentially powerful technique for assessing infant information processing as a way of identifying infants who are likely to be mentally retarded. A number of dependent variables may be derived from this paradigm including number of trials to reach the habituation criterion, the amount of attentional recovery, and other response classes such as affective responses to the novel stimulus. Here again, a certain amount of focused research must be conducted in order to develop this paradigm to the point where it will be easily adapted to being an assessment tool.

All three varieties of attentional paradigms described have been used in studies to predict later childhood intelligence. The predictive validity from early infant assessment of attentional measures has been very promising with the possibility of accounting for as much as 50% of the variation in childhood IQ using these measures. Of the three paradigms, the least is known about the habituation paradigm, although there are some data extant that show fairly strong relationships between infant habituation and IQ at age four. Fagan has provided the strongest evidence for a predictive relationship between attentional measures in infancy and IQ in later childhood.[25] The aggregate of these data strongly suggest that an attentional component of an assessment device that accounts for information processing is essential to identification of infants who will be retarded or developmentally disabled.

MANUAL EXPLORATION PARADIGMS

Piaget[29] assumed that infants learned about the characteristics of objects in their environments through direct contact with them. The implication of this was that object exploration—visually guided manual examination—was necessary for the development of competent cognitive growth. Although we now accept the possibility that manual exploration may not be a necessary condition for normal cognitive development, there seems to be some importance in how long, and how, infants manually explore the objects they have available to them. Some studies have found that there is a direct and strong correlation between object exploration and infant and childhood intelligence as measured by traditional assessment instruments. Two different paradigms will be discussed here, the technique where infants are given objects for a set period of time and their activities with these objects recorded, and the cross-modal perception paradigm where infants are presented with objects to explore manually without visual cues and then expected to recognize them visually from an array of objects.

Exploration of Objects

This technique is analogous to the visual-fixation paradigm described above except that here the infant is given objects to manipulate. The fact that many more behaviors are possible with the hands than with the eyes and that these behaviors are more easily observed makes this paradigm a potentially rich one for learning which aspects of objects infants find interesting. In this paradigm, the particular objects selected contribute to the information derived from coding the manual exploration. Ruff[30] has used this paradigm extensively, presenting simple objects to infants and charting the variety of their manual responses. She has found strong relationships between manual exploration during infancy and IQ later in childhood.[31] This was especially dramatic with samples of low-birth-weight infants who were studied longitudinally from infancy to early childhood. Another group of researchers using a slightly different way of coding manual exploration found strong contemporaneous relationships between exploratory behavior and performance on the Bayley Scales.[32,33] In addition, this second group also described the sequencing of behaviors during exploration and compared normal and Down's syndrome infants. They found slightly different patterns of exploration and discovered that these related differentially to developmental assessments.[23] A number of dependent measures are available from the assessment of object manual exploration. These include duration of exploration, frequency of exploration, and sequencing of behaviors expressed as vectors of conditional probabilities. This paradigm is simple to administer though somewhat difficult to score due to the extensive observational coding that is necessary. Further research using this paradigm could reduce the coding complexity considerably.

Cross-Modal Perception

Up to this point the discussion has been limited to presentation of stimuli and testing of responses in the same sensory modality. If a visual stimulus was presented initially, the test stimulus was also presented visually. The assumption that the infant has had to process the information to produce a response must be taken on faith since there are other possible explanations that often fall short of information processing per se. In this last paradigm, however, a stimulus is presented to the infant either haptically—without visual cues—or orally, and the test is presented visually. In the work of Susan Rose,[34] the major contributor to this literature, objects are placed in the infant's hand or in the infant's mouth and all visual cues are eliminated. Following an opportunity for haptic or oral exploration, the infant is presented with pairs of objects including the one already experienced and novel objects. These test objects are presented visually, the infant's visual attention being monitored for length of visual fixation. This is similar to the novelty preference method since the assumption is that the novel stimulus will be fixated longer than the familiar object although the infant may have never seen the familiar object before.[25] Rose and her colleagues have conducted extensive investigations using this paradigm and have demonstrated recently that this cross-modal transfer technique predicts childhood IQ better than "intramodal" transfer.[34] This has been shown in samples of normal children as well as in samples of preterm and low-birth-weight infants. It is clear from the data from Rose's laboratory that this is a highly sensitive and powerful assessment tool and should not be excluded from a test battery.

These last two paradigms are unique in that they provide information about how the infant responds to opportunities for extensive manual exploration of objects. Although manual exploration techniques would not be useful with infants having limited use of upper extremities, for other infants they may provide invaluable information regarding their potential for learning. Oral exploration would be extremely useful with infants having motor problems in the upper extremities. Adaptation of these techniques will require some continued efforts since an assessment battery must be simple to administer and simple to score if it is to enjoy widespread use. Nevertheless, it is reasonable to expect that these obstacles could be overcome in the near future. None of these techniques is so technical as to preclude relative ease of administration and scoring once the "bugs" are eliminated.

CONSTRUCTING THE INFORMATION-PROCESSING TEST

An overview of seven behavioral techniques that may be applied to the identification of infants at risk for mental retardation has been presented. Although each of these procedures has been used in basic experimental studies of young infants, none of them has yet been successfully developed as a test for screening infants. (Fagan has been developing an infant memory test using the novelty-preference paradigm.[24]) In order for these procedures to be further adapted for use as screening techniques, a great deal of research will be needed. Nevertheless, the adaptation of these procedures to facilitate the screening of infants is a manageable task. A number of steps must be undertaken in order to complete the development of a test or tests based on the information-processing procedures just outlined. These steps constitute a research program that would take a number of years to complete but would result in a set of procedures that would have superior power to predict which infants are most likely to develop disabilities.

The first step is to select items and procedures that can be used to screen the population at risk. The seven procedures that have been reviewed above are among the best candidates for inclusion. For each technique, however, specific items must be selected and alternative forms of each probably should be included in order to insure generalizability. Some of this selection process can be accomplished by reviewing the research literature for each technique.

The next step is to use the procedures and items selected as a prototype with clinical populations, for example, preterm infants, low-birth-weight infants, infants with intrauterine growth retardation, motor-impaired infants, and infants with Down's syndrome and other genetic anomalies. These clinical studies will help to define the limits of the different procedures.

A third step is to follow these clinical populations with standardized IQ tests at age three and also information-processing measures. This will allow estimates of predictive validity to be made and permit further refinement of the prototype test.

Next, items that do not discriminate well among different groups and between infants at risk and normal infants must be dropped. This will produce a test that is the penultimate version of the test. This version can be used to establish age norms for the different items and techniques.

The last step is to field test the final instrument in sites that would be comparable to settings in which the test is likely to be used. At the same time, age norms can be established in the field sites or where the test was developed. Age norms are important since some of the responses will change with age of the infants and developmental disabilites are based on age-appropriate behaviors.

These steps constitute the minimum set of procedures that would be necessary to adapt experimental techniques such as the information-processing methods described above and turn them into a valid and reliable test. Such a test could be used for predicting which infants from a population of infants at risk for mental retardation and developmental disabilities are most likely to have developmental problems. Although specific diagnoses may not be possible, decisions regarding the infant's ability to process information efficiently and quickly will allow more accurate screening of the at-risk population than is presently possible. At the same time, it should be noted that such a screening device would also identify at-risk infants who are likely to have no developmental problems related to cognitive processing. This new information-processing test should be much more sensitive and specific in identifying infants with mental retardation or developmental disabilities than existing traditional infant tests.

REFERENCES

1. BINET, A. & T. SIMON. 1914. Mentally Defective Children. W. B. Drummond (trans.). E. Arnold. London.
2. BINET, A. & V. HENRI. 1895. La memoire des phrases. L'Annee Psycholog. 1: 24.
3. TERMAN, L. M. 1916. The Measurement of Intelligence. H. Mifflin. Boston.
4. GESELL, A. 1926. The Mental Growth of the Preschool Child. Macmillan. New York.
5. GESELL, A. 1954. The ontogenesis of infant behavior. In Manual of Child Psychology. D. Carmichael, Ed. Wiley. New York.
6. GRIFFITHS, R. 1954. The Abilities of Babies. University of London Press, Ltd. London.
7. Cattell, P. 1940, 1960, 1966. The Measurement of Intelligence of Infants and Young Children. The Psychological Corporation. New York.
8. Bayley, N. 1969. Bayley Scales of Infant Development. Psychological Corporation. New York.
9. BAYLEY, N. The California First-Year Mental Scale. University of California. Berkeley, CA.

10. LESTER, B. 1986. Neurobehavioral assessment of the infant at risk. *In* Early Identification of Handicapped Infants. P. Vietze & H. Vaughan, Eds. Grune & Stratton. Orlando, FL. In press.
11. BRAZELTON, T. B. 1973. Neonatal Behavioral Assessment Scale. J. B. Lipincott Co. Philadelphia, PA.
12. HONZIK M. 1976. Value and Limitations of Infant Tests: An Overview, in Origins of Intelligence. M. Lewis, Ed. Plenum. New York.
13. ZELAZO, P. 1986. Information-Processing Approach to Infant–Toddler Assessment, in Early Identification of Handicapped Infants. P. Vietze & H. Vaughan, Eds. Grune & Stratton. Orlando, FL. In press.
14. MARQUIS, D. P. 1941. J. Exp. Psychol. **29:** 263–282.
15. SAMEROFF, A. J. 1972. Learning and adaptation in infancy: A comparison of models. *In* Advances in Child Development. H. W. Reese, Ed. Academic Press. New York.
16. BLASS, E. M., J. R. GANCHROW & J. E. STEINER. 1984. Infant Behav. Dev. **7:** 223–235.
17. FITZGERALD, H. E. & Y. BRACKBILL. 1976. Psychol. Bull. **83:** 353–376.
18. HULSEBUS, R. C. 1973. Operant conditioning of infant behavior: A review. *In* Advances in Child Development and Behavior. H. W. Reese, Ed. Academic Press. New York.
19. McCALL, R. B. 1986. Identifying developmental disabilities: Resume and future directions. *In* Early Identification of Handicapped Infants. P. Vietze & H. Vaughan, Eds. Grune & Stratton. Orlando, FL. In press.
20. SIGMAN, M. 1977. Child Dev. **47:** 606–612.
21. VIETZE, P., M. McCARTHY, S. McQUISTON, R. MacTURK & L. YARROW. 1983. Attention and exploratory behavior in infants with Down syndrome. *In* Infants Born at Risk: Perceptual and Physical Processes. R. Field & A. Sostek, Eds. Grune and Stratton. New York.
22. MacTURK, R. H., P. M. VIETZE, M. E. McCARTHY, S. McQUISTON & L. J. YARROW. 1985. Child Dev. **56:** 573–581.
23. MIRANDA, S. B. & R. L. FANTZ. 1973. Child Dev. **45:** 651-660.
24. FANTZ, R. 1964. Science **146:** 668–670.
25. FAGAN, J. F. 1986. Screening infants for later mental retardation. *In* Early Identification of Handicapped Infants. P. Vietze & H. Vaughan, Eds. Grune & Stratton. Orlando, FL. In press.
26. Cohen, L. B. 1976. Habituation of infant visual attention. *In* Habituation: Perspectives from Child Development, Animal Behavior, and Neurophysiology. T. Tighe & R. N. Leaton, Eds. Lawrence Erlbaum Associates. Hillsdale, NJ.
27. COHEN, L. B. & B. A. YOUNGER. 1983. Perceptual categorization in the infant. *In* New Trends in Conceptual Representation. E. Scholnick, Ed. Lawrence Erlbaum Associates. Hillsdale, NJ.
28. COHEN, L. B. 1986. Information-processing constraints on infant memory and categorization. *In* Early Identification of Handicapped Infants. P. Vietze & H. Vaughan, Eds. Grune & Stratton. Orlando, FL. In press.
29. PIAGET, J. 1952. The Origins of Intelligence in Children. (Trans. by M. Cook.). International Universities Press. New York. p. 30. H. Ruff. 1984. Dev. Psych. **20:** 9–20.
31. RUFF, H. 1986. The measurement of attention in high-risk infants. *In* Early Identification of Handicapped Infants. P. Vietze & H. Vaughan, Eds. Grune & Stratton. Orlando, FL. In press.
32. YARROW, L. J., G. A. MORGAN, K. D. JENNINGS, R. J. HARMON & J. L. GAITER. 1982. Infant Behav. Dev. **5:** 131–141.
33. YARROW, L. J., S. McQUISTON, R. H. MacTURK, M. E. McCARTHY, R. P. KLEIN & P. M. VIETZE. 1983. Dev. Psych. **19:** 159–171.
34. ROSE, S. 1986. Predicting cognitive development from infancy measures. *In* Early Identification of Handicapped Infants. P. Vietze & H. Vaughan, Eds. Grune & Stratton. Orlando, FL. In press.

Early Identification of Mental Retardation

Reactions and Comments

MARCEL KINSBOURNE

Department of Behavioral Neurology
Eunice Kennedy Shriver Center
for Mental Retardation
Waltham, Massachusetts 02254

In his lecture Dr. Vietze presented to us the figures that Fagan gave for the predictive value of his infant test procedure. It generated minimal Type I errors, but quite a few Type II errors; thus, very few children who are predicted to be normal will in fact be retarded. The reverse was quite frequent. In contrast, when a clinician examines a child for possible retardation, he is more apt to judge children to be normal who turn out to be retarded. So the difficulty with which we are clinically faced is of Type II error—failing to pick up retardation when in fact it is going to occur.

The statistics have to be viewed against the backdrop of how clinicians fare in attempting to predict outcomes for these children. If one makes one's sample heterogeneous enough by sprinkling into it a number of children much at risk for biological reasons of being retarded, then one can obtain impressive predictive validity; but still not do better than the clinician would have done, using more time-honored skills.

The question, however, is not only empirical, as to whether one can devise tests that make it possible to predict intelligence. It is also theoretical, because if it does turn out to be possible to predict intelligence years later by tests in infancy, that tells us what kind of a brain the brain is. Only a certain design for a brain will permit that to be done. If it proves to be impossible, this does not have to be because we have not found the right method. It could be because the brain by its very nature and constitution is not conducive to the approach. The issue addresses the major issues of continuity versus discontinuity in development and general versus special-purpose machinery in the brain.

If we evaluate early so as to predict for much later, we are assuming a continuous developmental sequence. But if the type of behaviors that are crucial for intelligence later on in life arise discontinuously during childhood, obviously they cannot be measured ahead of their time. It is an empirical question whether one is able to locate the origins of developmental sequences early in infancy; however, if a procedure yields results that do predict later intelligence, that could be for one of two reasons. It could be that intelligence both early and subsequently involves a major general factor, that there is general-purpose machinery that works well, or not so well, or poorly, and stays that way over time, and the procedure that is chosen by the experimenter has tapped that general-purpose machinery early. Alternatively, intellect comprises numerous special-purpose modules and any one test will tap only one of those, but for nonspecific biological reasons they nevertheless develop in parallel, so that by measuring any one of them, one effectively taps the others for the price of the one measure.

277

In mental retardation, which is a general delay, one cannot distinguish these two possibilities, but in selective learning problems, one can. I am going to make some comments about selective learning problems.

Our ability to prescreen for learning disability is something less than miraculous. By learning disability I am referring to the situation wherein a child achieves in school at a level less than that one would expect from his age, intellectual status, and apparent motivation. There are two major, broad-band domains of learning disability. One domain comprises disabilities that are selective, in which the child seems to be biologically ill-equipped to think in certain ways that are adaptive to particular school subjects. Reading is the subject matter most discussed, and the disability is often labeled dyslexia. The other type is an attention deficit, on account of which the child fails to focus on his or her schoolwork to the extent required for normal achievement. The attention-deficit type of learning disability is not selective to one school subject, but covers a range of subjects with perhaps occasional exceptions where the child is particularly motivated to succeed for idiosyncratic reasons. Yet, heterogeneous though they are, learning-disabled children, who are in the millions, have in common that they are not well served by the regular form of group instruction in the public schools, nor by the regular form of group instruction for educable mentally retarded children. They would not even be well served by classrooms for the learning disabled, because, unlike normal children who can be taught as a group, and unlike retarded children who can be taught as a group, the learning disabled are heterogeneous. What distinguishes them in the practical sense from the other children is the need for individualized educational and sometimes medical prescription. So it would be useful to be able to pick out these children who will require individualized education before they have spent several years failing and by virtue of that failure to come to notice. Certainly there have been many complaints about the definition of learning disability that calls for a child to be a certain amount, say two years, behind his chronological age in academic achievement, because that usually necessitates at least three years of failure before the child is admitted into the ranks of those qualifying for services. It would be better to be able to predict who, though intelligent, will fail in particular academic areas. The question is how to do it.

The most straightforward, pragmatic way of predicting failure in any subject is to show that the individual is already failing in that subject. This can often be done. After all, learning does not nowadays necessarily begin as late as grade one; children in increasing numbers attend preschool and kindergarten and their learning experiences, somewhat approximating those of the grade school classroom, are introduced to a varying extent. Therefore the teachers are in a position to observe how a child adapts to such instruction. Of course instruction proceeds more slowly for these younger children than in grade school, and the groupings will be different. The children in kindergarten and preschool tend to be taught in smaller groups or individually, so in that respect the experience is not comparable to that in grade school. But anything a child can be taught in first grade he or she can be taught in preschool, though more slowly. Specifically, early-reading programs are successful in teaching reading when begun even as early as age three, as our own research has shown. Of course, the programs have to be executed differently from the way they are done in grade school. One cannot just sit the child down with twenty-nine others and harangue him on something or other, because the child will not be socialized enough to that kind of experience to benefit from the instruction. I think the reason why age six is more suitable than, say, age four for beginning reading instruction is not that the four-year-old necessarily lacks the cognitive prerequisites for reading but rather that he lacks the cognitive requisites for learning to read in the particular way and context in which it is taught in first grade. This is not a condemnation of how reading is taught in grade one; it is simply an observation that that kind of teaching does not succeed with less-mature children who tend to get lost in a group and are uncomfortable with a fairly structured situation. At any rate, the facility with which an individually taught child acquires a limited sight vocabulary at age four or five is indeed a predictor, the best we have, of how that child will learn to read more words later. Knowledge of letters of the alphabet, for example, has been shown to be an excellent predictor of success in reading in grade one. So these rather mundane measures, when available, can be useful. Of course, one cannot count on them being available; that will vary greatly with the kind of instruction, if any, that a child has had in the preschool. If the particular preschool the child is in believes in enriching perceptual motor exercises, then the child will have learned nothing useful for years and nothing that one can test.

A second type of observation in the classroom that can be quite useful relates not so much to the child's ability to master a particular subject content as to the child's general task orientation, the child's readiness to focus on a task—that is, the task the teacher wants him to focus on, rather than the one he would choose, and to stay with that task till he succeeds. That task orientation can be readily observed in the preschool, and there is good informal reason for supposing it is quite a good predictor of how the child will fare. Certainly a shrewd kindergarten teacher faced with a child who has trouble with task orientation and perseverance may justifiably recommend holding him back a year while he matures further. This can be a useful step to take. Lack of task perseverance is striking in attention-deficit disorder. A child with an attention deficit will almost always manifest it before grade-school entry. The failure to maintain concentration; the restless, disorganized, scatterbrained, day-dreamy behavior, even the aggressive behavior, will all be only too apparent in the preschool. Perhaps that can be corrected adequately with stimulant medication, but only in the minority of cases is that correction so effective that the child's demeanor in the classroom ceases to be a problem. So one can predict a school problem rather readily from task orientation in the preschool. Here again, we are not talking about screening in any formal sense. We are not giving a particular test to children at large but making sensible observations. In my opinion, this is the best we have.

Can learning problems be predicted even before preschool entry? Clinical evidence, albeit much of it retrospective, suggests that the behavior of hyperactive children early in life, in about two-thirds of cases, was distinctly abnormal. So it should be possible to diagnose the kind of learning deficit that will arise from an attentional problem quite early on, sometimes even in infancy, not necessarily by special tests but by the general demeanor of the child.

The other domain of language is the one that underlies the most severe and perhaps the majority of the selective-learning problems, those that relate to reading and writing and sometimes also arithmetic. Many of those are the "tip of a language iceberg," and the underlying language deficit will have become apparent early on already by virtue of slow development of language skills in the second year of life.

A variety of precursors of language behavior can be identified in infants well before the infant's first words. These include interactional synchrony between the caretaker and the baby, a shared rhythm of movements when the caretaker speaks to the baby, and joint regard of caretaker and baby to a point in space, the ability of the baby to detect which sound comes from which visual source, and the baby's ability to point, while babbling, at a specific object as part of attending to it.

These are all potentially measurable functions, worth testing for predictive value for future language development. However, there are many learning-disabled children who do not show problems in the broad-band domain of attention or the domain of language, but in particular kinds of processing, for example, in spelling or writing, which are not to our knowledge represented much before school entry. So for predicting learning disability (which would have a momentous public-health impact, greater than that of predicting general mental retardation), there arises the interesting problem of predicting something that has not yet happened at all and for which even a precursor might not be identifiable. The readiness tests that exist, in my view, have not mastered that problem.

How may school readiness testing be justified? Ideally, the school will teach the child relevant material as soon as he is cognitively ready for it. It follows that if one tests that child at an earlier age, that must be too early for testing him on that mental operation. So, if possible, one finds a cognitive precursor, an earlier stage in the developmental sequence that will then lead the child into the key cognitive operations at the appropriate age and that, if it is delayed, will result in readiness also being delayed. In practice it is very hard to identify these cognitive precursors. Let us consider reading acquisition because it has been most studied in this connection.

What are the cognitive precursors to learning to read? Everybody who investigates reading feels he knows what they are, and everybody has a different list. I have mine,[1] and I can even boast of a 0.7 correlation between my tests at age four and subsequent reading achievement at age six,[2] which is probably a point or two better in predictive validity than somebody else's test. However, the fact is that we have no model of reading acquisition and thus what we use is of necessity arbitrarily rather than rationally selected. We do not know what the components are, worse still we do not know what the performance-limiting factors are, that have to be mastered to learn to read; therefore necessarily we cannot measure them with criterion-referenced tests. A criterion-referenced test enables one to determine whether a child's skill on the relevant variable is sufficient for a particular purpose. Being better than that is not necessary or even useful. In a convoy, the criterion to meet is to be able to sail at the rate of the convoy. The ability to sail twice as fast does not help in a convoy. So we do not need tests that evaluate the child relatively to norms. We need criterion-referenced tests. Nobody has those. So, realistically, we do not have readiness tests. But should we consider tests on an empirical basis? Maybe if we use a large arbitrary set of tests, we can get predictive value anyway.

There have been a number of attempts at predicting reading problems by giving a great miscellany of tests. These tests have had some limited success with respect to group data. The better the group does on those tests, the better the group performs subsequently in reading acquisition. The problem is that these measures cannot be applied with sufficient reliability to individuals. Even if, statistically speaking, the measures used do have predictive validity for a group, for the individual there is such a high incidence of Type I error, false identification, and Type II error, false rejection, that in terms of actually guiding the individual's education, the tests are not useful. Particularly undesirable is the risk of false-positive identification; incorrectly predicting that the child is bound to fail. Such predictions are apt to sow seeds of expectation in the minds of adults, notably teachers but also parents, by what is called the Rosenthal effect. This leads the teachers to underestimate the potential of those children and therefore to instruct them in ways that are at best patronizing and at worst impoverishing of their mental life, rendering self-fulfilling the prophesy that the child will fail. False-negative attribution is not as serious. To fail to predict that a child will have learning disability would be more serious if we had an effective intervention, which we do not. In fact, prediction itself is not really justified, even if it can be done, if we cannot then diagnose and treat. Even if one were to predict correctly at age three or four that a child would experience a learning difficulty at age six or seven, there is no consensus and no convincing data on what steps should be taken to prevent that happening. Obviously one can teach a child better, one can nonspecifically "enrich" the child's experience, but any child would benefit from that, not just one that is learning disabled. We do not need to do studies to convince ourselves that better teaching by more skilled or more motivated teachers is going to produce a better product. Beyond this, however, we do not have specific maneuvers that are known to be therapeutic. Maneuvers have been suggested at the level of behavior and the level of the brain. At the level of behavior, the particular forms of instruction range from Montessori to very methodical, progressive phonics curricula. Practitioners of every system that has been presented (and whenever there is no solution, many solutions are offered) have maintained that their system is almost always effective. The people who still in spite of all these systems are illiterate and cannot read should be ashamed of themselves, because they must really not have made proper use of all these good ideas. Yet their numbers are very great. That is not to say that the methodologies should not be used, but certainly we cannot at this point justify a nationwide effort for early identification even if we had the where-with-all to do it, given that the steps we can take to avert learning disability are limited at best.

The problem in learning disability is a problem of learning. Not one of the predictive tests of learning disabilities, including mine, is in fact a test of learning. None are tests of the rate at which the child is capable of acquiring particular classes of information. Like conventional IQ tests, they are all tests of what the child can currently do. It is an obvious commonplace that what one can do is at the mercy of a number of factors, situational as well as constitutional, whereas the rate at which one can pick up something one could not do at first is much more indicative of the biological potential of the organism. So in my opinion there is a bright future for a whole new set of procedures of prediction, if not in infancy, at least by age four or five, by which the ability of individuals to acquire certain types of information, representative of what they will have to do in reading and writing, will be quantified. Some years from now, we shall have such tests.

At the level of brain, it is much worse than that. The theorizing has been that the learning disability represents a failure to develop crucial parts of the nervous system, which no doubt is so in many cases, and that particular maneuvers can foster that development. The latter is the incorrect assumption. There is a range of futile offerings all the way up from "patterning" which purport to foster the development of cerebral hemispheric function, and I cannot review all of these. But I take as an example the notion of fostering the lateralization of language to the left hemisphere, which is a thread running through many programs, some not at all disreputable. The assumption is that somehow establishing peripheral laterality on the right will induce central laterality for language on the left. There is a statistical association between right-handedness/footedness (and to a miniscule extent right-eyedness) and left language lateralization and language in people who are ambidextrous and left-handed may be more bilaterally represented. But peripheral maneuvers to change that hand preference, for instance to switch it from left to right or to make it more consistent or to change eye preference—for instance, putting a patch over the left eye—have no effect on brain organization. Even if they had such an effect, it is not clear that it would help because although most of us have language lateralized in the left hemisphere, it is not clear that the minority that have bilateral language are any less well off with regard to language for that reason. There have been many studies, including some by myself and my colleagues,[3] comparing the cognitive abilities of right-handers and nonright-handers. If they are different, they are not much different. The few studies that do show differences, show different differences from study to study. So it is amazing but apparently true that as dramatic and ostensibly specific to the human species a phenomenon as left language lateralization should be violated in some five percent of us with such impunity. Why is language on the left in most of us if it might as well be on both sides, as it is in some of the rest of us? The answer may be that it is on one side because it does not need to be on both (but works no less well if it is). So even if it were possible by peripheral maneuvers to shift language from hither-to-yon, there is no current evidence that there would be any benefit in so doing.

To summarize, our attempts to predict learning disability lack validity in that what we measure does not necessarily qualify as a precursor of the academic skill in question, lack generality insofar as at best they cover only a small territory within that domain, lack test-retest reliability, with respect to components of these tests, and lack utility, because even if the results were valid, one would not know what to do next.

What then should one do? If a child has in the preschool years the obvious signs of delayed language development, that child undoubtedly is going to have a school problem, and one does not have to do special predictive tests. Undoubtedly if a child has a severe visuospatial problem and clumsiness, which occurs more rarely but certainly occurs, the child will have difficulties, again obvious, in writing, drawing, and visuospatially related activities. Importantly, if a child has an attention deficit—hyperactivity, underfocused attention—that child is likely to run into trouble. He may even achieve within the normal range, but he will not achieve up to his intellectual potential because if he is superior in intellect but has an attention deficit, it may degrade his performance to the normal level for the country, but not to the normal level for him. So the attention deficit is to be taken seriously.

There is another type of attention deficit, which I have called overfocused,[4] as opposed to underfocused, which is the opposite in that the child takes a long time to establish task orientation and when he has it will not give it up—only is willing to deal with one issue at a time, greatly resents interruption for other task demands, finds even the least irrelevant external change highly distracting, and at the end of the school period will not put his pen down, but in a compulsive way has to finish what he is doing. This overfocused attention, which incidentally sometimes comes with certain speech delays and deviant speech expressions, reminiscent of the autistic, is itself a quite common variant, which does have implications for school performance but that mysteriously has eluded observation until recently. Nevertheless, the overfocused style, just like the underfocused style, is worth picking up in the preschool, and it too calls for a certain amount of individualized understanding, sympathy, and management.

Finally, if one wants to predict school failure in first grade relative to a practical issue, such as whether the child should enter regular first grade or a school with additional resources for special needs, and a good teacher-pupil ratio, obviously the best thing to do is to test the child's incipient academic abilities as late as possible before school entry. The correlations between such test results and what happens in the first grade would be better the shorter the time interval between the test and the first-grade entry, because the longer that time interval, the more time there is for the biological variability of human development to inject noise into the correlation. My recommendation in that kind of situation is to do the tests just before one has to make a decision, usually the summer before or the late spring before school entry, and at that point obtain the best sense one can on what to do with the child come the next fall. But with respect to prescreening for learning disabilities in any way comparable to the accurate prescreening that is now possible for many metabolic disorders, that is totally impossible at this time.

REFERENCES

1. KINSBOURNE, M. 1976. Looking and listening strategies and beginning reading. *In* Aspects of Reading Acquisition. J. T. Guthrie, Ed. Johns Hopkins University Press. Baltimore, MD.
2. Light, M. C. 1980. A longitudinal study of the effect of a kindergarten early reading program. Doctoral dissertation. University of Toronto.
3. BRIGGS, G. G., R. D. NEBES & M. KINSBOURNE. 1976. Intellectual differences in relation to personal and family handedness. Quant. J. Exper. Psychol. **28:** 591–601.
4. KINSBOURNE, M. & P. J. CAPLAN. Children's Learning and Attention Problems. Little, Brown. Boston, MA.

DISCUSSION

PETER VIETZE: Dr. Kinsbourne, you ended with a comment about preschool tests that assess learning potential versus learning outcomes. There are some researchers who have been developing procedures that assess learning potential. These include Ann Brown and Joe Campione, and Carl Haywood, as well as other information-processing researchers who have been studying normal and retarded children and who are moving closer to preschool tests that assess learning. In addition, I would point out that Reuven Feuerstein has been saying that it is important to measure learning processes rather than learning outcomes for 20–30 years. Since this work has been going on for so long now, I think the development of tests that assess potential learning is going to take less time than you said.

MARCEL KINSBOURNE: Even earlier Milton Budoff used his Koh's Blocks test in very much the sense I am advocating.

SUSAN SKLOWER: Please comment on the self-fulfilling prophesies of using the tests, Dr. Kinsbourne.

MARCEL KINSBOURNE: One reason why screening tests are not used is an ethical one. You should not screen unless you have available a management procedure that can correct what is picked up by the screening. Now, in spite of many claims, there are no such proven procedures in developmental disabilities. Lots of people say lots of things. So when a child has been identified by a test, rightly or wrongly, as one who is going to have learning disabilities, we are not in a position to do anything more than watch him have it. In fact, as your question implies, the harder we look at him the more likely he is to have it, which is the well-known Rosenthal effect. When a child seems to be at risk for doing something poorly either because somebody says so or because he doesn't have a very attractive face, another inexpedient thing to have, then the teacher's attempts to teach are muted, pessimistic, and ineffective. So the Type I error of false identification is serious, whereas the Type II error of not picking up the impending learning problem matters less.

The Diagnostic Value of Ultrastructural Studies of Skin-Punch Biopsies and Buffy Coat for the Early Diagnosis of Some Neurodegenerative Diseases

KRYSTYNA E. WISNIEWSKI

Department of Pathological Neurobiology
Institute for Basic Research in Developmental Disabilities
New York State Office of Mental Retardation &
Developmental Disabilities
Staten Island, New York 10314

INTRODUCTION

Neurodegenerative diseases (NDDs) are disorders in which there is neuronal dysfunction often due to progressive central nervous system (CNS) structural abnormalities which may, or may not, involve the peripheral nervous system, and/or other organs or systems. Well over 600 varieties of neurodegenerative diseases have been reported to affect infants and children, while fewer than 100 have been described in adults.[1,2] The majority of these diseases are often genetically determined, frequently belonging to the subgroups of inborn errors of metabolism. These diseases may also be caused by nongenetic factors such as persistent viral infections, disturbances in host–immune interactions, or by chronic toxic and metabolic encephalopathies. NDD are divided etiologically into three major groups:

1. genetically determined (e.g., Tay-Sachs disease, neuronal ceroid lipofuscinosis);
2. nongenetically determined (e.g., subacute sclerotic panencephalitis, chronic toxic lead encephalopathy); and
3. unknown.

At the present time the NDDs of unknown cause constitute the largest group of disorders and further biochemical, molecular genetic, virological, and immunological studies are required to characterize these diseases.

The NDDs are classified topographically or anatomically according to whether the predominant locations involved are in the gray or white matter of the CNS (with or without PNS), or whether they are in other parts of the nervous system. The classification is as follows:

1. polioencephalopathies (mainly gray matter);
2. leukoencephalopathies (mainly white matter);
3. corencephalopathies (deep telencephalon, diencephalon, mesencephalon);
4. spinocerebellopathies (mainly brain stem, cerebellum);
5. diffuse encephalopathies (involvement of the whole neuroaxis); and
6. diffuse encephalopathies with involvement of other tissues or organs (multisystemic).

In the early stages of many of these diseases, only some of the symptoms may be present. Which symptoms are seen (seizures, dementia, pyramidal, extrapyramidal, cerebellar, and/or suprabulbar signs) depends on the site of the initial lesions in the CNS. As the disease progresses and more of the susceptible areas of the neuroaxis become involved, the entire group of symptoms becomes apparent. The onset of clinical symptoms of NDD may be in infancy, early or late childhood, juvenile or adult stages of life.

In recent years, ultrastructural studies of punch biopsies from tissues of skin, conjunctiva, buffy coat, peripheral nerve, rectum, and liver have been used increasingly as an aid in the early diagnosis of those neurodegenerative disorders of the CNS that are associated with lysosomal storage diseases, enzyme deficiencies, or hereditary neurometabolic diseases.[3-10] The categories of lysosomal storage diseases that can be diagnosed by the above method are neuronal ceroid lipofuscinosis, sphingolipidosis, mucopolysaccharidosis, oligosaccharidosis, mucolipidosis, and glycogenosis.

In the storage diseases, although the storage compounds are normal, the excessive accumulation of nontoxic, cytoplasmic constituents interferes with normal cell function. In lipid storage diseases, the accumulated lipids are poorly soluble in aqueous solutions, and, therefore, they precipitate within the cells in which they are synthesized. This occurs mainly within the cells' lysosomal compartment; under the electron microscope they appear as abnormal, cytoplasmic inclusion bodies. This excessive storage in the cell membrane may disturb the function of normal membrane-associated processes.

Recently, it has been shown that in different brain cells, individual neurons may have quite different metabolic pathways of transmitter biosynthesis, as well as different receptor proteins.[11] This heterogeneity of neurons accounts for the varying clinical pictures presented where the same type of enzyme defect or differing degrees of enzyme deficiency produces different levels of CNS involvement as well as differing clinical onsets.

In this paper, findings from our ultrastructural studies of skin biopsies and buffy coat (lymphocytes) obtained from patients with chronic neurological disorders are discussed. These patients were often developmentally disabled (DD), and present either clear-cut evidence or a hint of progressive encephalopathies—neurodegenerative diseases. These ultrastructural studies are valuable diagnostic tools in cases in which the enzymatic defect is poorly understood; where it has not yet been demonstrated by biochemical means; or, in cases where it has been demonstrated, but the unusual clinical presentation and biochemical variants have made diagnosis difficult. In cases with unusual clinical manifestations of the disease, skin biopsy is also taken for culture of fibroblasts for lysosomal enzyme studies, or used for further biochemical studies. The indications for and the limitations of using skin-punch biopsies as a diagnostic tool are also discussed.

MATERIALS AND METHODS

Between April 1978 and April 1985, 158 skin-punch biopsies and 40 buffy-coat fractions were obtained from 158 patients referred to the clinic as being suspected of having lysosomal storage or neurometabolic disorders associated or with chronic progressive encephalopathies. The ages of these patients ranged from 3 months to 32 years (TABLE 1).

TABLE 1. Age Distribution of Cases Studied Using Two Types of
Ultrastructural Studies

Skin-Punch Biopsy		Buffy Coat	
Age	No. of Patients	Age	No. of Patients
1–12 mo	21	1–12 mo	5
2–5 yr	62	2–5 yr	15
6–10 yr	36	6–10 yr	12
11–20 yr	22	11–20 yr	8
>20 yr	17	>20 yr	0
Total	158	Total	40

The 158 cases were examined and can be divided into three clinical groups:

Group I, definitive chronic progressive encephalopathy: Comprised of 65 patients with clear-cut clinical signs of progressive, neurological dysfunction and where a diagnosis of lysosomal storage disease was apparent.

Group II, definite chronic progressive encephalopathy: This group was comprised of 53 patients diagnosed clinically as having a NDD and who showed some clinical signs indicating a clinical variant or an atypical form of lysosomal storage disease, a partial enzyme deficiency, or a specific neurological syndrome.

Group III, undefined chronic encephalopathy: In this group, all 40 patients were initially suspected of having a NDD but, after a neurologic follow-up, 29 of the patients were diagnosed as cases of static encephalopathy.

All patients from clinical groups I, II, and III had neurological and psychological evaluations, and where indicated, neuroradiological (CTT scan), electrophysiological, for example, electroencephalogram (EEG), electroretinogram (ERG), nerve conduction (NC), electromyogram (EMG), brainstem auditory-evoked response (BAER), visual-evoked response (VER), and biochemical studies, for example, lysosomal enzyme, gene mapping, high-pressure liquid chromatography (HPLC), urine oligosaccharides, glycolipids, mucopolysaccharides, dolichol levels. When possible, cases have been reevaluated once a year. Some of these laboratory studies have been repeated.

METHOD FOR ULTRASTRUCTURAL STUDIES

Skin-punch biopsy specimens measuring 3 mm in diameter and 3 mm deep were taken with minimal trauma from the supine, middle aspect of the forearm, 10 cm below the elbow. In the majority of the patients, skin samples were taken from the left upper extremities. The patients required only a local infiltration of anesthetic and an adhesive strip dressing afterwards.

The skin fragment for the ultrastructural study was immediately fixed in a solution of 2.5% glutaraldehyde and after two to six hours was transferred to 0.1 M phosphate-buffered pH 7.2, 1% osmium tetroxide, followed by rapid dehydration using a series of graduated concentrations of alcohol and acetone and embedded in Epon. Thick sections (1.0 μm), stained with toluidine blue, were observed under light microscope to select the areas containing subcutaneous nerve bundles. Ultrathin sections were cut and mounted on copper grids and stained with uranyl acetate and lead citrate.

When possible, blood samples were obtained by venous puncture (40/158 cases). Following centrifugation (2,000 g, 10 min) of the citrated blood, the plasma was aspirated and replaced by a solution of 2% glutaraldehyde. After two hours, the solidified buffy coat was cut into small pieces that were postfixed in 1% osmium tetroxide for two hours at 4°C. Following dehydration in graduated concentrations of alcohol and acetone, the blocks were embedded in Epon. The ultrathin sections were double-stained with uranyl acetate and lead citrate. Micrographs were taken on a Hitachi HV 11 or Philips 300. At least three blocks were cut and examined for each clinical specimen. We looked for abnormalities in endothelial cells, pericytes, axons, Schwann cells, perineurium, fibroblasts, glands, and epithelial cells of the skin biopsies. Electron micrographs were made of each cell type and care was taken to obtain pictures of each kind of abnormal material in at least one cell of each cell type involved. At least 100 cells of the buffy coat were studied and inclusions counted.

The presence, in skin-biopsy and buffy-coat samples, of ultrastructural features characteristic of specific diseases, was considered as a positive observation only when intracellular inclusion bodies not present in normal skin were encountered in sufficient numbers, the minimum was 5% inclusion bodies although the minimum varied according to the age of the patients. An equivocal observation was one with nonspecific changes not related to lysosomal storage disease. The negative observation was associated with cases where lysosomal inclusion bodies were not found. Mast cells, melanocytes, and sweat glands were not included in the analysis because of the possibility of confusing their granules with abnormal inclusion bodies. Detailed analysis was undertaken to avoid misinterpretation of artifacts in normal skin cells, artifacts that could be mistaken for abnormal inclusions.[12,13] The number of micrographs taken per biopsy averaged 20 but was considerably greater in unusual cases. Abnormal inclusions were counted, and the periodicities of lamellar structures were measured, when present. The evaluation of the biopsies included consideration of clinical and other laboratory data.

RESULTS

The results are presented in TABLES 2–5. In TABLE 2, the three groups of patients are shown further divided into three subgroups according to ultrastruc-

TABLE 2. Results of Ultrastructural Study of Skin-Punch Biopsy in 158 Cases with NDD

Group	Totals	Subgroups		
		A Positive- Abnormal	B Equivocal- Nonspecific	C Negative- Normal
I—Definitive chronic encephalopathy—storage disease strongly suspected	65	65	0	0
II—Definitive chronic encephalopathy—storage disease only suspected	53	0	20	33
III—Undefined chronic progressive encephalopathy—storage disease not suspected	40	0	9	31
Total	158	65	29	64

TABLE 3. Results of Ultrastructural and Lysosomal Enzyme Studies of Cases With a Definitive Chronic Encephalopathy Strongly Suspected

A6 Neurometabolic Disorders	No.	Ultrastructural Diagnosis		Lysosomal Enzymes	
		Positive	Negative	Positive	Negative
Category 1					
Neuronal ceroid lipofuscinosis					
Early infantile	2	2	0	0	2
Late infantile	2	2	0	0	2
Juvenile	16	16	0	0	16
Category 2					
Mucopolysaccharidosis					
Hurler (1), Hunter (2), Sanfilippo (7)	10	10	0	10	0
Mucolipidosis-type II (2), IV (2)	4	4	0	2	2
Disorders of lipoprotein degradation					
Fucosidosis (1), sialidosis (1)	2	2	0	2	0
Category 3					
Lipidosis					
Gangliosidosis (GM_2, GM_1)	4	4	0	4	0
Niemann-Pick disease	2	2	0	1	1
Krabbe's disease	2	2	0	2	0
Fabry's disease	1	1	0	1	0
Metachromatic leukodystrophy	3	3	0	3	0
Category 4					
Glycogenosis					
Pompe's disease	4	4	0	4	0
Category 5					
Other, unclassified					
Adrenoleukodystrophy	4	4	0	0	4
Infantile neuroaxonal dystrophy	4	4	0	0	4
Lowe's syndrome	3	3	0	0	3
Suspected new type of lipid storage disease	2	2	0	0	2
Total	65	65	0	29	36

TABLE 4. Group II: Results of Ultrastructural and Lysosomal Enzyme Studies of Cases with Chronic Encephalopathy Storage Disease Suspected

Disease	No.	Ultrastructural		Lysosomal Enzyme	
		Equivocal	Negative	Positive	Negative
Atypical spinocerebellar degeneration	18	9	9	0	18
Alexander's disease	3	1	2	0	3
Politzeus Merzbacher	2	0	2	0	2
Olivo-ponto-cerebellar degeneration	2	0	2	0	2
Cerebro-oculomuscular syndrome	3	3	0	0	3
Cockayne's syndrome	3	0	3	0	3
Canavan's disease	5	0	5	0	5
Myotonic dystrophy	2	0	2	0	2
Charcot-Marie-Tooth disease	4	0	4	0	4
Sjögren-Larssen syndrome	2	2	0	0	2
Unspecified leukodystrophy	4	0	4	0	4
Acquired autoimmune deficiency syndrome	3	3	0	0	3
Rett's syndrome	2	2	0	0	2
Total	53	20	33	0	53

TABLE 5. Group III: Results of Ultrastructural and Lysosomal Enzyme Studies

Diagnosis	No.	Ultrastructural		Lysosomal Enzyme	
		Equivocal	Negative	Positive	Negative
Static encephalopathies	29	0	29	0	29
Atypical spinocerebellar degeneration	7	7	0	0	7
Familial Alzheimer's disease	2	2	0	0	2
No definite diagnosis	2	2	0	0	2
Total	40	9	31	0	40

tural abnormalities. Of the three subgroups, Subgroup A exhibited specific ultra-structural features of various types of storage or neurometabolic diseases and corroborated the clinical diagnosis in 100% of the cases. The fifty-three patients in Subgroup B were suspected of having a NDD, but the ultrastructural results were negative in 33 cases and equivocal in 20 cases. The other 40 patients (Subgroup C) examined and tested were originally diagnosed as having undefined chronic, pro-gressive neurological signs. The ultrastructural studies showed negative in 31 of these cases and the other nine were equivocal.

Group I, Category One

Group I, which comprised 65 patients, was further divided into five categories of disorders (TABLE 3) according to the following morphological similarities.

Neuronal Ceroid Lipofuscinosis (NCL)

This category contained 20 patients (FIG. 1 a,b,c,d). Two patients had the infantile form, two had the late infantile form, and 16 had the juvenile form of the disorder. Each of the various forms of the disorder is distinguished by the pre-dominance of one or more of the disorder's conventional cytosome types. The infantile form is distinguished by granular inclusions (FIG. 1a); the late infantile form by curvilinear bodies (FIG. 1b), and the juvenile form is distinguished by fingerprint and rectilinear bodies (FIG. 1c), and, in some cases, by lipofuscin inclusions (FIG. 1d).

Among the 16 patients with the juvenile form of the disorder, two presented clinically with spinocerebellar degeneration without clear-cut dementia, a history of seizures, or blindness. Osmiophilic and granular inclusion bodies, as well as vacuoles, were observed in samples taken from four of the remaining patients in this group, and a great variety of intracytoplasmic inclusions were detected in the cells from two of these four individuals. These inclusions were found in some fibroblasts, pericytes, endothelial and Schwann cells, smooth muscle of sweat glands, and occasionally, in axons. In the lymphocytes, vacuoles appeared, to-gether with fingerprint profiles, particularly in the juvenile form. In the infantile and the late-infantile form of the disorder, while vacuoles were absent from the lymphocytes, they did contain membrane-bound inclusion bodies. The clinical picture that emerges at these two stages of the disorder is expressed through such classical symptoms as dementia, seizures, retinal pigmentary degeneration with

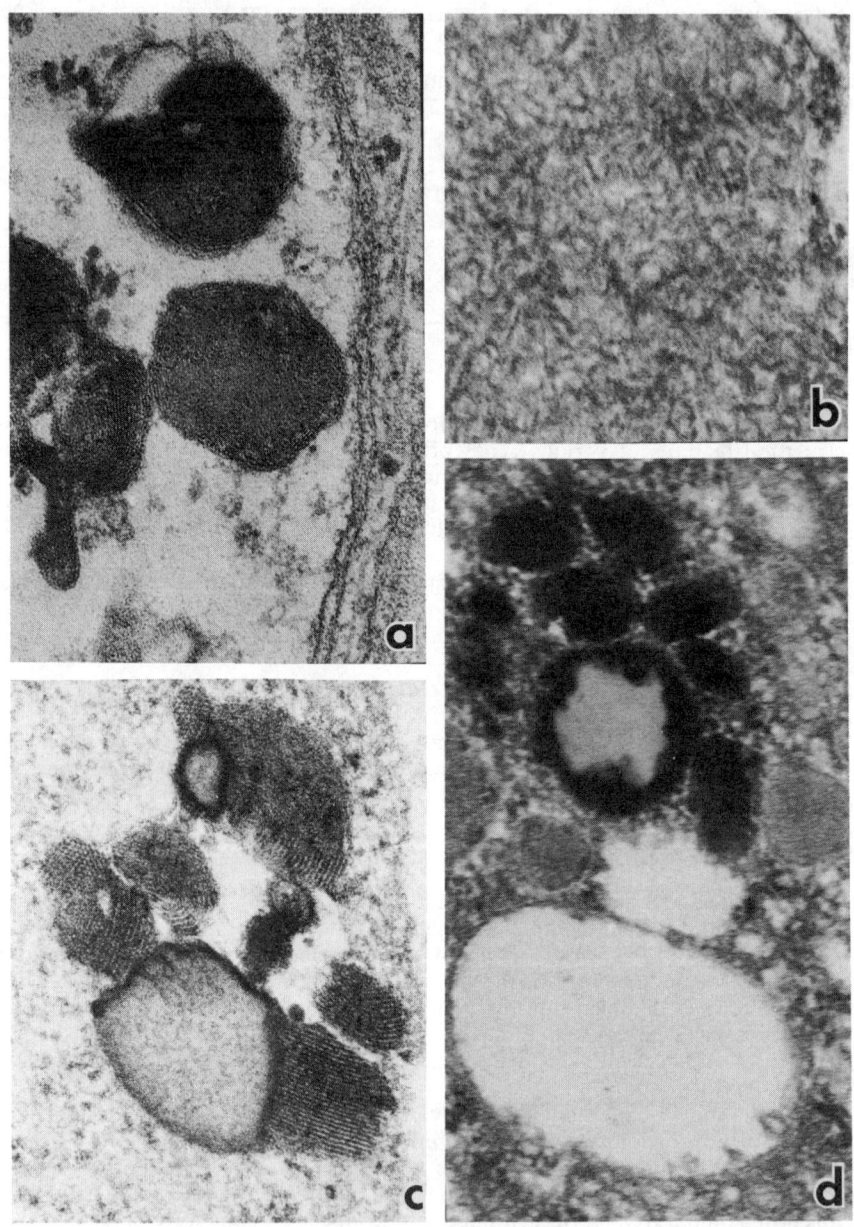

FIGURE 1. Neuronal ceroid lipofuscinosis lysosome inclusions: **(a)** granular (×50,000) in the infantile form; **(b)** curvilinear cytosomes (×120,000) in the fibroblasts in the late-infantile form; **(c)** fingerprints (cytosomes) with islands of lipid of electron lucent and of electron-dense bodies (×50,000) in the juvenile form; **(d)** vacuoles, lipofuscin, and electron-lucent bodies (×85,500) in a juvenile form. Original magnifications shown in parentheses; reproduced here at 90% original size.

292

optic atrophy, pyramidal, extrapyramidal, cerebellar, suprabulbar signs, and so forth. Urinary dolichol excretion was elevated in all of the 20 patients (TABLE 2), and electroretinography (ERG) was abnormal in only nine of the 20 patients.

Category Two

The second category includes mucopolysaccharidoses (MPS), mucolipidoses (ML), and disorders of glycoprotein degradation (LP). This subgroup comprises 16 patients, 10 of whom had MPS, four of whom had ML, and two of whom had LP.

Mucopolysaccharidoses (MPS)

Positive results were obtained from skin biopsies performed on 10 patients who had either MPS I, II, or III. This group of patients included one case of Hurler's syndrome (MPS I), two cases of Hunter's syndrome (MPS II), and seven cases of Sanfilippo's syndrome (MPS III). In this subgroup, all of the 10 patients exhibited various degrees of dementia, mental retardation, coarse features, dysostosis multiplex, and organomegaly. Electron microscopy of skin biopsies revealed extensive vacuolization in most fibroblasts, in smooth muscle of sweat glands, epithelium, perineurium, pericytes, endothelium, and Schwann cells, but rarely in axons. Vacuolization appeared to be most extensive particularly in MPS I and II. It was slightly less extensive in MPS III (FIG. 2a). Indentation of epidermal cell nuclei was seen in MPS I–III. The lesions in MPS III may be severe as a result of laminated lipid particles filling the vacuoles of the affected cells, which may also contain zebra-like bodies (FIG. 2b). The lymphocytes were affected in MPS I and II but rarely in MPS III. A diagnosis of MPS has been confirmed for the 10 patients in this sample by results obtained from analysis of urine quantities of mucopolysaccharide (MPS) and from studies of specific lysosomal enzymes.

Mucolipidoses (ML)

There are three known forms of ML in this subgroup (ML II, III, and IV). ML II is also known as I-cell or inclusion cell disease. The four ML patients in our study were of the ML II and IV varieties. Skin biopsies from the two patients with ML II revealed a deficiency of several acid hydrolases. This disorder was further characterized by the existence of membrane-bounded cytoplasmatic vacuoles in fibroblasts, perineurium, Schwann cells, pericytes, and endothelial cells. Zebra bodies were not in evidence in any of these cells (FIG. 3a and b). Inclusions with closely packed dense rings were particularly apparent in the endothelial cells. Vacuolization was also found in lymphocytes. Lysosomal enzyme studies confirmed the diagnosis of ML II.

Skin biopsies obtained from the two ML IV patients revealed, besides cytoplasmic vacuoles, small dense lipid membranous cytoplasmic and zebra bodies that appeared in all the different skin cells and also in the lymphocytes. Although some lysosomal enzyme studies were performed in these two cases, a new test, that is, catalytically defective ganglioside neuroaminidase GD_{1a}, was not performed.

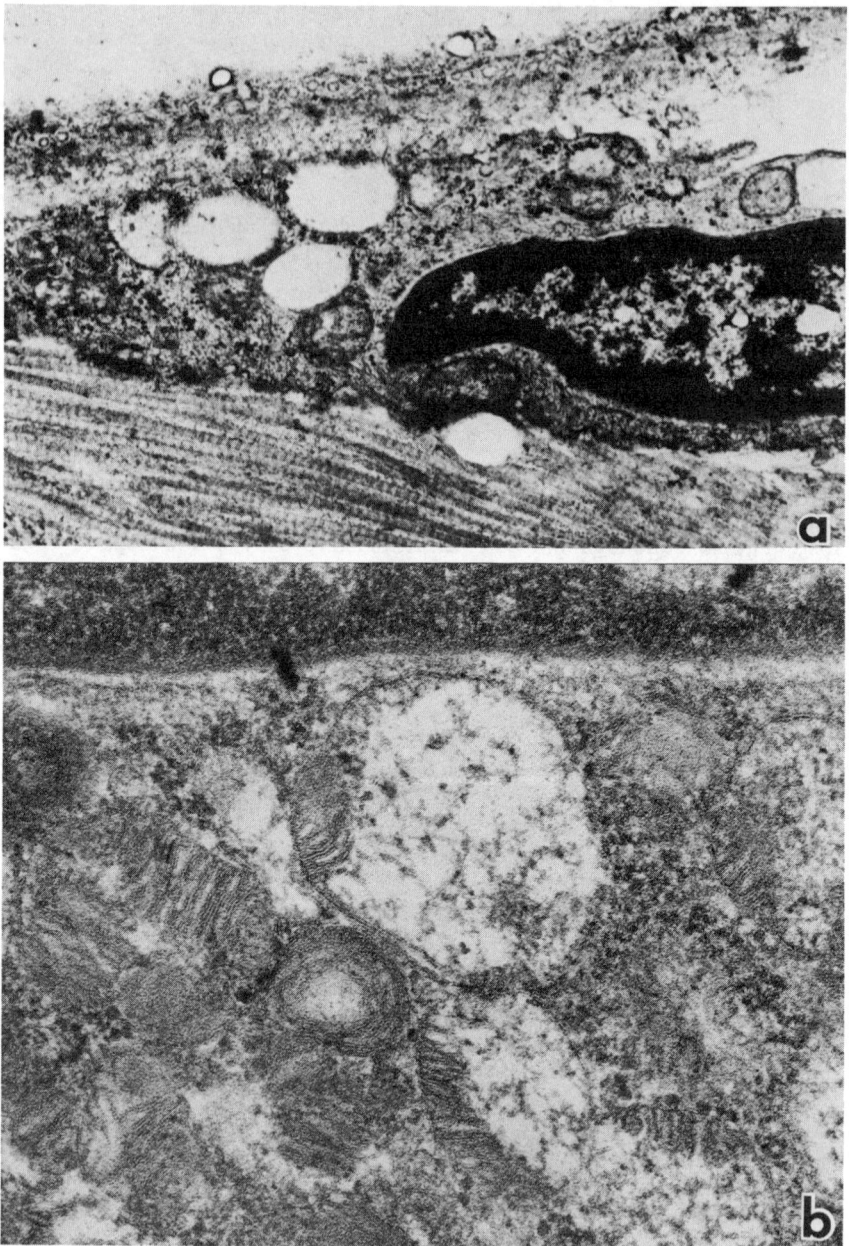

FIGURE 2. (a) MPS III, fibroblast with membrane-bound vacuoles (×19,400); (b) MPS II, membrane-bound cytoplasmatic vacuole and zebra bodies (×55,000). Original magnifications shown in parentheses; reproduced here at 95% original size.

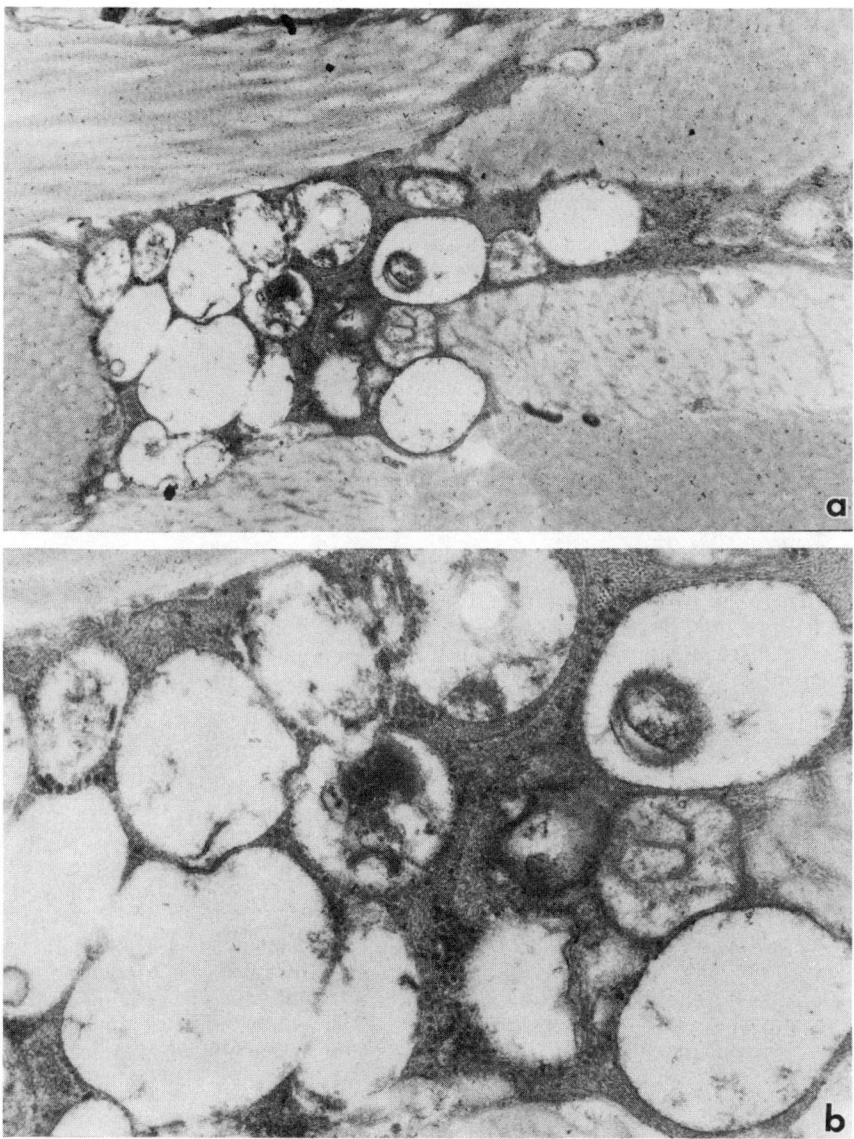

FIGURE 3. I-cell disease. **(a)** fibroblast with many membrane-bound vacuoles (×22,000); **(b)** higher magnification (×50,000). Original magnifications shown in parentheses; reproduced here at 65% original size.

Disorders of Glycoprotein Degradation (Oligosaccharidosis)

This subgroup contains the following disorders: mannosidosis, fucosidosis, sialidosis, and aspartyl-glycosammine. Only fucosidosis (one patient) and sialidosis (one patient) were observed in our patient sample. In the course of examining the skin cells of these patients, we found not only membrane-bounded vacuoles, similar to those observed in FIG. 3, but also lamellar inclusions and lipofuscins. Lysosomal membrane abnormalities were identified only in the fucosidosis patient.

Category Three

The third category includes gangliosidosis (G), Niemann-Pick disease (NP), Krabbe's disease (KD), Fabry's disease (FD), and metachromatic leukodystrophy (MLD). This subgroup includes 12 patients, four of whom have G, two of whom have NP, two of whom have KD, one of whom has FD, and three of whom have MLD.

Gangliosidosis (G)

There are two distinct storage diseases: Tay-Sachs, due to hexosaminidase A deficiency, and Sandhoff's, due to deficiencies of both hexosaminidase A and B. These diseases have several variants that are well identified biochemically or clinically. Abnormalities observed in skin samples obtained from the two Tay Sach's patients were confined to axons (FIG. 4a, b). An examination of the lymphocytes did not reveal any inclusion bodies. In skin cells obtained from the one patient with Sandhoff's disorder (which leads to a general visceral storage), small laminated zebra bodies appeared in fibroblasts, endothelial cells, glands, and Schwann cells (FIG. 5a–e) while dense bodies filled the axons.

The clinical presentations of these two disorders, in this study, were atypical, and therefore, the information derived from the electron-microscopic studies aided in making the initial diagnoses. Lysosomal enzyme studies also helped to confirm the diagnoses.

GM_1 gangliosidosis, due to β-galactosidase deficiency, is a disorder that involves not only neuronal tissue but also mesenchymal tissue containing widespread visceral deposits of keratan sulfate. In the infantile form of this disorder, the so-called pseudo-Hurler's syndrome, the skin and lymphocyte samples taken from one patient revealed vacuoles similar to those that appear in the subgroups of MPS, ML, and LP. Furthermore, abnormalities were also detected in the subcutaneous nerve bundles showed (FIG. 6).

Niemann-Pick Disease (NP)

In NP, the storage of sphingomyelin in the cells may, or may not, be accompanied by sphingomyelinase deficiency. The most common form of Niemann-Pick disease is the infantile form A. Forms B, C and D, E are less common. Skin biopsies obtained from the two patients type B and C showed small zebra bodies, with wavy parallel line of lipid, which often appear as washed-out inclusions (FIG. 7) stored often in endothelial cells, pericytes, fibroblasts, histocytes, Schwann

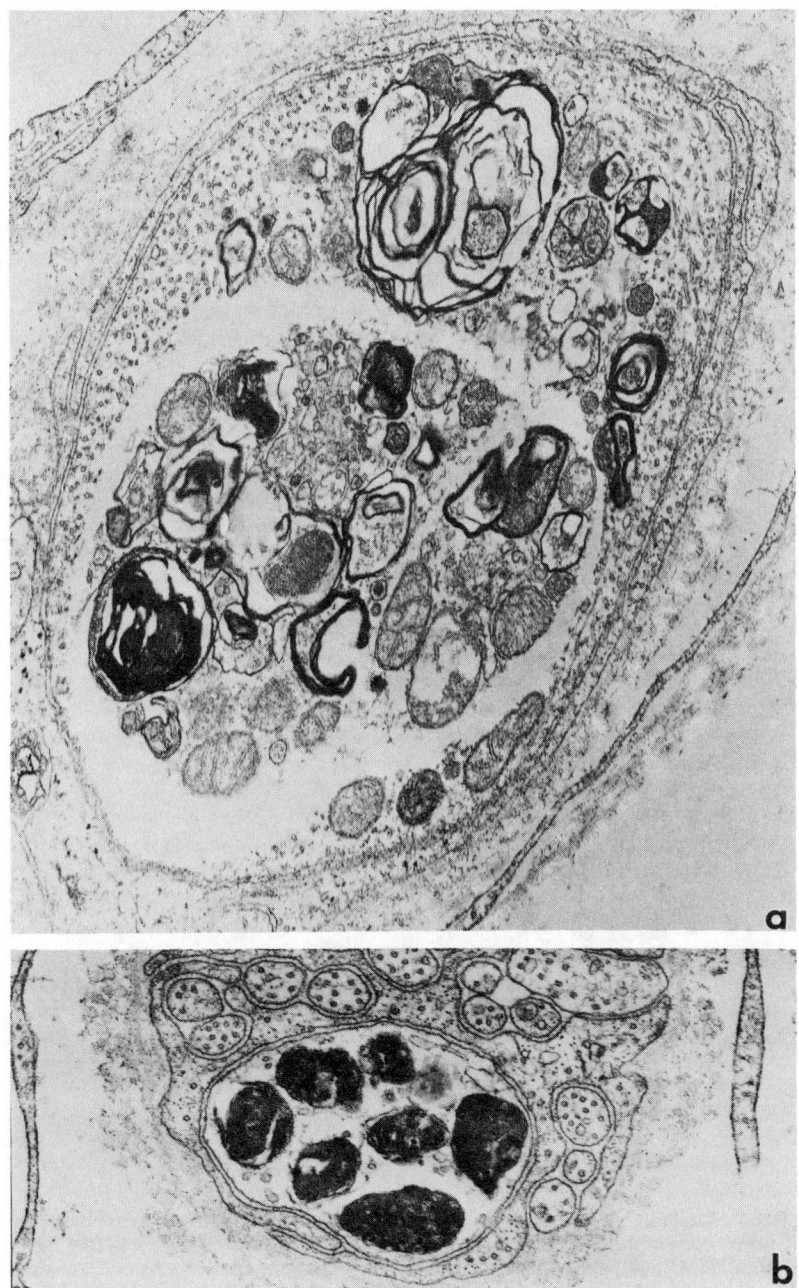

FIGURE 4. Tay-Sachs disease: **(a)** membranous inclusion bodies in the axons of the subcutaneous nerve bundles (×35,000); **(b)** axons with electron-dense bodies (×50,000). Original magnifications shown in parentheses; reproduced here at 70% original size.

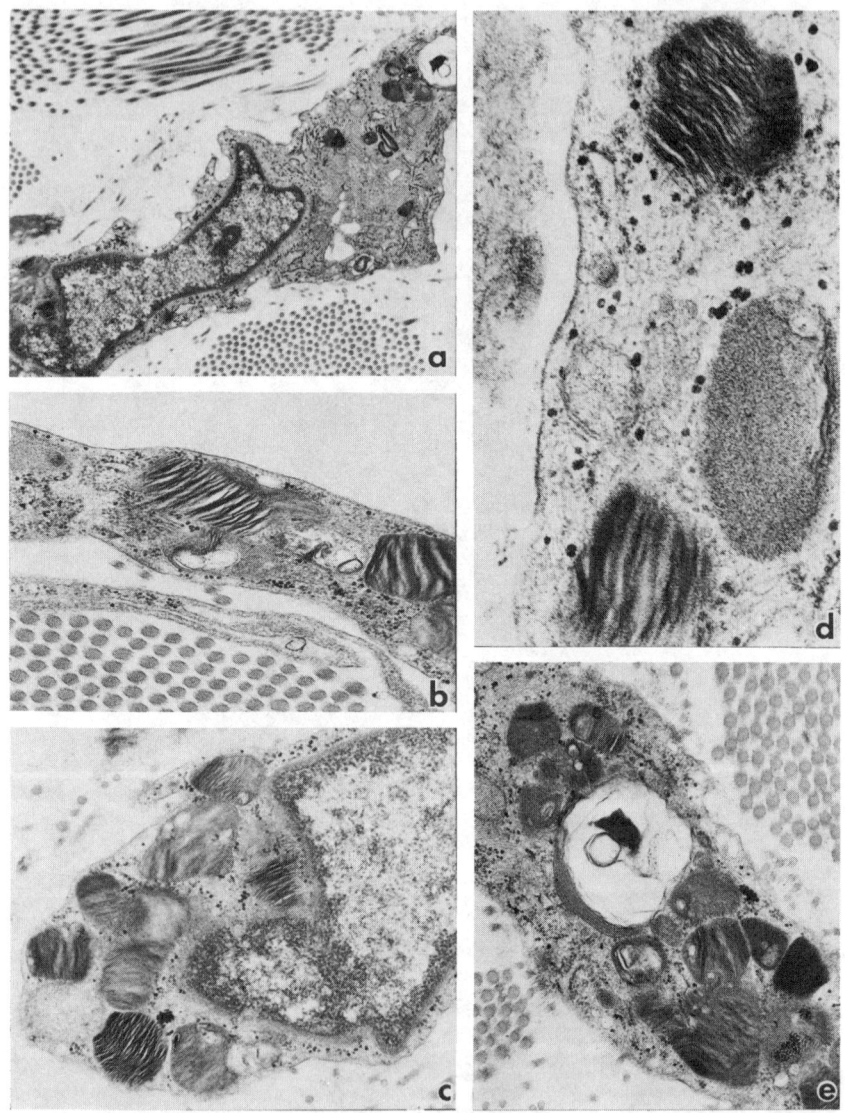

FIGURE 5. Sandhoff's disease, zebra body–type of inclusions with different sizes, direction of stripes seen in the cytoplasm of five different fibroblasts. **a,** (×9,700); **b,** (×35,000); **c,** (×19,400); **d,** (×85,500); **e,** (×35,000). Original magnifications shown in parentheses; reproduced here at 60% original size.

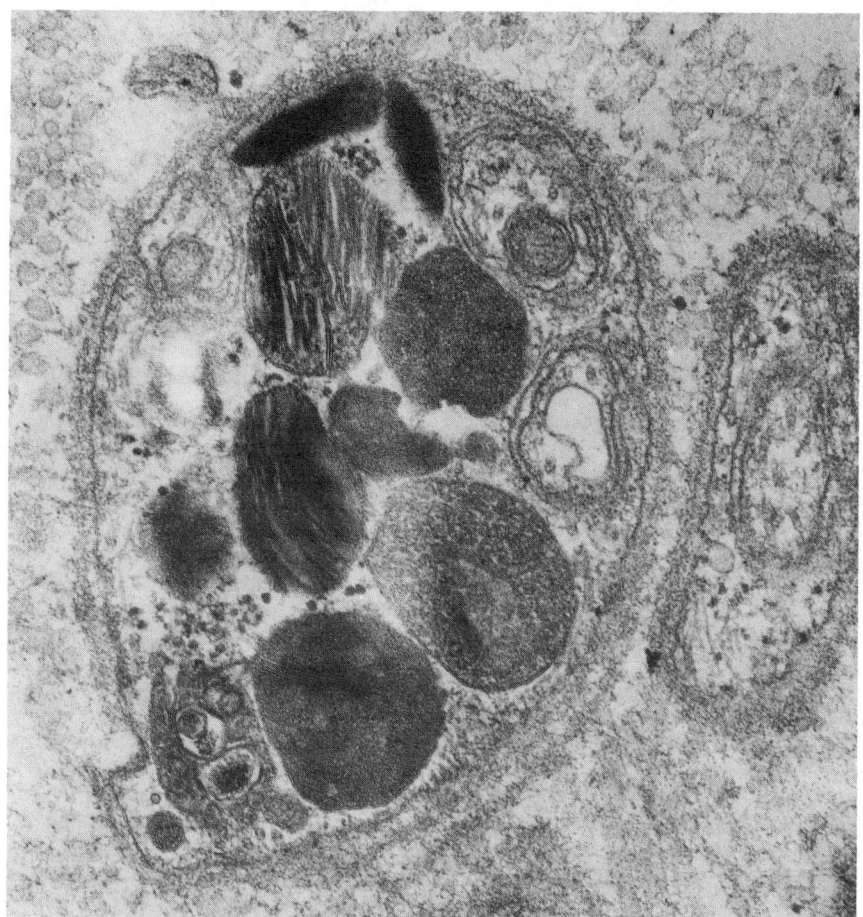

FIGURE 6. GM$_1$ gangliosidosis, subcutaneous nerve bundles filled with zebra bodies, granular, esmophilic inclusion bodies and vacuoles. Original magnification ×75,000; reproduced here at 75% original size.

cells, and smooth muscle of sweat glands. The axons contained dense, as well as zebra-like bodies. Histocytes contained large electron-dense inclusions within which numerous tiny, membrane-bounded vacuoles, multivesicular bodies and washed-out, unrecognizable inclusions were found.

Krabbe's Disease (KD) or Globoid Leukodystrophy

KD, which is caused by β-galactosidase deficiency, afflicted two of the patients in our study. Inclusions consisting of electron-lucent, needle-shaped irregular crystals were observed sometimes in Schwann cells and in dystrophic axons of skin biopsies; however, the lymphocytes were found to be normal.

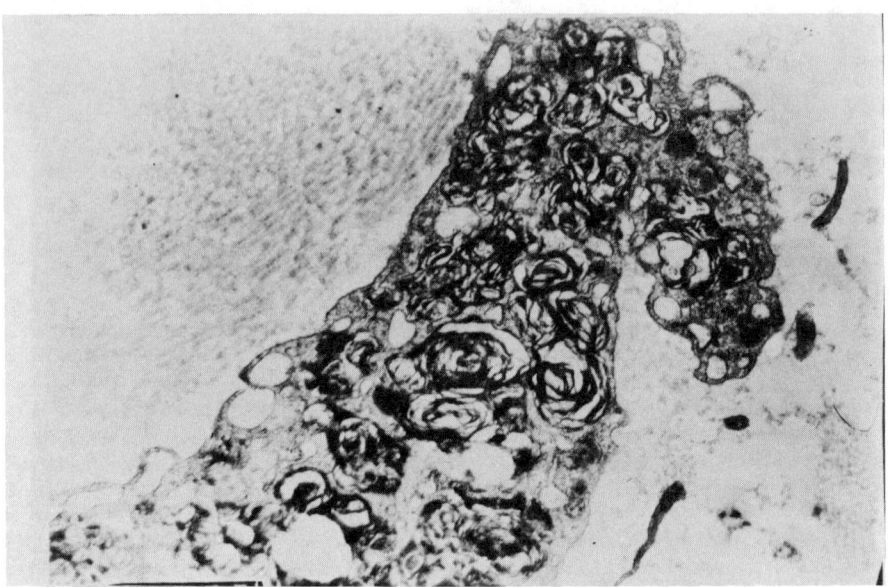

FIGURE 7. Niemann-Pick disease, type C multivesicular bodies and washout membranous inclusion bodies. Original magnification ×80,000; reproduced here at 70% original size.

Fabry's Disease (FD)

FD, the angiokeratoma corporis diffusum, is a sex-linked disorder where trihexosyl ceramide (TC) storage results from TC galactosyl hydrolase deficiency. In this one case study, small, laminated lipid bodies appeared in endothelial cells, pericytes, smooth muscle, fibroblasts, and perineurium. The lymphocytes did not reveal any abnormalities.

Metachromatic leukodystrophy (MLD)

In MLD, Tuffstone bodies or "herringbone" deposits or their derivatives (FIG. 8) were observed in Schwann cells and in histocytes of nerve fasicles of tissue from three patients (FIG. 8a,b,c). These Tuffstone bodies are the accumulation of sulfatide-linked lipid resulting from arylsulfatase A deficiency. Myelin degradation products could be seen in macrophages and Schwann cells. The lymphocytes, in three cases, were normal.

Category Four

Category four includes glycogenosis and its variants.

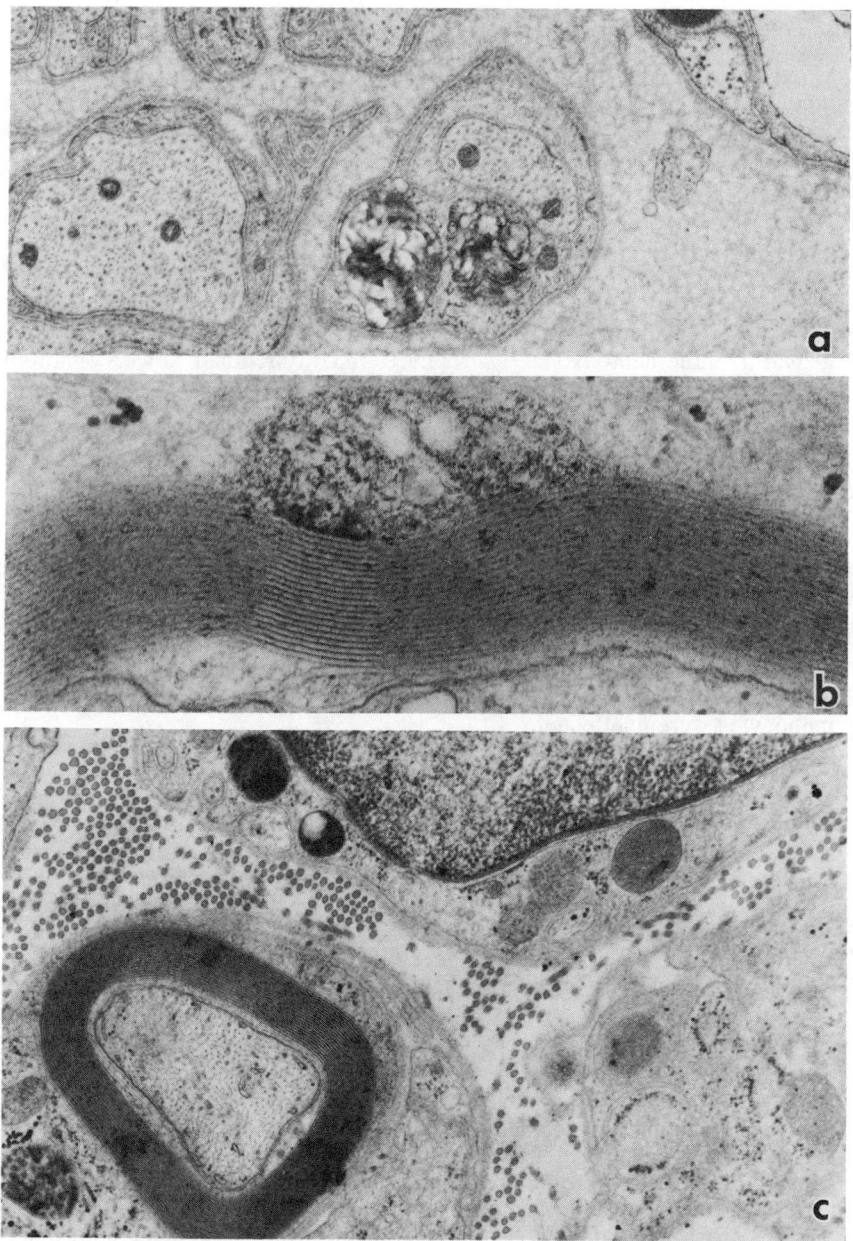

FIGURE 8. Metachromatic leukodystrophy: **(a)** subcutaneous nerve bundles with Tuff-stone–like inclusion bodies (×20,400); **(b)** bubble inclusion in the myelinated axons and Schwann cells (×84,000); **(c)** lipofuscin or dense bodies in the Schwann cells (×28,000). Original magnifications shown in parentheses; reproduced here at 75% original size.

Glycogenosis II—Pompe's Disease (P)

Four patients with this disorder were included in the patient sample used in this study. Acid maltase deficiency was detected in the leukocytes. Furthermore, extensive storage of membrane-bounded glycogen was observed in all of the different skin cells and in lymphocytes (FIG. 9a,b).

Category Five

Category five includes an unclassified, miscellaneous group of storage diseases. Of the 13 patients in this category, four had adrenoleukodystrophy (ALD), four had infantile neuroaxonal dystrophy (INAD), three had Lowe's syndrome (LS), and two were suspected to have new lipid storage disease (NLSD).

Adrenoleukodystrophy (ALD)

Two of the four patients with this disorder were clinically presented as having adrenomyeloneuropathy (AMN). One of these patients was a developmentally disabled child with initial cerebellar signs. Since the missing lysosomal enzyme is unknown in these cases, diagnosis was based on the abnormal volume of the long-chain fatty acids[23-26] in the plasma and fibroblasts. Needle-like inclusions were seen in the Schwann cells of only two of the four patients' skin biopsies. An increase of melanin pigment was observed in three of these tissue samples. The lymphocytes were unaffected.

Infantile Neuroaxonal Dystrophy (INAD)

In all four patients' biopsies, spheroids were filled with filamentous material (FIG. 10), and giant mitochondria were observed in some dystrophic axons. Other skin cells were unaffected. The lymphocytes were not affected. Clinical signs of involvement of the entire neuroaxis were present. A biochemical marker for the disorder has not yet been identified.

Lowe's Syndrome (LS)

Membrane-bounded vacuoles similar to those seen in the MPS group were observed in the fibroblasts of biopsies from the three LS patients. The lymphocytes, in these cases, were unaffected. These patients had had classical signs of progressive neurological dysfunction since late childhood. The skin-cell samples from these patients contained vacuoles, some electron-dense membranous inclusions, as well as axonal and vascular changes. The lysosomal enzyme studies were negative. Biochemical diagnosis could not be established, but urine MPS was found to be elevated four to five times the normal controls. The increased MPS was a low-sulfated chondroitin-4-sulfate.

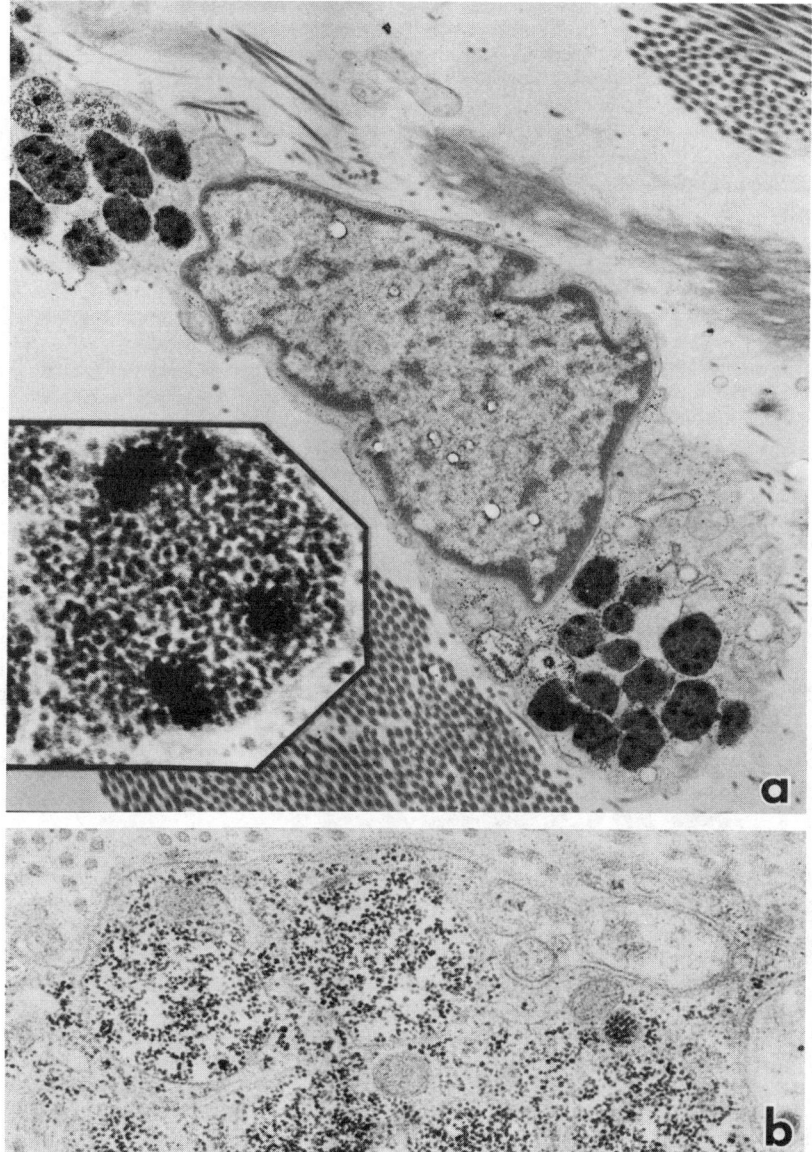

FIGURE 9. Pompe's disease: **(a)** Accumulation of glycogen in the fibroblasts (×9,700), higher magnification (×55,000); **(b)** in the subcutaneous nerve bundles (×20,000). Original magnifications shown in parentheses; reproduced here at 90% original size.

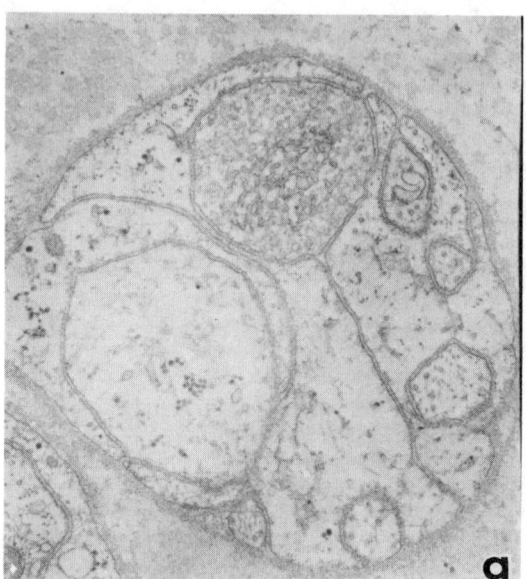

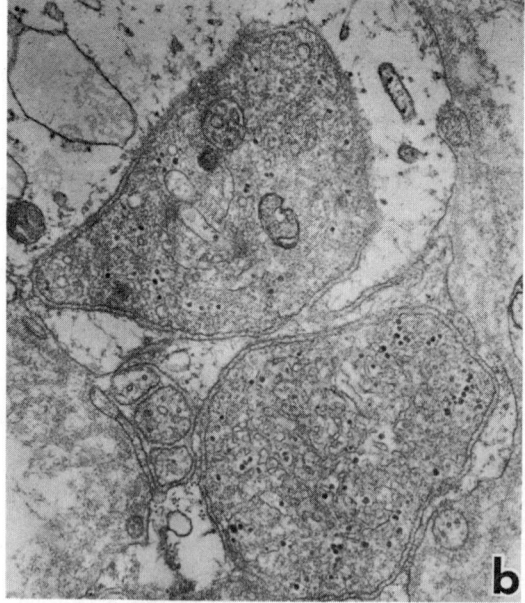

FIGURE 10. Spheroids in INAD. Unmyelinated nerve bundles. **(a)** Early accumulation of vesiculotubular profiles (×30,000); **(b)** more advanced changes in the spheroids (×35,000). Original magnifications shown in parentheses; reproduced here at 85% original size.

Suspected New Type of Lipid Storage Disease (NLSD)

Of the two patients in this subset of disorders, one seems to have a disorder of the Gaucher's disease variety, while the other has an unspecified variety of this new type of lipid storage disease. In the skin biopsy from the latter case, membranous and lipofuscin types of inclusion bodies were found in the cytoplasm of several types of cells. Both of these cases were studied intensively biochemically. In summary, ultrastructural studies corroborated the clinical diagnosis in all 65 of the patients in this group. The lysosomal enzyme studies were positive in 29 of the 65 (45%) of the cases. This would appear to demonstrate the value of ultrastructural analyses in this group of cases. A summary of the ultrastructural findings are given in TABLE 6.

TABLE 6. Cytoplasmic Inclusion Bodies Observed in Ultrastructural Studies of Skin-Punch Biopsies of 65 Cases of Neurodegenerative Disease

Lysosomal Storage Disease	Type of Inclusion Bodies
Neuronal ceroid lipofuscinosis	
Early infantile	Granular
Late infantile	Curvilinear
Juvenile	Fingerprint, rectilinear, lipofuscin
Mucopolysaccharidoses	
MPS I, II	Membrane-bound vacuoles and indentation of epidermal cell
MPS III (Sanfilippo's disease)	Membrne-bound vacuoles and indentation of epidermal cell, zebra bodies
Mucolipidosis II (I-cell disease)	Membrane-bound vacuoles, closely packed dense rings
Mucolipidosis IV	Small dense lipid membranous cytoplasmic and zebra bodies
Fucosidosis	Membrane-bound vacuoles, lamellar inclusions, lipofuscins
Sialidosis	Membrane-bound vacuoles, lamellar inclusions, lipofuscins
Lipidoses	
Tay-Sachs disease	Membranous inclusions in the subcutaneous nerve bundles
Sandhoff's disorder	Small laminated zebra bodies
GM, gangliosidosis infantile form (pseudo-Hurler's disease)	Membrane-bound vacuoles Zebra bodies
Niemann-Pick disease Types B and C	Small zebra bodies, wavy parallel lines of lipid with a washed-out appearance
Krabbe's disease (Globoid leukodystrophy)	Needle-shaped irregular crystals
Fabry's disease	Small, laminated lipid bodies
Metachromatic leukodystrophy	Tuffstone bodies or "herringbone" deposits
Glycogenosis	
Pompe's disease	Membrane-bound glycogen deposits
Other, unclassified	
Adrenoleukodystrophy	Needle-shaped, nerve bundles
Infantile neuroaxonal dystrophy	Spheroids filled with filamentous material; giant mitochondria, nerve bundles
Lowe's syndrome	Membrane-bound vacuoles, electron-dense membranes
New types of lipid storage disease	Membranous and electron-dense lucent inclusions

Group II

Group II included 53 patients suspected of having lipid storage disease and were categorized according to specific clinical syndromes or diseases (TABLE 4). Ultrastructural studies were nonspecific or equivocal in 20 of the 53 patients and were negative in the remainder. Lysosomal enzyme studies, performed to identify different variants of lipid storage diseases, were negative in all 53 patients.

Group III

Group III included 40 patients who initially were diagnosed as having undefined chronic encephalopathy (TABLE 5). After an intensive neurological follow-

TABLE 7. Markers of Neurodegenerative Diseases

Case history	Regression most common in first decade. Often similar problems in some other members of the family. Consanguinity may be (+).
Examination	Progressive neurological and psychological dysfunction. Signs of involvement: upper motor, lower motor neurons, or one of these, or whole neuroaxis or multisystemic.
Cause	Genetic Nongenetic Unknown
Topography	Nervous system involvement. A. Gray matter. B. White matter. C. Peripheral system. D. Extraneuronal involvement—coarse features, hepatosplenomegaly, dysostosis multiplex
Ultrastructural	Cytoplasmic inclusion bodies. See TABLE 6 for specifics.
Neurophysiological	Abnormalities increase with the duration of the disease. A. Electroencephalogram B. Evoked potentials C. Nerve conduction velocities D. Electromyogram E. Electroretinogram F. Visual evoked responses
Neuroradiological	Progressive brain atrophy, white-matter degeneration shown on CTT, NMR images that increase with duration of the disease.
Radiological	Skeletal survey including spine and skull.

Biochemical	A. Urine studies:
	  1. Quantitative study for mucopolysaccharidoses (MPS).
	  2. Dolichol excretion.
	  3. Thin-layer chromatography (for oligosaccharidoses).
	  4. High-pressure liquid chromatography for lipid analysis.
	B. Lysosomal enzyme studies with substrates:
	  1. Fluorigenic, chromagenic, and artificial.
	  2. Natural, radioactive labeled.

up, 29 were diagnosed as having static encephalopathy, reducing the number of NDD patients who could be characterized as having progressive neurological disease from the original 158 to 129. Ultrastructural studies were negative in 31 of the 40 cases, and were equivocal in the remaining nine. Lysosomal enzyme studies were negative in all 40 cases.

In summary, all the 129 cases with NDD had the characteristic clinical histories, neuropsychological, electrophysiological, neuroradiological, and biochemical findings (see TABLE 7).

COMMENTS

Our studies show that ultrastructural examination of skin-punch biopsies and buffy-coat fractions are valuable methods for diagnosing NDD resulting from lysosomal storage diseases but not for nonlysosomal storage diseases.

In this study we examined 158 patients suspected of having NDD. Intensive clinical evaluations led us to classify 65 of the cases (Group I), suspected strongly for lysosomal storage disease, as manifesting a definitive chronic progressive encephalopathy (CPE), 53 (Group II) as having a definite but atypical form of lysosomal storage disease CPE, and 40 (Group III) as having an undefined CPE. Characteristic abnormal inclusions were observed in biopsy samples in all 65 CPE cases (TABLES 3 and 6). Biochemical, lysosomal abnormalities were found only in 29/65 cases in tissue samples of patients from Group I. Neither ultrastructural nor biochemical studies of skin biopsies, taken from patients from Group II (53 patients) and Group III (40 patients) provided any useful diagnostic information.

The largest group of cases with specific ultrastructural abnormalities within Group I, category one, was NCL (constituting 20 of the 65 patients in this group). Our studies, as well as others,[14–22] point to the presence of pathogenic lysosomal inclusion bodies in the biopsy material.

Clinical and ultrastructural studies may not be specific for some of the diseases mentioned above, not even for some of the storage diseases that are well defined biochemically. For instance, in three siblings in their late 30s, we found electrophysiological and ultrastructural abnormalities resembling those of NCL, while biochemical assays indicated MPS III.[24] In yet another case, two siblings were diagnosed as having NCL while the clinical presentation suggested spinocerebellar degeneration, without dementia or visual problems.[25] A clinical presentation of spinocerebellar degeneration was described previously in cases that were eventually diagnosed as adrenomyeloneuropathy, gangliosidosis, glutamate dehydrogenase deficiency, and Krabbe's disease.[26–31] Lowe's syndrome, previously thought to be due to aminoacidopathy, now appears to be due to abnormal metabolism of MPS.[32,33]

Different clinical presentations associated with different CNS topographies may have the same lysosomal enzyme deficiency. Among the different lysosomal storage diseases, some of the factors described for gangliosidosis may play an important role in explaining why disorders with the same lysosomal enzyme deficiency have different clinical presentations. There is speculation that gangliosidosis and other lysosomal storage diseases have different receptor proteins and different neurotransmitter biosynthetic pathways in different nerve cells.[11] That may explain why the same enzyme deficiency may be responsible for different clinical onsets and symptoms related to different CNS topography in affected individuals.

The fifth category of disorders in Group II, the unclassified storage diseases, have clearly defined ultrastructural abnormalities and, although the particular enzyme deficiency is unknown, some biochemical markers do exist.[34] For instance, long-chain fatty-acid accumulation in ALD; increased levels of MPS in the urine in Lowe's syndrome, increased GD_{1a} ganglioside neuraminidase in ML IV,[35] and elevated urinary levels of dolichol in NCL.[23] Ultrastructural examination of skin biopsies were instrumental in identifying the abnormalities characteristic of INAD.[36,37] The NLSDs, the suspected new lipid storage diseases, require further biochemical studies to determine their specific abnormalities.

For some of the NDDs (Groups I–III, TABLES 3–5), where the lysosomal enzyme or the biochemical marker is unknown, the development of molecular biological techniques—using DNA probes to quantify studies of restriction-fragment-length polymorphisms closely linked to gene abnormalities—may better define their cause, as we noted above.[38,39] A better definition for the cause of NDD also requires a multidisciplinary approach provided by the cooperation of clinicians, pathologists, molecular geneticists, and biochemists.

Our investigations indicate that ultrastructural studies of skin-punch biopsy are useful for the screening of children and adults suffering from progressive encephalopathy. This method should be applied only in cases where an inborn error of metabolism, especially storage disease, is suspected. Although biochemical analysis of some NDDs is the simplest technique for diagnosing most of the progressive neurological diseases, because a molecular marker is not available for many of these disorders, ultrastructural studies of skin biopsy and buffy-coat fractions provide an important diagnostic tool, particularly where the clinical presentation is also atypical. The data in this study show that ultrastructural studies corroborated the diagnosis in all 65 cases where a definitive clinical diagnosis had been made, whereas biochemical confirmation occurred in 29 of the 65 (45%) cases. It should also be pointed out that where a definitive clinical diagnosis cannot be made, the ultrastructural studies were at best equivocal, and all of the biochemical assays are negative; in other words, we have much more to learn about diagnosing neurodegenerative diseases.

ACKNOWLEDGMENTS

We wish to thank Fred Connell for technical support, Lawrence Black for bibliographical assistance, Dr. Donald Snider and Lea Soifer for editorial assistance, and Patricia Codoner for secretarial assistance.

REFERENCES

1. DYKEN, P. & N. KRAWLECKI. 1983. Neurodegenerative diseases of infancy and childhood. Ann. Neurol. **13:** 351–364.
2. KATZMAN, R. 1981. Early detection of senile dementia. Hosp. Pract. **16:** 61–76.
3. O'BRIEN J. S., J. BERNETT, M. L. VEATH & D. PAA. 1975. Lysosomal storage diseases. Arch. Neurol. **32:** 592–599.
4. DOLMAN C. L., P. M. MACLEOD &. E. CHANG. 1975. Skin-punch biopsies and lymphocytes in the diagnosis of lipidoses. Can. J. Neurol. Sci. **2:** 67–73.
5. MARTIN, J. J., C. CEUTERICK & J. G. LEROY. 1976. Contribution de la biopsie cutanee au diagnostic des encephalopathies metaboliques. Rev. Neurol. Paris **132**(9): 639–651.
6. MARTIN, J. J. & C. CEUTERICK. 1978. Morphological study of skin biopsy specimens:

A contribution to the diagnosis of metabolic disorders with involvement of the nervous system. J. Neurol. Neurosurg. Psychiatry **41:** 232–248.

7. CHAMOLES, N. A., J. E. MANZITTI & A. L. TARATUTO. 1979. Diagnostico de enfermedades metabolicas mediante el estudio enzimatico ultraestructural de la conjuntiva. Palestra Oftalmol. Panamer. **3**(2): 83–89.

8. YAMANO, T., M. SHIMADA, S. OKADA, T. YUTAKA, H. YABUUCHI & Y. NAKAO. 1979. Electron microscopic examination of skin and conjunctival biopsy specimens in neuronal storage diseases. Brain Dev. **1:** 16–25.

9. ARSENIO-NUNES, M. L., F. GOUTIERES & J. AICARDI. 1981. An ultramicroscopic study of skin and conjunctival biopsies in chronic neurological disorders of childhood. Ann. Neurol. **9:** 163–173.

10. DOLMAN, C. L. 1984. Diagnosis of neurometabolic disorders by examination of skin biopsies and lymphocytes. Semin. Diagn. Pathol. **1:** 82–97.

11. SANDHOFF, K. & E. COTZELMAN. 1984. The biochemical basis of gangliosidoses. Neuropediatrics **15:** 85–92.

12. ZELICKSON, A. S. 1967. Ultrastructure of Normal and Abnormal Skin. Lea and Febiger. Philadelphia.

13. SIPE, J. C. & J. S. O'BRIEN. 1979. Ultrastructure of skin biopsy specimens in lysosomal storage diseases: Common sources of error in diagnosis. Clin. Genet. **15:** 118–125.

14. CARPENTER, S., G. KARPATI & F. ANDERMANN. 1972. Specific involvement of muscle, nerve, and skin in late-infantile and juvenile amaurotic idiocy. Neurology **22:** 170–186.

15. WITZLEBEN, C. L., K. SMITH, J. S. NELSON, P. L. MONTELEONE & D. LIVINGSTON. 1971. Ultrastructural studies in late-onset amaurotic idiocy: Lymphocyte inclusions as a diagnostic marker. J. Pediatr. **79:** 285–293.

16. WITZLEBEN, C. L. 1972. Lymphocyte inclusions in late-onset amaurotic idiocy. Value as diagnostic test and genetic marker. Neurology **22:** 1075–1078.

17. MACLEOD, P. M., C. L. DOLMAN, E. CHANG & S. PANG. 1981. Ultrastructural studies of skin biopsies and lymphocytes in neurodegenerative disease. J. Neuropathol. Exp. Neurol. **40:** 360. (Abstract)

18. ZEMAN, W. 1976. The neuronal ceroid lipofuscinoses. In Progress in Neuropathology, Vol. III. H. M. Zimmerman, Ed.: 203–223. Grune and Stratton. New York.

19. GOEBEL, H. H., W. ZEMAN & H. PILZ. 1975. Significance of muscle biopsies in neuronal ceroid lipofuscinosis. J. Neurol. Neurosurg. Psychiatry **38:** 985–993.

20. STEKHOVEN, J. H., U. J. VAN HAELST, E. M. JOOSTEN & M. C. LOONEN. 1977. Ultrastructural study of the vacuoles in the peripheral lymphocytes in juvenile amaurotic idiocy. Juvenile form of generalized ceroid lipofuscinosis. Acta Neuropathol. **38:** 137–142.

21. ARSENIO-NUNES, M. L. & F. GOUTIERES. 1975. An ultramicroscopic study of the skin in the diagnosis of the infantile and late-infantile types of ceroid-lipofuscinosis. J. Neurol. Neurosurg. Psychiatry **38:** 994–999.

22. CEUTERICK, C., J. J. MARTIN, P. CASAER & G. W. EDGAR. 1976. The diagnosis of infantile generalized ceroid lipofuscinosis (type Hagberg Santavuori) using skin biopsy. Neuropaediatrie **7:** 250–260.

23. WOLFE, L. S., N. M. NG YING KIN, J. PALO & M. HALTIA. 1983. Dolichols in brain and urinary sediment in neuronal ceroid lipofuscinosis. Neurology **33:** 103–106.

24. WISNIEWSKI, K., R. RUDELLI, M. LAURE-KAMIONOWSKA, S. SKLOWER, G. E. HOUCK, JR., F. KIERAS, P. RAMOS, H. M. WISNIEWSKI & H. BRAAK. 1985. Sanfilippo disease, type A with some features of ceroid lipofuscinosis. Neuropediatrics **16:** 98–105.

25. WISNIEWSKI, K., M. DAMBSKA, I. RAPIN, R. PULLARKAT, S. SKLOWER, R. MADRID & H. M. WISNIEWSKI. 1985. Two cases of slowly progressive degeneration with polyneuropathy associated with ceroid lipofuscinosis in one family. Ann. Neurol. **18:** 400. (Abstract)

26. MARSDEN, C. D., J. A. OBESO & A. E. LANG. 1982. Adrenoleukomyeloneuropathy presenting as spinocerebellar degeneration. Neurology **32:** 1031–1032.

27. ROSEN, N. L., R. LECHTENBERG, K. WISNIEWSKI & R. PULLARKAT. 1985. Adreno-leukomyeloneuropathy with onset in early childhood. Ann. Neurol. **17:** 311–312.
28. RAPIN, I., K. SUZUKI, K. SUZUKI & M. P. VALSAMIS. 1976. Adult (chronic) GM_2 gangliosidosis. Atypical spinocerebellar degeneragion in a Jewish sibship. Arch. Neurol. **33:** 120–130.
29. WILLNER, J. P., G. A. GRABOWSKI, R. E. GORDON, A. N. BENDER & R. J. DESNICK. 1981. Chronic GM_2 gangliosidosis masquerading as atypical Friedreich ataxia: Clinical, morphologic, and biochemical studies of nine cases. Neurology **31:** 787–798.
30. PLAITAKIS, A., W. J. NICKLAS & R. J. DESNICK. 1980: Glutamate dehydrogenase deficiency in three patients with spinocerebellar syndrome. Ann. Neurol. **7:** 297–303.
31. THOMAS, P. K., J. P. HALPERN, R. H. M. KING & D. PATRICK. 1984. Galactosylceramide lipidosis: Novel presentation as a slowly progressive spinocerebellar degeneration. Ann. Neurol. **16:** 618–620.
32. WISNIESKI, K. E., F. J. KIERAS, J. H. FRENCH, G. E., HOUCK, JR. & P. L. RAMOS. 1984. Ultrastructural, neurological and glycosaminoglycan abnormalities in Lowe's syndrome Ann Neurol **16:** 140–49.
33. KIERAS, F. J., G. E. HOUCK, JR., J. H. FRENCH & K. E. WISNIEWSKI. 1984. Low-sulfated glycosaminoglycans are excreted in patients with the Lowe syndrome. Biochem. Med. **31:** 201–210.
34. MOSER, H. W., A. B. MOSER, K. K. FRAYER, W. CHEN, J. D. SCHULMAN, B. P. O'NEILL & Y. KISHIMOTO. 1981. Adrenoleukodystrophy-increased plasma content of saturated very long chain fatty acids. Neurology **31:** 1241–1249.
35. BEN-YOSEPH, Y., T. MOMOI, L. C. HAHN & H. L. NADLER. 1982. Catalytically defective ganglioside neuraminidase in mucolipidosis IV. Clin. Genet. **21:** 374–381.
36. WISNIEWSKI, K. & H. M. WISNIEWSKI. 1980. Diagnosis of infantile neuroaxonal dystrophy by skin biopsy. Ann. Neurol. **7:** 377–379.
37. MARTIN, J. J., J. G. LEROY, J. LIBERT, M. VAN EYGEN & N. LOGGHE. 1979. Skin and conjunctival biopsies in infantile neuroaxonal dystrophy. Acta Neuropathol. (Berlin) **45:** 247–251.
38. ROWLAND, L. P. 1983. Molecular genetics, pseudogenetics, and clinical neurology. The Robert Wartenberg Lecture. Neurology **33:** 1179–1195.
39. KOLODNY, E. H. & S. YATZIV. 1985. Laboratory approaches for inherited neurometabolic diseases. Dev. Med. Child. Neurol. **27:** 252–257.

DISCUSSION

W. TED BROWN: How often do you see pathologic change in the mitochondria in skin-punch biopsies and buffy coat?

K. E. WISNIEWSKI: Mitochondrial pathology is commonly seen in dystrophic axons, especially in neurometabolic disease when the whole neuroaxis is involved. We see mitochondrial pathology in lysosomal storage disease, such as neuronal ceroid lipofuscinosis.

PHILLIP SWENDER: A couple of years ago, we supplied the clinical material for a research project studying the heterozygote state with cystic fibrosis using skin-punch biopsies—a technique of pyrolysis/gas chromatography of cultured skin fibroblasts. That might be a logical technique to apply to these conditions, but I'm not aware whether it has ever been done. This method involves "pyrolyzing" the cultured fibroblasts and making a chromatograph and then analyzing curve differences in the chromatographs. This has been done with Tay-Sachs disease, but I do not know whether it has been done with any of these other conditions.

KRYSTYNA WISNIEWSKI: We have not performed this type of chromatography with cultured skin fibroblasts. Our ultrastructural studies mainly look, not at the skin fibroblast culture, but at the fibroblast immediately after embedding; this is done to exclude nonspecific changes forming after artificial nutrition. We use ultrastructural studies to help to diagnose suspected lysosomal storage disease with atypical clinical presentation which has many biochemical variants, for example, Lowe's syndrome and some types of MPS.

MARIA MICHEJDA: Important advances have been made in antenatal diagnosis. Fetal skin biopsy via fetoscopy has become a very useful technique in detecting many fetal problems. The technique is very similar to skin-punch biopsies. Was it ever done for early antenatal diagnosis of neuronal axonal dystrophy?

KRYSTYNA WISNIEWSKI: Neuronal ceroid lipofuscinosis has been diagnosed prenatally, but infantile neuroaxonal dystrophy has not. In both conditions the biochemical markers are still unknown, which is why ultrastructural studies might be very helpful in prenatal diagnosis.

Early Detection of Lysosomal Storage Diseases

EDWIN H. KOLODNY[a]

Biochemistry Department
Eunice Kennedy Shriver Center
for Mental Retardation, Inc.
Waltham, Massachusetts 02254

Department of Neurology
Massachusetts General Hospital
Boston, Massachusetts 02114

The lysosomal storage diseases have attracted considerable interest in the field of developmental disabilities. They comprise more than 40 diseases with a combined incidence of at least 1 per 5,000 births.[1] The majority of these disorders seriously impair central nervous system functioning. Their biochemical basis, with only a few exceptions, has been clearly elucidated. Their genetics are well understood and prevention is feasible. Possibilities for treatment exist through enzyme therapy, organ transplantation, and gene surgery; however, definitive treatment is not yet available for any of these diseases so that they remain the focus of an intense and widespread research effort.

Early detection of the lysosomal storage diseases clarifies the prognosis, defines the measures that can be taken for habilitation and, most importantly, enables the patient's family to take preventative action in order to avert the birth of subsequent children who are similarly affected. However, the diagnosis of these diseases is by no means straightforward because of considerable clinical and biochemical heterogeneity and the need for specialized laboratory testing. Their recognition at an early stage is difficult because the time of onset is variable and different for each of the lysosomal storage diseases and the progressive and degenerative nature of the condition is only recognized after a period of time has elapsed. This presentation will discuss measures the clinician may take to confirm the impression of a lysosomal storage disease, some of the newer techniques used for early detection, and the problems and pitfalls encountered in arriving at an exact diagnosis.

The concept of the lysosome was introduced in 1955 by deDuve and his coworkers. They found, by tissue fractionation, that they could isolate a membrane-bound intracellular organelle that was rich in lytic enzymes with maximal activity in a pH range of 4–5.[2] Hers, in 1963, identified the cause of Pompe's disease as a deficiency of the lysosomal enzyme acid α-glucosidase.[3] This first description of a human lysosomal enzyme defect established the criteria for a new category of inherited metabolic disease. Common features are (1) the accumulation of one or several related types of macromolecules within lysosomes, (2) distortion of the cellular architecture by the stored material, and (3) a severe deficiency in the activity of a specific catabolic enzyme normally present within the lysosomal membrane.

[a] Address for correspondence: Eunice Kennedy Shriver Center, 200 Trapelo Road, Waltham, MA 02254.

Lysosomal enzymes act as glycosidases, proteases, phosphatases, sulfatases, and lipases. Their natural substrates are glycolipids, glycoproteins, mucopolysaccharides, mucolipids, oligosaccharides, glycogen, and fatty-acid esters of cholesterol and triglyceride. The enzymes are glycoproteins composed of one or more polypeptide chains of 25–100K daltons. Synthesis of the lysosomal enzyme takes place within the endoplasmic reticulum where the nascent peptide acquires short chains of sugar. It then passes on to the Golgi apparatus where additional post-translational processing occurs including further glycosylation as well as phosphorylation and proteolysis. In cultured cells, there are specific receptor sites that internalize lysosomal enzyme present in the culture media. Lysosomal enzymes containing the recognition marker, mannose-6-phosphate, are able to bind to the receptor site and reenter the lysosome.

TABLE 1 lists some examples of different types of molecular defects causing deficient activity of lysosomal enzymes. Enzyme protein that is abnormal tends to be rapidly removed by proteolysis so that whether or not any enzyme is produced, measurements of enzyme activity show a uniformly low level. Enzymatic hydrolysis of certain of the sphingolipids depends upon the presence of an additional factor, a nonenzymatic helper protein that combines with the lipophilic substrate to improve its binding with the enzyme.[4] In the case of activator deficiencies, test-tube assays for the activity of these enzymes may be normal; thus, precision in the diagnosis of these diseases depends ultimately upon demonstrating the exact nature of the molecular defect. However, for most lysosomal diseases, we can provide adequate clinical management and genetic counseling without completely defining the molecular basis for the enzyme deficiency.

We know from examining fetuses with various forms of lysosomal storage disease that the process of intracellular storage may begin early in fetal life. The organs affected are those that normally have the largest concentration of the nondegradable material. For example, brain, particularly gray matter, is rich in gangliosides while only trace amounts of these compounds are found outside the central nervous system; therefore, defective hydrolysis of ganglioside such as occurs in the G_{M1}- and G_{M2}-gangliosidoses results primarily in signs of central nervous system dysfunction. On the other hand, sphingomyelin, an important membrane lipid, is present in cells throughout the body so that when its hydrolysis is blocked, as in Niemann-Pick disease, it accumulates not only in brain but in many different organs of the reticuloendothelial system.

Clinical signs may not appear until the amount of stored material reaches a critical threshold within the cells of a particular region of the brain or other vulnerable organs. At this point, the cells no longer function properly and may be destroyed. Outside the nervous system, new parenchymal cells may for a time be generated but neurons that are lost to the storage process cannot be replaced.

TABLE 1. Molecular Defects in Lysosomal Enzymes

1. Complete absence of enzyme protein (CRM[a] negative, no mRNA).
2. Enzyme protein present but catalytic activity deficient (CRM positive, mRNA present).
   a. abnormal K_m or V_{max}
   b. common subunit defective (ex. Sandhoff variant of G_{M2}-gangliosidosis)
   c. subunit association defective (ex. adult-onset G_{M2}-gangliosidosis)
   d. processing defect in recognition marker (ex. I-cell disease)
   e. sphingolipid activator protein deficient (ex. AB variant of G_{M2}-gangliosidosis)

[a] CRM = cross-reacting material.

Consequently, the brain cortex shrinks and is replaced by gliosis and the subtending white matter loses its axons. Similarly, in a leukodystrophy, axons are laid bare of their myelin and undergo secondary degeneration. As a result, connections are severed between the brain cortex and structures in deeper gray matter, the brain stem and spinal cord.

We become suspicious of a lysosomal storage disease when the child fails to develop or loses milestones after a period of early normal development. This is a diagnosis we also consider when we encounter progressive coarsening of facial features, enlargement of the liver and spleen, or myoclonic seizures. Very often the early signs are subtle and there is a natural inclination to discount them, but at some point an experienced parent, grandparent, babysitter, teacher, or other caretaker becomes concerned because of the much more rapid progress of a younger sibling or playmate of the same age. Sometimes the family is prompted to seek a workup because of a contemplated or actual pregnancy and the desire to prevent the birth of another child with the same disease.

The essential clinical features upon which we rely are listed in TABLE 2. The onset is generally earlier and the progression more rapid in diseases primarily affecting gray matter such as the gangliosidoses and Hurler's disease than in metachromatic leukodystrophy or the oligosaccharidoses. Seizures and early loss of cognitive skills incline one's thinking toward a neuronal storage disease, whereas microcephaly and peripheral neuropathy direct attention to a leukodystrophy.

If examination of the eye discloses corneal clouding, Hurler's disease, Maroteaux-Lamy disease, and mucolipidosis IV must be seriously considered. Atrophy of the optic disc occurs early in the leukodystrophies and later in the course of a gangliosidosis and sialidosis. A cherry-red spot and surrounding white halo in the macula is another feature of these latter two diseases as well as of Niemann-Pick disease type A and some cases of type B. Pigmentary and macula retinal degeneration commonly occur in the neuronal ceroid lipofuscinoses.

Facial dysmorphism and skeletal dysplasia will suggest a mucopolysaccharidosis, oligosaccharidosis, or mucolipidosis. Patients with these diseases may also have an enlarged liver but hepatomegaly and splenomegaly without facial dysmorphism is more suggestive of a sphingolipidosis such as Farber's disease, Gaucher's disease, or Niemann-Pick disease.

TABLE 2. Clinical Features

1. Age of onset
2. Rate of progression
3. Nervous system involvement
gray matter
white matter
peripheral nervous system
4. Eye signs
corneal clouding
optic atrophy
cherry-red macula
retinal pigmentary degeneration
5. Extraneural involvement
facial dysmorphism
hepatosplenomegaly
skeletal dysplasia

Clinical laboratory tests useful in the diagnosis of lysosomal storage diseases are presented in TABLE 3. Biochemical analyses that measure substrate concentrations and levels of enzyme activity are more specific than hematological, radiological, neurophysiological, and pathological studies, but the results of these latter tests can confirm whether a storage disease exists and help to localize the pathology. We have found examination of the bone marrow and biopsy of the skin to be especially useful. The presence of foam cells in these tissues immediately directs our attention to certain sphingolipidoses, that is, Sandhoff's disease, G_{M1}-gangliosidosis, Farber's disease, and Niemann-Pick disease and to the oligosaccharidoses, mannosidosis and fucosidosis. The presence or absence of foam cells in the bone marrow helps to differentiate disorders that are otherwise similar such as Tay-Sachs and Sandhoff's disease, and Gaucher's and Niemann-Pick disease. It is particularly useful in separating out from among cases with hepatosplenomegaly those children with a lysosomal storage disease. Electron microscopic examination of the skin, as described elsewhere in this symposium by Dr. K. Wisniewski, is, in our hands, also particularly helpful in differentiating lysosomal storage diseases from other causes of degenerative nervous system disease in children. Only very rarely do we find it necessary to recommend a brain biopsy for such children. We do this when the child's diagnosis remains enigmatic, the pathology appears confined to the central nervous system, and the parents are young and desire additional children.

Skin biopsy provides tissue not only for histologic studies but also for cell culture. The cultured cells can be grown on cover slips, fixed, and stained for examination by light and electron microscopy. With this technique, we have been

TABLE 3. Clinical Tests

  I. Hematological
    A. peripheral smear
        1. vacuolization of lymphocytes and monocytes (ex. mannosidosis)
        2. hypergranulation of neutrophils (ex. neuronal ceroid lipofuscinoses)
        3. autofluorescent granules in leukocytes (ex. neuronal ceroid lipofuscinoses)
    B. bone marrow
        1. foam cells (ex. Niemann-Pick disease)
        2. Gaucher cells (ex. Gaucher's disease)
        3. Reilly-Alder bodies (ex. Hurler's disease)
 II. Radiological
    A. skeletal survey
        1. J-shaped sella, thickened cortices, spatula-like ribs, anterior beaking of T_{12} & L_1 vertebrae, pelvic dysplasia (ex. mucopolysaccharidoses)
        2. Erlenmeyer-flask deformity of humerus & femur (ex. Gaucher's disease)
    B. neuroimaging (brain CT and MRI scans)
        1. cortical atrophy (ex. neuronal ceroid lipofuscinoses)
        2. leukomalacia (ex. metachromatic leukodystrophy)
III. Neurophysiological
    A. evoked potentials
        1. brainstem auditory evoked responses (BAER) (ex. leukodystrophies)
        2. visual evoked responses (VER) (ex. neuronal ceroid lipofuscinoses)
    B. nerve conduction velocities (ex. leukodystrophies)
    C. electromyogram (ex. Pompe's disease)
IV. Morphological
    A. leukocyte/fibroblast pellet (ex. mucolipidosis IV)
    B. biopsy of skin/conjunctiva (ex. fucosidosis)
    C. rectal biopsy (ex. late-onset G_{M2}-gangliosidosis)

able to show the excessive accumulation of lipids in various sphingolipidoses and in mucolipidosis IV. In an unknown lipidosis, the cells can be stained with various lectins to analyze the chemical nature of the terminal nonreducing bond in the stored material.[5] Their content of neutral sphingolipids and monosialo-gangliosides can also be determined. The quantitation, done by HPLC on cells harvested from a single 60-mm petri dish, can confirm the presence of a storage disease in situations where the usual *in vitro* enzyme assay would not be definitive as, for example, in the case of the AB variant of G_{M2}-gangliosidosis.[6] Cultured cells are also used for turnover studies to differentiate among clinical variants of a particular disease with different ages of onset, as reported for metachromatic leukodystrophy.[7] For genetic complementation studies, cultured cells from two different variants with the same enzyme defect are fused and the hybrid cells examined for cross-correction of the enzyme defect. Skin biopsy and cell culture are, therefore, important avenues for the early detection of lysosomal storage diseases.

The biochemical approach to lysosomal diseases involves both substrate studies and enzyme assays and uses those fluids and tissues that are most readily available (TABLE 4). In addition to cultured skin cells, we rely heavily upon tests of blood and urine. The Berry spot test screens for an excess of mucopolysaccharide in the urine. Those dysmorphic patients who are Berry spot test negative, that is, do not have a mucopolysaccharidosis, can be tested for an oligosaccharidosis by thin-layer chromatography (TLC) of their urine. After staining the TLC plate for carbohydrate, banding patterns are obtained that are characteristic for each of the oligosaccharidoses. Plasma may be used in conjunction with HPLC to quantitate the abnormal composition of neutral glycolipids present in Gaucher's disease, Fabry's disease, and Sandhoff's disease.[8] Occasionally, tissue lipid analyses may be required as in the case of Niemann-Pick disease type C where there is no discernable enzyme defect. HPLC is also employed for urine sulfatide quantitation in cases of metachromatic leukodystrophy.

Enzyme assays have been the most effective and precise way to diagnose the lysosomal storage diseases. Artificial chromogenic and fluorigenic compounds are available that conveniently substitute for the glycolipids, mucopolysaccharides, or other natural components accumulating in these diseases; however, for some reactions, the natural substrate is preferred with a radioactive label inserted to increase the sensitivity of the detection system.

TABLE 4. Biochemical Analyses

  I. Substrate accumulation
      A. Berry spot test—urine (ex. mucopolysaccharidoses)
      B. Thin-layer chromatography—urine (ex. oligosaccharidoses)
      C. High-pressure liquid chromatography
              urine (ex. metachromatic leukodystrophy, Fabry heterozygotes)
              plasma (ex. sphingolipidoses)
 II. Enzyme assays
      A. Substrates
          1. artificial, chromogenic, and fluorigenic
          2. natural, radioactivity labeled
      B. Problems
          1. enzymatic basis for some storage diseases unknown
          2. clinical heterogeneity
          3. activator protein deficiencies
          4. pseudodeficiency

In the case of Tay-Sachs disease, a substantial decline in cases has come about through the use of an artificial substrate assay for the prepregnancy detection of Tay-Sachs carriers. The widespread acceptance of the carrier test could not have been accomplished were it not for the development of a relatively simple and inexpensive test for hexosaminidase A (hex A); thus, in the case of this particular lysosomal disease, the identification of couples at risk for having an affected child allows prenatal diagnosis to be carried out in a family before the birth of even one affected child.

Numerous problems, however, presently limit the usefulness of this type of testing for the early detection of lysosomal storage diseases:

(1) The enzymatic bases for several lysosomal storage diseases, such as the neuronal ceroid lipofuscinoses, Niemann-Pick disease type C, and mucolipidosis IV, are not yet known.

(2) Enzyme assays do not explain the clinical heterogeneity seen in many of the lysosomal diseases. For example, the enzyme deficiency appears to be the same in infantile and adult Pompe's disease, in late-infantile and adult-onset metachromatic leukodystrophy, in type 1 and type 2 G_{M1}-gangliosidosis and in Hurler's disease and Scheie's disease. Moreover, storage diseases that first appear late in childhood or in adult life would have to be diagnosed preclinically in order for this information to aid the patient's family in the prenatal analysis of subsequent siblings.

(3) Certain diseases such as the AB variant of G_{M2}-gangliosidosis and one variant of metachromatic leukodystrophy result from deficiency of a nonenzymatic activator protein that cannot be detected in the usual *in vitro* assay system.

(4) Very low levels of enzyme activity have been detected in some healthy people. These pseudodeficient individuals have no evidence of lysosomal storage and can competently dispose of the relevant natural substrate when challenged in tissue-culture turnover studies.[9]

The molecular differences causing clinical heterogeneity are beginning to be understood. Three different forms of late-infantile G_{M2}-gangliosidosis are known, each of which has a subvariant with an onset later in childhood or in adult life. Hex A is missing in Tay-Sachs disease, hex B in Sandhoff's disease, and the activator of G_{M2}-ganglioside hydrolysis is deficient in the AB variant. The plasma and leukocyte hex A deficiency in classical Tay-Sachs disease and in adult-onset G_{M2}-gangliosidosis are similar. However, with the use of an immunochemical technique that examines the processing of the enzyme molecule, hex A, and its component parts, the alpha and beta chains, the different forms of these diseases can be differentiated. Neufeld and others in her lab have shown that the α-chain precursor is completely absent in classical Tay-Sachs disease and that the β-chain precursor is missing in classical Sandhoff's disease. In some forms of adult-onset G_{M2}-gangliosidosis, the α-chain preprotein appears but is defective. It does not progress to mature β-chain and does not associate with α-chains to form hex A.[10]

Where are things headed in the field of lysosomal enzymology? Finer and finer resolution is being sought and one of the pathways is through molecular biology. The mapping of genes for the various lysosomal acid hydrolases has proceeded rather rapidly so that the chromosomal localization of quite a few of these enzymes is now known. Two are on the X-chromosome—the Fabry's disease gene and the Hunter's disease gene. The others are scattered throughout the genome (cf. TABLE 5).

During the last year the isolation of cDNA clones for at least five lysosomal enzymes has been reported—α-fucosidase,[11] β-glucosidase,[12] β-glucuronidase,[13] and the α- and β-chains of hexosaminidase.[14,15] It is entirely reasonable to sup-

TABLE 5. Chromosome Assignments of Genes Coding for Some of the
Lysosomal Enzymes

Enzyme	Chromosome No.
α-L-Fucosidase	1
Acid β-glucosidase	1
β-Galactosidase	3
Aspartylglucosaminidase	4
Arylsulfatase B	5
Hexosaminidase B	5
β-Glucuronidase	7
Acid lipase A	10
Hexosaminidase A (α-chain only)	15
α-Glucosidase	17
α-Mannosidase	19
Arylsulfatase A	22
α-Galactosidase A	X
Iduronate sulfatase	X

pose that, using these cDNAs as probes, it will be possible to demonstrate molecular heterogeneity among individuals who are carrying different forms of each lysosomal enzyme deficiency state. To do this, we will draw blood, isolate the DNA, use specific restriction endonucleases to degrade the leukocyte DNA in a known fashion, separate the pieces of DNA by agarose gel electrophoresis and then transfer a blot onto nitrocellulose paper. We will identify from the great mass of DNA fragments the specific fragments of interest and determine their molecular weight using specific probes obtained from a plasmid vector and labeled with phosphorus-32. The cDNA clones could be used to analyze genomic DNA, mRNA, and familial inheritance patterns by the way the probe anneals with specific fragments of DNA obtained from the patient. DNA errors such as deletions, duplications, and substitutions of the type now known to occur in the hemoglobinopathies will undoubtedly be found. At this juncture, lysosomal enzymes assays would become more of a screening tool and a prelude to more definitive diagnostic study using the tools of molecular biology. We are perhaps five years away from the realization of this goal.

REFERENCES

1. KORNFELD, S. & W. S. SLY. 1985. Lysosomal Storage defects. Hosp. Pract. **15:** 71–82. For details of each disease see also: J. B. Stanbury, J. B. Wyngaarden, D. S. Fredrickson, L. Goldstein & M. S. Brown, Eds. 1983. The Metabolic Basis of Inherited Disease. 5th Edition. McGraw-Hill. New York.
2. DEDUVE, C., B. C. PRESSMAN, R. GIANETTO, R. WATTIAUX & F. APPELMANS. 1955. Tissue fractionation studies. Intracellular distribution patterns of enzymes in rat-liver tissue. Biochem. J. **60:** 604–617.
3. HERS, H. G. 1963. Alpha-glucosidase deficiency in generalized glycogen-storage disease (Pompe's disease). Biochem. J. **86:** 11–16.
4. LI, Y-T., I. A. MUHIUDEEN, R. DE GASPERI, Y. HIRABAYASHI & S-C. LI. 1983. Presence of activator proteins for the enzymic hydrolysis of G_{M1} and G_{M2} gangliosides in normal human urine. Am. J. Hum. Genet. **35:** 629–634.
5. ALROY, J., V. ORGAD, A. A. UCCI & M. E. A. PEREIRA. 1984. Identification of glycoprotein storage diseases by lectins: A new diagnostic method. J. Histochem. Cytochem. **32:** 1280–1284.

6. RAGHAVAN, S., A. KRUSELL, T. A. LYERLA, E. G. BREWER & E. H. KOLODNY. 1985. G_{M2}-ganglioside metabolism in cultured human skin fibroblasts: Unambiguous diagnosis of G_{M2}-gangliosidosis. Biochim. Biophys. Acta **834:** 238–248.
7. PORTER, M. T., A. FLUHARTY, J. TRAMMELL & H. KIHARA. 1971. A correlation of intracellular cerebroside sulfatase activity in fibroblasts with latency in metachromatic leukodystrophy. Biochem. Biophys. Res. Commun. **44:** 660–666.
8. ULLMAN, M. D. & R. H. MCCLUER. 1977. Quantitative analysis of plasma neutral glycosphingolipids by high performance liquid chromatography of their perbenzoyl derivatives. J. Lipid Res. **18:** 371–378.
9. CHANG, P. L. & R. G. DAVIDSON. 1983. Pseudo-arylsulfatase-A deficiency in healthy individuals: Genetic and biochemical relationship to metachromatic leukodystrophy. Proc. Natl. Acad. Sci. USA **80:** 7323–7327.
10. D'AZZO, A., R. L. PROIA, E. H. KOLODNY, M. M. KABACK & E. F. NEUFELD. 1984. Faulty association of α- and β-subunits in some forms of β-hexosaminidase A deficiency. J. Biol. Chem. **259:** 11070–11074.
11. FUKUSHIMA, H., J. DEWAT & J. S. O'BRIEN. 1984. Molecular cloning of a human fucosidase cDNA. Am. J. Hum. Genet. **40:** 137s.
12. GINNS, E. I., P. V. CHOUDARY, B. M. MARTIN, S. WINFIELD, B. STUBBLEFIELD, J. MAYER, D. MERKLE-LEHMAN, G. J. MURRAY, L. A. BOWERS & J. A. BARRANGER. 1984. Isolation of cDNA clones for human β-glucocerebrosidase using the $\Lambda_{gt}11$ expression system. Biochem. Biophys. Res. Commun. **123:** 574–580.
13. GUISE, K. S., R. G. KORNELUK, J. WAYE, A. M. LAMHONWAH, F. QUAN & R. A. GRAVEL. 1984. Characterization of cDNA clones encoding human β-glucuronidase. Am. J. Hum. Genet. **40:** 139s.
14. MYEROWITZ, R., R. L. PROIA & E. F. NEUFELD. 1984. Deficiency of α-chain mRNA in Ashkenazi Tay-Sachs fibroblasts. Am. J. Hum. Genet. **36:** 148s.
15. O'DOWD, B., F. QUAN, H. F. WILLARD, A. M. LAMHONWAH, R. G. KORNELUK, J. A. LOWDEN, R. A. GRAVEL & D. J. MAHURAN. 1985. Isolation of cDNA clones coding for the β subunit of human β-hexosaminidase. Proc. Natl. Acad. Sci. USA **82:** 1184–1188.

DISCUSSION

JOSEPH FRENCH: When some of these normal cellular constituents accumulate in excess in lysosomes, why do neuronal tissues not function well and prematurely expire?

EDWIN KOLODNY: This is the unique problem with the lysosomal storage diseases, and why they are of great importance to the neuroscientist. We are dealing with an element, the nerve cell, that cannot regenerate. The progressive accumulation of stored material can mechanically disrupt the cell. It can cause, as we know, aberrant formation of synaptic contacts, in regions of the nerve cell membrane where they do not belong. This causes meganeurite formation and leads to physiological consequences of disastrous proportions. Once the nerve cell is disrupted, it cannot be replaced by mitosis of other cells, and we lose the function of that cell.

EDMUND JENKINS: You mentioned a cost savings by using bright-field microscopy rather than electron microscopy. What is the relative precision and accuracy, in terms of false-positive and -negative diagnosis, between bright-field and electron microscopy?

EDWIN KOLODNY: Electron microscopy is very important, but as one works one's way to electron microscopy one may effect a cost savings. Autofluorescent granules accumulate in the neuronal ceroid lipidfuscinoses and are an example of this. We can collect white cells, via Fiacoll gradient sedimentation separation of monocytes and lymphocytes in a 10-milliliter sample of blood, smear them out on a glass slide, and then look at them under the light microscope using ultraviolet light illumination. The autofluorescent granules of neuronal ceroid lipidfuscinosis are obvious. When they are not obvious, electron microscopy of a skin biopsy is almost always negative. It is a useful screening procedure costing approximately $50 per specimen, which compares with about $800, the cost at our institution for the electron microscopic study of a skin biopsy specimen.

Another advance that I wanted to tell you about was the use of lectin staining. Dr. Joseph Alroy of the Tufts New England Medical Center has been able to demonstrate specific patterns of staining for the different storage diseases using specific lectins. With this method one can identify an unknown disease using paraffin-embedded sections. Dr. Alroy, who is a veterinarian, has been able to do lectin staining with specimens from animals that have died 10–20 years ago of previously undiagnosed forms of storage diseases.

Introduction

DONALD A. SNIDER

Institute for Basic Research in Developmental Disabilities
New York State Office of Mental Retardation &
Developmental Disabilities
Staten Island, New York 10314

In this session Dr. Robert Guthrie of the State University of New York at Buffalo leads with a description of the extent of low-level lead exposure in the United States. His paper presents data that challenge popular assumptions about exposure to lead.

Dr. Allen Crocker, Director of the Developmental Evaluation Clinic at the Children's Hospital in Boston, provides an overview of prevention efforts in selected states and describes successful prevention programs.

Two presentations address medical education: Dr. Elena Lesser of the Downstate Medical Center in Brooklyn describes curriculum trends in medical schools. She compares the progress made in introducing basic science material related to MRDD and medical student contacts with mentally retarded persons during their clinical training.

Dr. Robert Kugel, Medical Director of the Terence Cardinal Cooke Health Care Center in New York, also describes the status of developmental disabilities in health and medical education. He describes how the practicum or clerkship can be used as a vehicle for addressing medical aspects of mental retardation and shaping students' attitudes toward retarded persons.

Lead Exposure in Children: The Need for Professional and Public Education

ROBERT GUTHRIE

Departments of Pediatrics and Microbiology
State University of New York at Buffalo
Buffalo, New York 14214

In discussing the subject of lead exposure in preschool children and the great need for more education of both the public and the medical profession on the subject, I would like to begin by pointing out that about 15 years ago, the President's Committee on Mental Retardation, Washington, DC, came to the conclusion that approximately 50% of mental retardation can be prevented. Assuming that this is true, then we have the difficult task of defining both mental retardation and its prevention. Mental retardation means to me the retardation of intellectual growth during childhood. When intellectual growth stops, as it does in everyone at age 14–18 years, retardation of the intellectual growth process means that an individual plateaus at a lower IQ. On this basis, all of us are retarded to some degree, since there are hundreds of environmental and genetic causes of mental retardation, and no one can escape the influence of at least a few of these. Of course, the medical, legal, and educational professions arbitrarily define the mentally retarded as those with an IQ of less than 70.

The challenge of prevention of lead exposure is an all-encompassing one. Since most of us are in an environment where we are exposed to lead, probably all of us have been affected to some degree by low-level lead exposure. For that reason, if we could eliminate lead from the environment, we should eliminate retardation of intellectual growth and other effects on the central nervous system of many, many more individuals than the two or three percent of the population we usually consider when we think of mental retardation.

The April 1985 issue of the *American Journal of Public Health*[1] has a photograph of New York City automobile traffic during the decade of the 1920s. In this issue, there is a historical account of the controversy surrounding the introduction of tetra-ethyl lead in gasoline for automobiles, which took place in the 1920s. The possibility that this would be a health problem was seriously considered, but was dismissed after a special Federal commission concluded that it would not cause any damage to health. The decision was made even though it was known that tetra-ethyl lead was causing serious illness in workers at the DuPont plant where it was being made. Now, 60 years later, we know that lead pollution of the environment from leaded gasoline is indeed a serious health problem.

In 1970, the Erie County Department of Health in New York, where I reside, began a program to prevent lead poisoning in children. This was two years before there was Federal money available to deal with the problem. Our county, which includes the city of Buffalo, was of course not the first; other cities in the East, such as Boston, New York, Philadelphia, and Chicago, had already begun programs. At that time, the major problem considered (and this is still of major importance) was the consumption by smaller children of lead-based paint in old, dilapidated urban houses. For the past 13 years, some Federal money has been made available. Because there has never been enough, this money has been distributed only to the major cities in the country for screening in urban centers. As a result, the impression of the public and the medical profession is that the problem exists only in the center of large cities; hence, the current misconception by the medical and public communities of the extent of this problem.

In 1979, the first publication, a way of measuring previous exposure to lead in children who were five to seven years of age, was based on the measurement of lead in baby teeth, or shed teeth. The publication by Needleman and his associates appeared in the *New England Journal of Medicine*.[2] This was followed by other, similar studies, also attempting to find if there was a difference in development of children who had a larger amount of lead accumulated in baby teeth, as compared to children with a smaller amount. The findings of Needleman, which were based on very careful studies, are conclusions with which I agree: that there is a statistically significant but small decrease in mean IQ in the group with high lead, a decrease in school achievement, and an increase in behavior problems and hyperactivity. However, other studies, particularly those in England, have not confirmed this; so there is controversy concerning this type of study.

A different approach that has impressed me is that of Otto and his associates at the University of North Carolina.[3] They have measured a significant alteration in the electroencephalographic pattern in the central nervous systems of young children two years after they had known elevated blood lead levels. They found a direct correlation between the increase in blood lead levels two years before and the extent of this alteration, and that this correlation extends down into the so-called "normal" range with no apparent threshold.

These studies have confirmed in the human species the findings that have been made in experimental animals: low-level lead exposure, below that which would cause immediate effects or apparent illness in a child, does cause permanent effects in the developing central nervous system.[4] As a result of these studies, the center of attention for the lead problem has shifted from concern with high levels of lead of 50 to 100 μg/dl of blood, down to lower levels. More recently, the Centers for Disease Control in Atlanta, Georgia, has advocated lowering the cut-off in screening from 30 μg of lead/dl to 25 μg/dl of blood.[5] During the past decade, we are no longer recognizing a "safe" lead level as a level of lead below which there are no immediate symptoms. We realize that lead can cause deleterious effects to development years later in a child who shows no symptoms at all during the period of exposure and does not appear to present a problem to either the parents or to the physician.

In addition to these studies showing the effect of low-level lead exposure on children, a five-year national survey was completed in 1980 by the National Health and Nutrition Examination Survey that indicated much wider exposure of children to lead than had been previously believed. These studies, which began to be published in 1981,[6] indicated that in preschool children under five years of age, as many as 4% (1/25) had levels of lead of 30 μg/dl of blood or greater. There was a wide variation in the incidence of children with these levels, depending on socioeconomic circumstances and where they lived, and also depending on whether they were black or white. The very lowest incidence was 0.7%, found in children of white families with an income of greater than $15,000/year; while the highest incidence, 18.5%, was found in children of black families below the poverty level of $6,000/year of family income, or black children living in urban centers. In fact, in every category, socioeconomic or urban–rural, the incidence in black children was much higher than in white children. This has not been explained, and may possibly be due to a biological difference in lead absorption between black children and white children.

In addition, the survey revealed that the mean blood lead level had dropped from 15 μg/dl to 10 μg/dl over a five-year period, and that this decrease in mean blood level was almost parallel to the decrease in the use of leaded gasoline in the United States during that period.[7]

Only a short time after the above findings had been published, the Environmental Protection Agency announced in 1982 that they were prepared to relax regulations on leaded gasoline so that more lead could be added rather than less. There was a tremendous adverse reaction to this disclosure by both the professional community and the public. As a result, within one year's time, the EPA reversed itself and announced that it was going to restrict the lead content of gasoline even more than it had previously. Recently, the EPA has repeatedly announced that it advocates the ban of the sale of all leaded gasoline in the United States.

Approximately 25 years ago, our laboratory became very interested in the possibility of developing simple screening methods to detect inborn errors of metabolism, particularly those that are preventable, such as phenylketonuria (PKU) and others. We introduced for this purpose the idea of collecting dried blood spots on filter paper and the use of an ordinary paper punch to produce a quantitative sample for testing.[8,9] This made it practical for the first time to screen every baby born before it leaves the hospital nursery by sending large numbers of these specimens to a regional laboratory where screening for a number of different rare conditions could take place. During the past two decades, programs including as many as six to eight different tests have been used on this filter-paper blood specimen.[10,11]

Approximately 15 years ago, we became interested in the possibility of using the same kind of specimen to detect lead exposure and iron deficiency, taking advantage of a method that had already been developed by Dr. Sergio Piomelli[12] for the extraction of a dried spot of blood or a small sample of blood in a test tube, and detecting the accumulation of erythrocytic protoporphyrin (EP). EP accumulates when there is a lack of iron for the terminal enzyme in the sequence of heme synthesis, ferrochelatase, or the presence of lead, which poisons this enzyme. Dr. William Murphey and Mr. Adam Orfanos, in our laboratory, succeeded in simplifying the extraction method of Piomelli and applying it to a disc punched from a spot of blood on filter paper.[13] Since there is no intention of using the blood specimens for a microlead determination, no special precautions are needed in collecting the sample to avoid contamination with lead, as is the case in practically all existing screening programs. This is the basis of a program that we consider to be a very successful demonstration of an approach for screening all children annually in the first few years of life.

During the past four years, we have succeeded in testing nearly 40,000 children. Of the first 30,000, approximately 25,000 were in the suburbs and rural areas outside the city of Buffalo. In these children, we have found an incidence of 0.6% for those with a blood lead level of 30 μg/dl or greater. An additional 1% of children have been found to have iron deficiency.[14] I believe it is important to point out that our incidence of 0.6% corresponds closely to the value of 0.7% in the national survey for children of white families with incomes above \$15,000 annually. This appears to be an accurate description of the socioeconomic status of the children that we have been screening outside the city of Buffalo. We feel, therefore, that we have confirmed the national survey. As a result, we conclude that it is very important to screen these children, who otherwise would not be screened at all.

When more and more children in the suburbs and rural areas are included in screening programs, the children detected will stimulate greater concern by the public. It is my feeling that this is the direction to proceed, not only for medical reasons, but because in the long run, this will provide the political and fiscal solution. As you may remember, there was little federal money available for the drug problem in this country until it was discovered that there was a problem in the suburbs. As long as people felt that the drug problem was only in the urban centers, nothing much was done about it. Once the public realizes that the national survey is correct, and that the results of Needleman, Otto, and others apply to children everywhere, not just in our urban centers, then I am sure that state and federal funds will become available.

Our federal government has recommended that all preschool children under age six years be screened periodically with the EP test for iron deficiency and lead. The agencies recommending this include the Centers for Disease Control (CDC),[15] and the Maternal and Child Health Service. In 1982, a memo was sent to every state from the Assistant Surgeon General's office, advocating that all children be screened;[16] therefore, I am actually only proposing what is already officially recommended. Yet, at the present time, according to the CDC, only approximately 3% of children under age six are screened for these conditions. The problem is: How do we screen for the other 97%? If we are going to have mass screening of children with iron deficiency and lead exposure, the approach of using a filter-paper blood spot that does not require skin preparation for a micro-lead test, would seem to be the most practical method of accomplishing this. Current methods of screening would continue to be used for the existing programs in the urban centers, where the higher risk justifies the greater expense.

A description of our own program would not be complete without giving you a short account of the results we have on specimens sent to us from Kuwait. Kuwait is an Arab country, and we feel the results on specimens sent to us from Kuwait are an indication of problems in other Arab countries. It appears that the major source of lead in the hundreds of children we have screened is the black eye cosmetic which is used from cradle to grave by women in Kuwait. It is known as *kohl* in Kuwait; the same eye cosmetic is called *surma* in India, Pakistan, and Bangladesh. There is a possibility that the eye cosmetic may be a source of lead exposure in children in a very large fraction of the population of the world, and the cause is unrecognized in those countries.

What we have found is that children who are born in Kuwait have a problem of congenital lead toxicity that is not found in the Western world.[17] This condition is due to the practice of women during pregnancy using kohl as an eye cosmetic and applying it to their infants and children as well. One of the most popular and frequent sources of kohl in Kuwait is galena, or crude lead found in the earth. This is ground into a black powder, which sparkles due to light reflected from the platelets of lead disulfide in the material. Kohl is commonly applied to the umbilicus of infants at birth, and to their eyes. As they grow older, little girls apply it to their own eyes, as do the mothers. For the rest of their lives, these women use black eye cosmetic. If the kohl is made with galena, it has become obvious that it may be absorbed and affect the fetus before birth and also, of course, affect the children. In Kuwait, I was shown x-rays of infants born with "lead lines" in their skeletons. Our first results from Kuwait on 96 children screened led to detection of 17 infants and young children with lead poisoning, with one death from lead encephalopathy.[18,19]

We know that this is also a problem in India and Pakistan because of reports in the British literature of young children who are hospitalized because of lead encephalopathy. In these cases, the lead source has been traced to the use of surma by the mother, as well as on the eyes of these very young infants. It is important to realize that there may be hundreds of thousands of people living in the United States who are from these countries, and that this may be a problem that pediatricians might not be aware of.

Finally, I should point out that we became interested in lead exposure and iron deficiency as detected by this test, not only because we feel that it is fully justified to screen every child periodically, but because this practice should lead to the same type of regional mass screening as was developed for newborn screening. There will be an opportunity to develop and use other tests on the same blood specimen; there must be many genetic and environmental problems in young children that can be detected with these same filter-paper blood specimens at the regional laboratories.

REFERENCES

1. ROSNER, D. & G. MARKOWITZ. 1985. A "gift of God?": The public health controversy over leaded gasoline during the 1920s. Am. J. Public Health 75(4): 344–352.
2. NEEDLEMAN, H. L., C. GUNNOE, A. LEVITON, R. REED, H. PERESIE, C. MAHER & P. BARRETT. 1979. Deficits in psychologic and classroom performance of children with elevated dentine lead levels. N. Engl. J. Med. 300(13): 689–695.
3. OTTO, D., V. BENIGNUS, K. MULLER, C. BATONN, K. SEIPLE, J. PRAH & S. SCHROEDER. 1982. Effects of low to moderate lead exposure on slow cortical potentials in young children: Two-year follow-up study. Neurobehav. Toxicol. Teratol. 4: 733–737.
4. NEEDLEMAN, H. L., ED. 1980. Low-Level Lead Exposure: The Clinical Implications of Current Research. Raven Press. New York.
5. CENTERS FOR DISEASE CONTROL. 1985. Preventing lead poisoning in young children: A statement by the center for disease control. January, 1985. Department of Health and Human Services. Atlanta, Georgia.
6. ANNEST, J. L., D. O'CONNELL, J. ROBERTS & R. S. MURPHY. 1981. Blood lead levels from the second national health and nutrition examination survey, 1976–1980. In Childhood Lead Poisoning Prevention and Control: A Public Health Approach to an Environmental Disease. F. F. Cherry, Ed.: 93–102. Department of Health and Human Resources. New Orleans, LA.
7. — 1982. Blood-lead levels in U.S. population. Morbid. Mortal. Wkly. Rep. 31:(1): 132–134.
8. GUTHRIE, R. 1961. Blood screening for phenylketonuria. J. Am. Med. Assoc. 178: 863.
9. GUTHRIE, R. & A. SUSI. 1963. A simple phenylalanine method for detecting phenylketonuria in large populations of newborn infants. Pediatrics 32: 338–343.
10. GUTHRIE, R. 1972. Mass screening for genetic disease. Hosp. Pract. 7: 93–100.
11. GUTHRIE, R. 1980. The organization of a regional newborn screening laboratory. In Neonatal Screening for Inborn Errors of Metabolism. H. Bickel, R. Guthrie & G. Hammersen, Eds.: 259–270. Springer-Verlag. Heidelberg.
12. PIOMELLI, S., B. DAVIDOW, V. F. GUINEE, P. YOUNG & G. GAY. 1973. A Screening Micromethod for Lead Poisoning. Pediatrics 51: 254–259.
13. ORFANOS, A. P., W. H. MURPHEY & R. GUTHRIE. 1977. A simple fluorometric assay of protoporphyrin in erythrocytes (EPP) as a screening test for lead poisoning. J. Lab. Clin. Med. 89: 659–665.
14. GUTHRIE, R., A. P. ORFANOS, K. WIDGER, D. NADLER & C. FRANCEMONE. 1985. Lead exposure in suburban and rural children: Results of a screening program. Am. J. Public Health. Submitted for publication.
15. HOUK, V. N. 1981. Lead poisoning prevention services. In Childhood Lead Poisoning Prevention and Control. A Public Health Approach to an Environmental Disease. F. F. Cherry, Ed.: 21–27. Department of Health and Human Resources. New Orleans, LA.

16. ASSISTANT SURGEON GENERAL. 1982. Erythrocyte protoporphyrin (EP) screening of children for undue lead exposure and iron deficiency. Bureau of Health Care Delivery and Assistance Regional Memorandum 82–15. November 10.
17. SHALTOUT, A., S. A. YAISH & N. FERNANDO. 1981. Lead encephalopathy in infants in Kuwait: A study of 20 infants with particular reference to clinical presentation and source of lead poisoning. Ann. Trop. Paediatr. 1: 209–215.
18. SHALTOUT, A. A., M. M. GHAWABY, M. C. M. HUNT & R. GUTHRIE. 1985. High incidence of lead poisoning detected by FEP screening in arabian children. In Science and Services in Mental Retardation. J. M. Berg, Ed. Methuen & Co., Ltd. London. In press.

DISCUSSION

DAVID SOIFER: Dr. Guthrie, in Kuwait, the whole Middle East, and other developing countries, to what extent does industrial pollution from lead battery and other kinds of factories where carry lead is involved contribute to the kinds of problems that you are talking about?

ROBERT GUTHRIE: I am sure it does not contribute as much as the eye cosmetics, kohl, or surma. Developing countries, as you can guess, are slow to recognize the dangers of pollutants as they industrialize.

ALLEN CROCKER: Could you mention the information on distribution of the problem by sex in the cultures where eye cosmetic use is preeminently, I would presume, a female problem. I imagine there must be spillover beyond the newborn situation. How does that work out?

ROBERT GUTHRIE: Of course, half of the infants who are exposed before birth are male and the other half female. The practice often is to begin applying the eye cosmetic on the infant right after birth, and there is also a practice of putting the same cosmetic on the umbilicus of the newborn baby. This would probably affect both sexes.

PEGGY BROOKS-BERTRAM: Is there now a major effort under way to establish a public health awareness program to change cultural practice regarding the use of the eye cosmetic?

ROBERT GUTHRIE: Some of these countries have sufficient resources to provide public health education. Kuwait is the wealthiest country in the world per capita, and I believe that the problem will be documented in that country, eventually leading to an educational campaign by the government. This will affect the other Arab countries in turn. The Indian authorities are just beginning to evince interest in an erythrocyte protoparphyrin screening program like that in Kuwait. I am intrigued by our EP test on filter-paper blood spots because the follow-up to screening is public education, rather than a special diet treatment, as with PKU.

IRVIN EMANUEL: As I recollect, in India in some of the villages there are community pots of this cosmetic. Older studies have shown that these pots were a source of transmission of traucoma.

ROBERT GUTHRIE: You might be interested in knowing that an ophthalmologist in Bombay was promoting use among his patients of an antibiotic added to lamp black as an eye cosmetic to combat traucoma. I have also been told that lumps of galena are used in cooking pots in villages, and little children in some parts of these countries sit around these pots and breathe the fumes to keep off the evil spirits.

THOMAS SHEPARD: Do you have any feel for the teratogenic effects of this?

ROBERT GUTHRIE: Yes, I have a suspicion, but I am not sure that kohl is going to turn out to be teratogenic. From what I saw and heard in Kuwait it seems that teratogenic effects exist, but the major effect seems to be afterwards. The children are born with lead lines in their skeletons, but then the blood levels become elevated as they use this material; that is when the effects are seen. The effects of lead as a teratogen and the effects of lead after birth have been studied extensively in animals, but not in humans, and the findings are just being confirmed now. The studies of Otto and others simply show that the human brain is affected subtly and asymptomatically, in much the same way as in animals. The major thing we are concerned about is what happens after birth.

Prevention of Mental Retardation: 1985[a]

ALLEN C. CROCKER

Developmental Evaluation Clinic
The Children's Hospital
Boston, Massachusetts 02115

It is appropriate in the closing or summary portion of this workshop to look at the status of efforts in the prevention of mental retardation and developmental disabilities. For this I will provide an overview with liberal use of quotations from current news articles for a commentary on our cultural commitment.

A public communication that arose from the Association for Retarded Citizens (FIG. 1) demonstrates the positive yearning that characterizes this field. The issue of what percentage of developmental disabilities is presently approachable by prevention techniques is an intriguing business. Such discussion has been part of many major conferences. There are a number of tactical handicaps in that the true incidence figures are incompletely known and the technology of counting effects is primitive at the moment.

What is the prevention message that is going to the public at the present moment? Rather appropriately a significant portion of it relates to concerns about the most dramatic time—the most relevant portion of human existence—namely the period of gestation. I have put together a potpourri of some of the kinds of publication materials that have been presented (FIG. 2).

"Take care of yourself."

"Take care of yourself for your baby's sake."

"Take good care of both of you."

This theme of the baby in your life comes from the states of Alabama, Indiana, and Illinois. Many times the attempt to recruit active concern uses conventional public health techniques—insinuating an element of personal responsibility. Participation is reflected in these pamphlets from Pennsylvania, California, and Massachusetts, where the "you" part is prominently featured in the message (FIG. 3).

The appeal for efforts in thoughtful attention to pregnancy and optimal nurturance for the baby is reflected in a couple of conference reports, one from California and one from New York. Specifically, the insinuation is that we are doing something that is central to the progress and value of our society as a whole. I don't know of any more commanding title for a conference than "The Right To Be Born Well" (FIG. 4).

It is of interest to take a look at where some of the islands of success are in 1985, and to use this as a kind of bellwether of where our investment has taken us to date. I am going to just touch on these and then drop back and look at one state's program in more detail.

In the area of primary prevention, rubella is clearly the flagship. I quote from a news report that emanated from the Centers for Disease Control: "the reported incidence of rubella and congenital rubella syndrome has decreased consistently

[a] This work was partially supported by Project 928 from the Department of Maternal and Child Health of the National Institutes of Health and by Grant 03DD0135 from A.D.D.

Last year 100,000 babies were born mentally retarded.

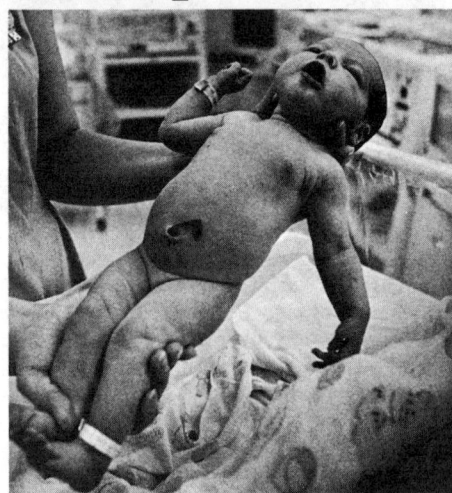

50,000 didn't have to be.

Every five minutes in the United States today, a baby is born mentally retarded. It's a true national tragedy because mental retardation can be prevented. In fact, half the time, it doesn't have to happen at all.

Few cases of mental retardation are inherited. And no case is an isolated natural phenomenon. The causes can occur before birth, during birth or in very early childhood. Some causes are unknown. 50% of the causes we know about are preventable today.

Prevention of mental retardation is one very big reason why the Association for Retarded Citizens needs your help. We need to inform pregnant women and women wanting to raise families about causes we've already isolated.

Please give to the ARC. And help 50,000 out of 100,000 babies to be born normal.

Association for Retarded Citizens
2501 Ave. J Arlington, Texas 76011

WHEN YOU GIVE HELP YOU GIVE HOPE.

FIGURE 1. Description in a magazine advertisement of the prevention program of the Association for Retarded Citizens of the United States.

since 1969." In the first thirty-three weeks of 1984, a provisional total of 500 cases of rubella were reported for the entire country, which represents a 33% decline from the comparable period in 1983. The extraordinary excitement accompanying the reduction in rubella is reflected in virtually every state. The figures in Massachusetts for example show that the total incidence of reported rubella for 1981 to 1984 has been 23, 2, 8, and 16, respectively. This in a state where the usual reported figures in the pre-1963–64 period were in the neighborhood of 3,000 to

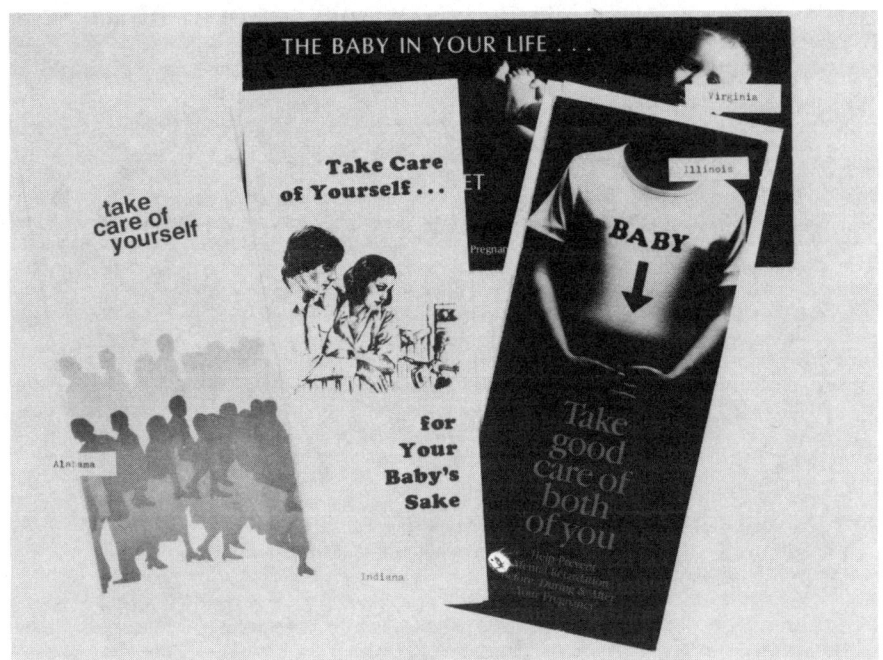

FIGURE 2. Public education pamphlets published by the ARCs of Alabama, Illinois, Indiana, and Virginia.

FIGURE 3. Appeals to young parents to inform themselves about the best care during pregnancy (California, Massachusetts, Pennsylvania).

FIGURE 4. The hopeful themes of prevention—conferences in California and New York.

4,000. Congenital rubella, by the way, has had no identified occurrence in Massachusetts for three or four years. The accomplishment represented by reduction of rubella, and thus of congenital rubella, is one in which we have reason to take pride.

Another area where a tremendous amount of devotion has occurred on an official level as well as in public education is in prenatal care. I was extremely heartened to find on May 21, 1985, an account in a Boston newspaper of a resolve that our Title V program in the Department of Public Health had introduced to the legislature. Efforts to remove all financial barriers to health care for pregnant women who have no health insurance or Medicaid coverage are gathering momentum in Massachusetts. The most concrete proposal is a $9.5-million-dollar item in the House Ways and Means budget that would provide universal entitlement to prenatal care and target "near-poor" women for special efforts to improve pregnancy results. This picks up the so-called medically indigent cluster of the population. It is extremely gratifying that Massachusetts is moving toward a nearly comprehensive provision of quality prenatal care.

The next item in primary prevention that is pertinent is the whole area of newborn intensive care. I have often commented in the past that this is the big one,[2] the one where there is the most clear-cut numerical accomplishment as a result of concrete efforts. FIGURE 5 presents data from Peter Budetti's work. It points out that as neonatology and the formation of newborn intensive-care units (NICUs) have grown, we have produced vastly increased numbers of normal children. Also, disabled children have decreased rather than increased while this

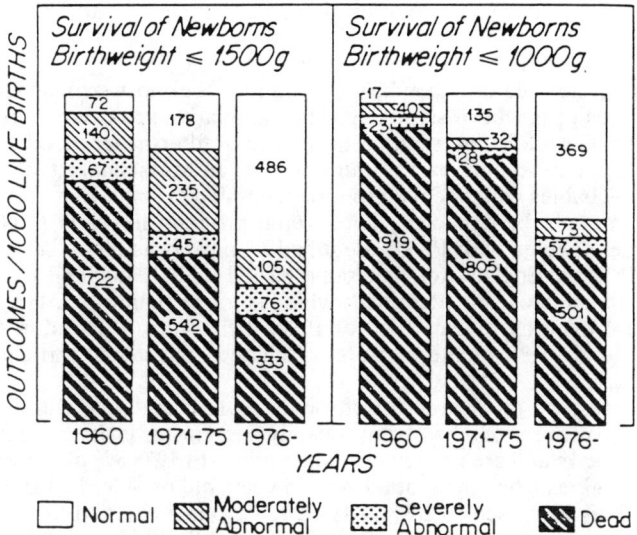

FIGURE 5. Composite data on the outcome for low-birth-weight babies, after Budetti. (From Thompson et al.[1] Reprinted by permission.)

astonishing improvement in mortality was occurring in NICUs. This is an extremely encouraging component of the prevention world.

Another effort in primary prevention is that of letting the consumer know well about the dangers of alcohol and pregnancy. From a recent article (April 14, 1985) in the *Boston Globe,* "Governor Michael S. Dukakis and the president of Consumer Value Stores (CVS) yesterday announced a joint project in which state-printed literature describing alcohol-induced problems during pregnancy will be available in all 111 CVS stores in Massachusetts." The article goes on to point out that they will be distributing 400,000 copies of a brochure, of which many will surely be read. In the state of Massachusetts, about 100 babies a year are born with fetal alcohol syndrome, plus many more with lesser effects. Reaching to the public with information about this risk is one of our current societal commitments.

Reduction of head injuries and direct cerebral damage from accidents is another concern. In Massachusetts a major program is underway based in the Division of Family Health Services to look at injury prevention in childhood. One of the first efforts by the persons administering the program has been to document the staggering number of head injuries and general trauma in childhood (e.g., one of every 50 teenagers is injured as a motor-vehicle occupant each year).

The final area of primary prevention I would like to mention is that of family-life education, with the notion that young people can be prepared for entering the responsibilities and opportunities of parenthood with a more informed perception of the issues. Some current newspaper articles illustrate this effort. One headline I saw recently in a Boston newspaper was "Human Development Class Produces a Battle in Athol" and another states "Family Life Course OK'ed" (in Essex). This is characteristic of the varying fortunes that school systems have experienced while attempting to introduce courses aimed at documenting the commitment of parenthood.

———————◆•••◆———————

Looking at the area of secondary prevention, the most striking element has been in the area of newborn screening. The cumulative achievement represented by Dr. Robert Guthrie's contribution in the use of filter-paper blood spots beginning in 1962, is now simply part of our heritage. In Massachusetts in 1984 there were nine new babies with phenylketonuria identified, a somewhat higher level for the year than might be expected. The overall program for New England documented 14 new babies with PKU, identified as newborns and eligible for direct treatment. In New England for that same period with 124,000 births, there was one child with galactosemia identified, which is lower than usual. If one looks at the congenital hypothyroidism data for 1984, there were 14 found in Massachusetts and a total of 39 in the five New England states that participate in the regional program.[3]

Amniocentesis in the search for chromosomal aberrations, with its potential for secondary prevention, has been steadily increasing in its penetrance. The figures in Rhode Island are particularly interesting. In 1978 8% of the women who gave birth to babies when the mother was 35 years old or older had amniocentesis performed. This increased to 20% in 1979, to 37% in 1980, and to 41% in 1981.[4] Rhode Island is now running somewhere in the middle 40s for use of amniocentesis for older women, which is consistent with most states in New England. I have the feeling that on a cultural and technologic base, we will level off at somewhere around 50% in the use of this kind of prevention opportunity.

When one looks at the involvement of obstetricians in a partnership to seek identification of pregnancies carrying babies who have neural-tube defects, namely the measurement of the maternal serum alpha-fetoprotein, similar kinds of numbers are emerging. In those states where a major effort has been undertaken to educate the professional and consumer communities, the figures are moving up. At the present moment in Maine, which is the state with the highest use of alpha-fetoprotein measurement in the country, somewhere around 45% of all pregnancies are employing this type of screening. In Great Britain it is running around 50%, and if you look at the area around Edinburgh, it is running at an astonishing 70% of all pregnancies.[5]

The maternal serum alpha-fetoprotein screening situation has undergone a revitalization with the realization in the last two years that low levels in a pregnant woman's serum can probably be used as an index of risk for the presence of Trisomy 21 in a fetus. One can construct charts of risk that use a combination of maternal age and alpha-fetoprotein measurement; these allow a significantly more accurate use of amniocentesis in the prevention of births of children with Down's syndrome. Several states in the New England area are underway with activities of this sort. When Connecticut launched such a program, the first effect was an approximate doubling of the requirement for cytogenetic technologic capacity, so the need for resolve includes that of governmental agencies.[6]

Finally, in the area of secondary prevention, a progress report on the pursuit of carrier screening in Tay-Sachs disease is appropriate. As of the end of 1984, there had been almost 500,000 people, virtually all Ashkenazi Jews, tested for absence of the enzyme hexosaminidase A, and in that time 19,000 carriers were identified. Also, 483 carrier/carrier marriages were documented of people who had not been aware of the direct risk to their children. If one looks at the cumulative experience between 1969 and 1984 in the 77 worldwide centers where Tay-Sachs studies are prominent, 259 fetuses were diagnosed with this disorder.[7]

—◆•••◆—

Going on into tertiary prevention, our efforts in the area of early intervention have been substantial. There are two major national studies under way at the moment that are looking finally to produce prospective documentation of the effects of standardized forms of early intervention. One hopes that these studies will be blessed with data that are convincing and transportable to legislatures.

When one looks over the usual territory of prevention, one sees that in this last decade we have made a significant degree of progress in areas or islands of pertinence.

—◆•••◆—

I think it would be useful if we are to try to make an accounting of all this to get down to specifics in one individual area. For this purpose, I would like to focus on the program that the state of Tennessee undertook in 1981. A Governor's Task Force on Mental Retardation Prevention was created and met for a year under the chairmanship of the governor's wife, Honey Alexander. They came up with a 27-point set of resolves as the state plan for Tennessee.[2] It came at a very fortuitous time because of the presence of a popular governor in the statehouse, the governor's wife giving this major effort, and a climate of achievement that had been heightened by the activities and publicity of this "blue ribbon" panel. Tennessee took off in an attempt to make it real.

A few months ago a study was made to see what could be documented after three years of a vigorous state plan (see Acknowledgments). I will give you a few highlights of this because Tennessee is to be commended for such a significant, well-thought-through and accounted activity.

The first item was in the area of prenatal care programs, the one that is ascendant in the country as a whole. In 1981 there were publicly funded prenatal-care programs in 34 out of 95 counties in Tennessee. In that year 2,088 women were served. In 1984, after $1.8 million had been pumped into the program from the state legislature, there were prenatal-care programs in all 95 counties. In that year 14,000 women were served, or seven times as many as three years previously. So prenatal care had been taken from a haphazard to a well-ordered base.

Services for family planning had been present in Tennessee at a moderate level for many years. By the end of 1983, there were programs available in all 95 counties, and in that year 157,000 people were served.

Family-life education had had no identifiable basis, but during this period a curriculum was written by the Department of Education and the Tennessee General Assembly passed a resolution urging its adoption. There has been slippage between the assembly's resolution and broad utilization, but this is an important first step.

Birth-certificate information was filled out by doctors in 1981. By 1984 medical records technicians were being trained and involved—a significant improvement. The Center for Health Statistics within the Department of Health and Environment had become vigorously concerned with the scope and accuracy of birth-certificate information.

A Genetics Network was formed in Tennessee. Utilization was increased in the regional perinatal-care centers, particularly in relation to maternal transports, and a fifth center was opened. Developmental screening was greatly increased as part of an infant follow-up program, and as part of a revitalized EPSDT. Immunization programs were enhanced. Newborn screening was made tighter. Early intervention programs were made a higher priority.

In summary, in Tennessee three years after the beginning of an inspired, statehouse-based prevention plan, one can already observe a large increase in services. I spent a number of days in Nashville recently trying to link the documentation of increased services to decreased incidence of mental retardation and related outcome measures. This is difficult territory, and we were unable at least at this point to make these connections to my satisfaction. I think if a state is going to make a substantial investment in prevention, the ability to record service levels and measure the effects is extremely important.

------◄•••►------

What Tennessee has now done, in 1984, is to shift from an initiative for prevention of mental retardation to an overall healthy child initiative and to incorporate the prevention plan right into it. This is a very sensible way to go because many of the things one would put into a prevention program are, in point of fact, simply relevant to a generally healthy outcome for children.

There are many areas left for our attention. Large ones include securing a better understanding of the background or causation of fetal malformation, chromosomal aberrations, altered behavior states, psychoses, and autism. We need to extend family-life education and public education, to encourage universal development of state plans (now present in only 15–20 states), to look at the effects of social disadvantage more broadly, and to work for better acceptance of human variation.[8]

I had a chance to speak in Iowa a few weeks ago, and Al Healy, who was introducing me, saw fit to use a few lines from Emily Dickinson[9] to indicate the motivation that characterizes prevention programs. I was very touched by that and offer the poem to you now:

> If I can stop one Heart from breaking
> I shall not live in vain
> If I can ease one Life the Aching
> Or cool one Pain
>
> Or help one fainting Robin
> Unto his Nest again
> I shall not live in Vain.

This indeed is the kind of aspiration for our children that rests behind the prevention movement.

ACKNOWLEDGMENTS

Special appreciation is expressed to Ms. Marguerite W. Sallee and Ms. Susie M. Baird, Department of Health and Environment, Nashville, TN.

REFERENCES

1. THOMPSON, G. H., I. L. RUBIN & R. M. BILENKER, Eds. 1983. Comprehensive Management of Cerebral Palsy.: 53. Grune & Stratton, Inc. New York.
2. CROCKER, A. C. 1982. Current strategies in prevention of mental retardation. Pediatr. Ann. **11:** 450–457.

3. LEVY, H. L. 1985. Personal communication.
4. CROCKER, A. C. 1986. Societal commitment to prevention of developmental disabilities. *In* Advances in Prevention of Developmental Disabilities. S. M. Pueschel & J. A. Mulick, Eds. Academic Guild Publishers. Cambridge, MA. In press.
5. HADDOW, J. E. 1985. Personal communication.
6. GREENSTEIN, R. M. 1985. Personal communication.
7. KABACK, M. M. 1985. Annual update of TSD carrier detection and prenatal diagnosis experience. Report to the National Tay-Sachs and Allied Diseases Association.
8. CROCKER, A. C. 1983. The thoughtful thirteen. *In* Assessment of the National Effort to Combat Mental Retardation from Biomedical Causes; Conference Proceedings. Department of Health and Human Services. Washington, DC.: 14–20.
9. JOHNSON, T. H., Ed. The Complete Poems of Emily Dickinson. Little, Brown. Boston, MA. p. 433.

DISCUSSION

EDMUND JENKINS: How did the Tennessee program originate?

ALLEN CROCKER: A blue-ribbon panel was put together that was called the Governor's Task Force on Mental Retardation Prevention. It was chaired by Honey Alexander, the governor's wife. When the task force's work was finished, it was assured that it would not die because the responsibility for carrying out the recommendations of the task force was assigned to the Children's Services Commission, an entity that reports directly to the governor. The program persisted as a visible and important element. Much can be learned from what Tennessee has achieved, for this state has done a particularly exemplary job of finding ways to accomplish prevention activities.

ROBERT GUTHRIE: Dr. Crocker is too modest to mention what Massachusetts has done on lead, but I have to mention it because it is so unusual. Only 3% of children under six in the United States are being screened every year according to the Centers for Disease Control. Of those 3% (about a half-a-million children), one-fifth of them, 100,000 children, live in Massachusetts. Massachusetts screens 85% of the children in Boston and one-fourth of all of the children across the state. Massachusetts spends as much state money as federal money; it is probably the only state to do so.

PETER VIETZE: Although the governor and his wife have been given credit for the Tennessee program, I think kudos should also go to Carl Haywood, who brought mental retardation to the attention of the governor's office and had a lot to do with organizing that commission.

ALLEN CROCKER: Carl Haywood has been a potentiating figure throughout; I am glad you mentioned him. I would also like to pay tribute to Marguerite Sallee, who has been the professional troubleshooter within the Department of Health and Environment and who has stayed on top of the data collection and helped keep public attention focused on the program.

IRVIN EMANUEL: I think that any discussion of the prevention of mental retardation should include the role of elective abortion. As we look at the national statistics on abortion, about three-quarters of elective abortions are performed on unmarried women, and about one-third on teenagers—both high-risk groups. Older women, thirty-five and above, also make disproportionate use of elective abortion. These groups are at high risk for poor pregnancy outcome. Needless to say, the increasing role that prenatal detection can have for prevention of developmental disabilities depends on abortion rights, which are now being threatened in this country. In addition, the right to other types of family planning is being threatened. I think we should all be aware of these trends, and do what we can to combat these negative forces.

An Overview of the Medical School Curriculum in Mental Retardation and Strategies for Change

ELENA K. LESSER

Assistant Dean for Education
Downstate Medical Center
Brooklyn, New York 11203

Three points pertaining to medical school curriculum in mental retardation and developmental disabilities will be discussed in this presentation.

(1) The current state of affairs with respect to the undergraduate education of physicians in mental retardation and developmental disabilities (mental retardation henceforth);

(2) How the current state of affairs came to be; and

(3) A direction and proposed strategies for curricular change from the perspective of a medical educator—a curriculum generalist outside of the field of mental retardation.

THE CURRENT STATE OF AFFAIRS

Let us begin with an event that came to mind when the author searched her life experience for a link to the field of mental retardation. She had a job for a short time as a hospital volunteer before going to college.

One day, the Director of Volunteer Services was speaking with a colleague. "You'll never believe what just happened," she said. "The Dean came marching down here himself. He was upset because Tommy delivered his package to the wrong office. The Dean asked who had picked up the package. I looked it up, and told him it was Tommy. The Dean began yelling at Tommy for making the mistake—he carried on like a maniac, and I just couldn't stand to hear it so I finally screwed up my courage and interrupted him. I said, 'Excuse me, sir, but the boy couldn't help it—he's retarded.' Now wouldn't you think they'd teach them about retardation in medical school? He's a doctor, isn't he?" It had never occurred to her that it was possible to graduate from medical school without being able to recognize such a common human condition. The implications were frightening.

Data in an article by Oster[1] appearing in the early seventies estimated the amount of curriculum time in California devoted to mental retardation to be approximately ten hours, one long day, out of four years, or only 0.1% of total time. According to Holt and Huntley,[2] most schools in Britain in the same period offered one week or less of the four years. Neither article described the specific content or experiences being counted, but the estimates are close enough to lend credence to one another.

Now the situation has improved. Most schools offer substantially more than a day or a week on subjects related to mental retardation. The necessary topics are "covered." The climate has changed. While not exactly as fashionable as medical ethics or geriatrics, mental retardation has achieved respectability. There is a growing recognition that physicians should devote more attention to mental retardation than they have in the past. The factors that have led to this point include:

• The Kennedy family's commitment;
• The Willowbrook exposé;
• The principle of normalization—physicians see more mentally retarded patients;
• Knowledge of the extent to which mental retardation is preventable;
• The existence of Kennedy Centers, which stand as symbols for unfinished business.

Yet despite the increased time and more favorable climate, a problem still exists. Even when physicians are in role and aware that they are dealing with a mentally retarded person, they have problems. Willer, Ross, and Intagliata[3] reviewed the literature and found physicians were still "criticized for their role and manner in dealing with retarded persons and their families." Physicians were called "abrupt," "unwilling to counsel family members," "unaware of many basic facts of mental retardation and its etiology," "uninformed about alternative services," and "much too willing to recommend institutionalization."

According to a study by Holt and Huntley,[2] physicians themselves feel unprepared to treat and work with mentally retarded patients. Apparently, whatever they are learning in medical school is not clinically useful enough. The solution, however, lies not with more curricular time, nor even with more course content. What needs to be changed is the use of time, the teaching process, methodology, and emphasis.

The Willer, Ross, and Intagliata paper is interesting because the authors surveyed all of the U.S. medical schools to determine the type of undergraduate curriculum offered in mental retardation. Approximately half of the schools responded, and of the responding schools, only half offered curricula the authors considered acceptable.

By the authors' own admission, the criteria established for acceptability were minimal. They were: "Some specific mental retardation topics in the core curriculum, a clinical component which could be either some offered lectures, or lectures coupled with a visit to a state institution." A school's mental retardation sequence was considered acceptable if an elective opportunity was offered, "even if only a few students availed themselves of it."

Yet even with this minimal goal, only half the schools were considered acceptable. Apparently a student could graduate from one of the accredited medical schools without ever having worked up a mentally retarded patient.

Let us examine the specifics about what is wrong.

(1) The search revealed no well-established goals or objectives for a program or sequence of educational experiences for medical students; no agreed-upon training standards. Standards have been published for pediatric and psychiatric residencies, and they undoubtedly exist for neurology and for the fellowships. But little seems to be available at the undergraduate medical student level.

(2) No core curriculum, nor specific curriculum content in mental retardation has been established or promulgated. A set of competencies was developed at the University of Rochester[4] and tested with good results for a unit within a fourth-year clerkship, but results were published after just twelve students had rotated through. For the most part, curriculum in mental retardation at the undergraduate level tends to be uncoordinated, disconnected bits and pieces instead of a coherent, unified whole.

Every reader knows the old chestnut about five men who can't agree on what an elephant looks like because each views the elephant from his own vantage point. None sees the whole, and each man winds up with a different definition of an elephant.

What faculty educators seem to be doing is giving each student the whole elephant, but only after it has been cut up into bits and pieces. Never having seen a whole elephant, the student can't reconstruct an image out of this jigsaw puzzle. When the elephant finally materializes in the form of a mentally retarded patient, the student may not recognize it. The bits and pieces don't automatically come together and students have trouble making them fit. Why not let the students see the whole elephant first?

(3) Much of what is taught in preclinical lecture courses or in lecture format within the clerkships, is detached from a meaningful clinical context. This violates what we know serves human memory and learning. The passive lecture characterizes much of medical education and is certainly not unique to mental retardation. But it is nevertheless a serious problem.

(4) Clinical contact with mentally retarded patients is neither mandated nor scheduled so it tends to be hit or miss. Even if a student does happen to encounter a mentally retarded patient, there is no guarantee that the student will be observed or supervised. A student's discomfort with the patient or negative attitude is likely to go unnoticed.

(5) The particular mental retardation curricular emphasis that has evolved is placed on biochemical–genetic causes, not on sociological–familial aspects—an insufficient emphasis on prevention, services, and treatment. The set of published goals for pediatric residency by Cohen and Diamond[5] makes no direct reference to prevention.

(6) Little or no information is disseminated, nor contact planned, with other professions that deal with mentally retarded persons—social work, occupational therapy, physical therapy, special education, psychology, neurological psychology, and so on. The role of the physician in this multidisciplinary constellation is unclear. There are insufficient distinctions in the functions of the various specialists within the field of medicine.

(7) New areas such as medical ethics and law in relation to mental retardation need to be incorporated.

HOW DID IT COME TO BE THIS WAY?

The reasons appear to be primarily psychological. We fear what is different from us. We fear an affliction *more* when we sense we *could have* had it. I think on some level I'm more comfortable looking at the elephant piece by piece. The whole elephant is a bit awesome and frightening. I had a little contact with mental retardation patients in a clinical psychology postdoctoral internship. I did testing and futile attempts at psychotherapy. I felt very insecure.

As a very verbal person, it is frustrating and disconcerting not to be able to communicate verbally and get an appropriate response. I count on my verbal skills to make contact with others. Some mentally retarded persons are embarrassing and unpleasant to look at. They call up my own fears of not being intelligent or attractive enough.

My children are intellectually normal (at least they act that way most of the time). I feel guilty when I think of what parents of mentally retarded children must experience. It makes me feel I have no right to complain about my offspring.

There are no dramatic turnarounds working with retarded children or adults. One must be satisfied with small changes. There are exceptions, especially when the retardation is secondary to psychological problems, but by and large there are no cures.

Finally, in a strange way I found that I envy the retarded. They seem to be free of life's responsibilities, are not as burdened as I often feel. It is probably a myth, but I suppose I envy their guilt-free dependency, their eternal childhood.

What an incredible amalgam of powerful and inconsistent feelings. I wouldn't know how to begin sorting them out. Putting them on the table is hard enough. If I lacked the opportunity to sort them out as medical students do, I think I would react with avoidance. It wouldn't really be a conscious decision. It would just happen.

WHAT TO DO ABOUT IT? SOME SOLUTIONS AND STRATEGIES FOR CHANGE

What can be done within the current system?

(1) We need to develop a set of agreed-upon goals and objectives for training of all physicians at the undergraduate level, and promulgate them. Since medical schools don't consider mental retardation one of their high priorities, they are not about to initiate action on their own. The associations and institutions devoted to service and research in mental retardation–developmental disability should take the initiative, with or without collaboration with medical school faculties. These groups should provide the medical schools a standard to react to, to modify if necessary, and challenge, but even if its value is only heuristic, provide *something*. This standard could be based on what is available for residency level and adapted for mental retardation training. The role that nonspecialist physicians should be prepared to play with mentally retarded patients and families needs to be conceptualized, agreed upon, and spelled out.

(2) A core content needs to be specified, with essential knowledge, skills, and attitudes to be learned. The current emphasis needs to shift toward prevention, and should include medical ethics and law.

(3) Although it may be necessary at this stage in medical curricular–mental retardation history to blend into the prevailing departmentally organized basic science course structure that characterize the first two years of study in most medical schools, some student contact with mentally retarded people very early in medical school curriculum would be valuable. I approve of appropriately conceived and guided clinical experience for all medical students from day one. Otherwise it does not appear to me that what is taught in the basic sciences is retained. Students should examine at least one mentally retarded patient in the physical diagnosis course. The clinical context is crucial. It is difficult to convince physicians and basic scientists of this method of education because most did not experience it.

(4) Not everything needs to be learned through direct clinical contact. Active small-group or individual problem-solving sessions are also valuable and should include as much clinical case material as possible. There are a number of problem-solving methods and formats being developed for basic science teaching in medicine, and I would urge you to explore them.

Dr. Howard Barrows, for example, who is a nationally recognized problem-solving expert, has developed forty Problem-Based Learning Modules (PBLM),[6] in use at McMaster University and currently at perhaps as many as half the medical schools in the country. It seems that none of these modules illustrate the problems of mentally retarded patients; clearly, some should be created.

Dr. Barrows, Dr. Paula Stillman at the University of Massachusetts, Worcester, and a number of other medical educators are using live-simulated patients for teaching and evaluative purposes. Are any of these "patients" mentally retarded or developmentally disabled? The newer, more active teaching methodologies have to be exploited and made available to medical school faculty. This is the direction in which the schools are moving.

(5) Willer, Ross, and Intagliata[3] recommend that medical schools require a chronic illness clerkship, and that each student work with a number of mentally retarded–developmentally disabled patients. This seems like an excellent idea. Such a clerkship should be longitudinal, rather than a single block, giving the student an opportunity to work with chronic patients over time, as happens in practice. The objective would be to teach students a sense of what work with chronic patients is like and to give students a chance to develop patient-management skills in this area. Every medical student should be given the opportunity to explain prevention of retardation to a parent of a first-born, preventably retarded child.

(6) Little will really change until students are properly evaluated on their knowledge, skills, and attitudes about mental retardation and the developmentally disabled. The first step would be to evaluate the knowledge of graduating seniors. There is a national trend toward a final comprehensive clinical evaluation for all medical students. Those concerned about appropriate training should be investigating opportunities to insure that at least some mental retardation–developmental disability problems are included.

(7) Nothing will change until faculty sensitivities are cultivated. Faculty development seminars, continuing education programs that are experiential in nature and confront attitudinal issues head-on, are probably necessary. Opportunities for faculty who have not had significant contact with MR patients should be arranged. Faculty members need to increase their own comfort level to help students to do so. If the faculty does not increase their own level of understanding, they will continue to offer students an elephant in pieces.

REFERENCES

1. OSTER, J. 1974. Training of medical and dental students in mental retardation. Acta Paediatr. Scand. (Suppl.) **246:** 7–37.
2. HOLT, K. S. & R. M. C. HUNTLEY. 1973. Mental subnormality: Medical training in the United Kingdom. Br. J. Med. Educ. **7:** 197–202.
3. WILLER, B., M. ROSS & J. INTAGLIATA. 1980. Medical school education in mental retardation. J. Med. Educ. **55**(7): 589–601.
4. SIMEONSSON, R. J., W. KENNEY & L. WALKER. 1976. Child development and disability: Competency-based clerkship. J. Med. Educ. **51:** 578–81.
5. COHEN, H. J. & D. L. DIAMOND. 1984. Training and preparation of physicians to care for mentally retarded and handicapped children. Appl. Res. Ment. Retard. **5:** 279–91.
6. BARROWS, HOWARD S. 1985. How to Design a Problem-Based Curriculum for the Preclinical Years. Springer Publishing Co. New York.

DISCUSSION

ALLEN CROCKER: Part of the dilemma is caused by the split between the hard science aspect of mental retardation and the philosophical and societal aspects of this problem. We have had pretty good success in engaging students and house staff in areas that are clearly molecular or genetic—these students can be moved into the area of hard science. Although some of these hard science topics may not stand out as mental retardation per se, it is sprinkled through the curriculum and in the later training experiences. The other half of the problem, however, is societal, in the sense that the mentally retarded are a group of individuals with very special needs. Here our medical schools do not have a particularly distinguished record. How would you look at the social meanings of human variation as a school responsibility?

ELENA LESSER: I feel very strongly that medical schools have a responsibility to consider the social implications of human variation. I do not think there is too much debate about that. The question is how. My own feeling is that the way to do it is to introduce as much medicine as possible—including basic science—in a meaningful, clinical context. I do not see how you can teach "human variation" in the sterile confines of the lecture hall and expect students to come away understanding how mental retardation affects mentally retarded persons and their families. It is a more convenient way to teach, but it does not work.

HENRY WISNIEWSKI: I was very happy to hear what you said because we have been struggling with this problem for quite awhile. As you know, the field of mental retardation has an identity crisis; it is a multidisciplinary in nature. I wonder whether a new discipline of mental retardation and developmental disabilities might be born. . . . Let us put the discussion outside the medical establishment in terms of medical schools, and see whether this discipline can be identified with a clear-cut curriculum. I'm not sure that many of our friends from pediatric neurology, developmental pediatrics, pediatrics, or child psychiatry will subscribe to this idea. It is one thing that probably needs in-depth analysis.

JOSEPH FRENCH: Apropos of that comment by Dr. Wisniewski, demography and economics may be on our side. Because of the success of modern medicine, the chronically ill may be the only patients left.

DAVID SOIFER: I'd like to comment on Dr. Crocker's comment and ask him a question. A student may learn how to *recognize* a person with Down's syndrome, but does he or she learn in medical school how to *serve* someone with Down's syndrome? Does he or she learn how to take a history or to do a physical examination on someone who cannot communicate? Does the student learn how to deal with the developmentally disabled individual as a medical patient, as well as learning how to identify a specific lesion that brings about a developmental disability? Hard science about mental retardation may be creeping into the curriculum, but the management of the mentally retarded patient is not covered.

Dr. Lesser, is there room to include the developmentally disabled in general medical courses, such as those dealing with physical diagnosis? Can you get patient populations that are available to students, so that they can learn to work with retarded persons while they are in medical school?

ELENA LESSER: At least one of the patients a medical student encounters in the physical-diagnosis/medical-interviewing portion of his or her training should be a mentally retarded person. This would be an ideal point to enter the curriculum.

There is a fair degree of acceptance for teaching medical students by an organ-system approach. I think what we are reaching for in this discussion is going a step further, which would mean a curriculum based on *whole* patients, that is, focused on whole patients' medical problems. This is how the curriculum is focused at McMaster University, for example. Students begin with a presenting problem, working first with paper and pencil, then with simulated "trained" patients, and finally with actual patients. The type of problem students encounter becomes increasingly complex as they gain knowledge and experience. The beauty of this approach is that it propels students into the physician-in-training role from the outset, and never allows them to lose sight of the whole person.

PEGGY BROOKS-BERTRAM: To increase the interest of doctors in the mentally retarded, we might go back to high-school career programs. These might be structured so that highly motivated young people could explore the field of mental retardation. Programs might allow students in the last years of high school to participate in special health career programs with university-affiliated faculties, so that they could have experiences with interdisciplinary teams looking at the problems of mental retardation.

ELENA LESSER: Our society segregates the retarded. We cannot reasonably expect medical students to bridge the gap singlehandedly. Judicious, carefully planned "mainstreaming" of mentally retarded persons into regular elementary school classrooms would also help to close the gap. Persons admitted to medical school would then be guaranteed to have had significant natural contact as children with persons who are mentally retarded. They would have had contact with mentally retarded children in the course of elementary school classes. There would be less of a barrier to overcome.

Another very interesting method would be to have medical students learn to "role-play" a retarded person, and spend a few hours being reacted to by strangers, by physicians. There are any number of approaches that could be effective.

Developmental Disabilities in Health and Medical Education: Status and Strategy for Change in the Curriculum

Practicum Experience

ROBERT B. KUGEL

Department of Pediatrics
Terence Cardinal Cooke Health Center
New York, New York 10029

Medical education today is buffeted by many conflicting forces. Some of the factors are entrenched interests; some are due to reluctance to alter what most people feel is adequate; and some are the competitive factors among medical school departments, all of which believe their subject matter is the most important. One factor, however, that has always been present is that of time. In almost all medical schools, the demand for new courses is greater than the time available in the standard four-year curriculum. In addition, existing courses continually add new subject material as new medical discoveries come into being, thereby making the demand for time even greater. Unfortunately subject matter that is thought to be of lesser importance is often especially difficult to place into the overcrowded medical curriculum.

DIFFICULTIES AND OBSTACLES

A major issue for the medical student or resident is to appreciate the contrast between acute care and chronic care. Medical teaching has been dominated for many decades by the emphasis given primarily to acute-care matters. Little time and attention have been given to discussions about the differences between acute and chronic care. What everyone likes about acute care is the brevity of the patient contact and the satisfaction of seeing a ready and presumably satisfactory solution to a problem. On the other hand chronic care often means dealing with situations for which there is no one answer and where problems often go on without discernable resolution or often without the possibility even of amelioration.

The hospital is frequently staffed by all manner of people who lack concern for chronically ill, long-term-care patients. Uncomplimentary terms such as "crock" and "gork" are frequently used to describe those patients who require long-term care. To a large extent the attitudes that physicians have are identical with those that predominate in society as a whole. While it is acceptable to have a three- to five-day illness, for which there is an ultimate recovery, it is often very disconcerting for the patient, his family, his school, or his employer to deal with the needs that have long-term aspects, requiring care of many years. The needs of the long-term patients are quite different and in the planning for these patients, the physician must take these needs into account.

When it comes to dealing with mental retardation or any of the developmental disabilities, these negative attitudes become especially pronounced. Not only is there negativism with reference to long-term care, but also there is negativism with reference to the subject matter itself. In addition, during the past 10–15 years, a lot of hostility has been directed toward medicine in general, especially in so far as how it relates to mental retardation is concerned. The negative statements about the medical model and the negative statements about mental retardation only make trying to entice the young student into this field all the more difficult.

An additional concern that comes to the attention of curriculum committees is the issue of prevention versus cure. When viewing the field of mental retardation, the issues about care seem so complex that many educators would prefer to talk about the possibility of prevention. While prevention is certainly an important feature, it should not eclipse the very real need for long-term care in all its dimensions and diversity.

THE PRACTICUM EXPERIENCE

The definition of practicum varies greatly, but generally embodies the idea of having an experience that deals with a particular subject or condition in a practical manner in contrast to a theoretical one. In medical education the clerkship is often equated with a practicum experience in which a student spends a specified period of time in one or another subspecialty area in medicine, seeing patients, doing various examinations and procedures, and learning from first-hand experiences.

Often the practicum experience or clerkship may be quite unstructured or unsupervised. While such an arrangement may be useful for some students, it is frequently not for others. Ideally the practicum experience should be one that combines opportunities for direct patient contact and at the same time offers opportunities for discussions so that the student has a chance to try out his thoughts and ideas about his patient, but under supervision. The practicum should be structured in such a way that it does not compete with other tasks and responsibilities.

If the medical school curriculum is to be modified so that it influences large numbers of students, then a rotation or practicum experience needs to be made mandatory as opposed to being an elective. Too often schools develop a whole series of elective clerkships that are seldom selected by any student. Though such listings are often used to impress survey groups with the comprehensiveness of the curriculum, they accomplish little in the matter of influencing the prevailing understanding of students and their later actions as physicians.

A PRACTICUM IN MENTAL RETARDATION

Ideally in a practicum experience concerned with mental retardation, there should be opportunities to see and understand what other agencies dealing with mental retardation have to offer. Such agencies should include social agencies and educational agencies, and there should be information provided about the different role of the public versus the voluntary-type agency.

One special consideration relates to the question of the amount of time that should be devoted to a practicum experience. The usual clerkship experience might vary from one week to one month. Although there is no set time, in the case of chronic disability, it takes at least a week to orient a student to the concept of chronicity with all of its implications and therefore, most authorities would think that a rotational time of one month would be appropriate.

A special problem peculiar to the field of developmental disabilities is the role being assigned to the physician. Throughout medical training physicians are taught that for the most part they operate as individuals. Even though with greater experience the student learns that this concept is not always appropriate, he may find the field of mental retardation particularly confining and confusing, with its layering of committees and other groups that are concerned about the welfare of people who are mentally retarded. There is a special requirement to appreciate and present the importance of the broad, most all-encompassing approach to health as opposed to a more narrowly defined concept of illness only. Both nursing and medicine may be included under the rubric of health, but the issues facing each are clearly not identical.

Incumbent upon any learning process that will take place during a practicum experience is to offer opportunities to understand the wishes and desires of various groups, such as, parents, planners, educators, and therapists. These factors are important features that should be made part of the practicum experience.

While there certainly is a need to have a particular focus on traditional clinical problems as seen in mental retardation, opportunities for presenting some research issues should not be overlooked. An example might be the epidemiology of lead poisoning in New York City. Various items of basic research as well as applied research would be appropriate.

OTHER EXPERIENCES

There are a variety of other opportunities for gaining new understanding about mental retardation. One such program is the Masters in Public Health Program that has been recently developed by New York Medical College. This program offers certain students the opportunity to have a special diploma designation of Masters of Public Health in Developmental Disabilities. This 45-hour program offers the student with advanced training an experience and opportunity to further improve his skills in the context of the MPH program. Students in the health sciences (medicine, nursing, physical therapy, occupational therapy, speech therapy, etc.) as well as others, such as education and psychology, might well find this approach appropriate to their needs.

Another special area akin to a practicum is the residency program. Pediatrics has long had an interest in the person who is developmentally disabled, but there has only recently been much interest in internal medicine and in neurology. These areas need to be strengthened. Clearly with the aging of the whole population, the need for special concerns that relate to the older person who is mentally retarded will become increasingly important. Time in the residency program for these considerations must be made available if the practitioners for the future are to accommodate their practices to include people with developmental disabilities.

DISCUSSION

ROBERT GUTHRIE: A popular program at our university medical school is one in which medical students attend meetings run by the Association for Retarded Citizens at which they listen to a panel of parents who describe their initial contact with physicians when their child was born. The parents also take questions from the students. This program has been an important, useful, and popular one with medical students.

TERRENCE DOLAN: Three years ago, there was not a significant component in the curriculum of the medical school at the university of Wisconsin addressing mental retardation and developmental disabilities. In the last three years, however, the university has developed new training opportunities for residents. In that newly developed program, the Waisman Center UAF has become a required month-long rotation for all pediatric residents, as well as for residents from the departments of pediatric neurology, psychiatry, orthopedics, and rehabilitative medicine. The most significant thing we can teach these students in a single-month program is to remind them that they are a part of a larger system in two senses. First, they are part of a team that involves other disciplines that all have an important role to play in the care and treatment of developmentally disabled children. Second, they are part of a larger system that involves other organizations, agencies, and support groups that also play important roles in the treatment of these children. At the termination of the training exercise, then, it will be more clear to the trainees that the practicing physician has much knowledgeable support available to him or her from other organizations and disciplines.

JAMES GALLAGHER: I just wanted to mention a training program that a pediatrician, Mike Sharp, has conducted at the University of North Carolina that involves community placement for pediatricians in training; it also involves their going through a videotape simulation that I found to be very useful. If you want to see something very painful, watch a physician in training try to explain, in simulation, to parents that they have a Down's syndrome child. It is a very traumatic experience for the trainee, but it is also a very valuable one.

Introduction

DONALD A. SNIDER

Institute for Basic Research in Developmental Disabilities
New York State Office of Mental Retardation &
Developmental Disabilities
Staten Island, New York 10314

In his remarks at the opening of this workshop, Arthur Y. Webb, the Commissioner of the New York State Office of Mental Retardation and Developmental Disabilities, spoke of the necessity for interaction between science, education, and service to create excellence in the field of mental retardation. This session on technology transfer embodies this idea.

Mr. James G. Hill, Chief of the Office of Planning and Evaluation of NICHD, leads with a description of the rationale and process for consensus development conferences sponsored by that institute. This process is used when there is a scientific controversy or a gap between current knowledge and practice on a medically important issue.

Dr. James Gallagher, Director of the Frank Porter Graham Child Development Center at the University of North Carolina, notes how few scientists concern themselves with how their discoveries become distributed, applied, transformed, or used in the larger society. Dr. Gallagher proposes a model for using knowledge to influence practice and suggests how this model could be tested empirically.

Finally, I attempt to examine technology transfer in the medical field and postulate several ideas that might enable those of us in the field of developmental disabilities to better capitalize on existing and future technological capabilities.

Technology Assessment at the National Institute of Child Health and Human Development: An Alternative Model

JAMES G. HILL

Chief Office of Planning and Evaluation
National Institute of Child Health and Human Development
National Institutes of Health
Bethesda, Maryland 20205

Technology assessment has been undertaken in response to concerns for cost, safety, and efficacy of medical devices and procedures[1] as medicine comes to rely more and more on technology for diagnosis, treatment, and rehabilitation. The National Institute of Child Health and Human Development (NICHD) has assumed responsibility for technology assessment in the areas of its responsibility. The NICHD, a noncategorical institute of the National Institutes of Health (NIH) is concerned with human development in all its aspects. The programs of the NICHD include one, Mental Retardation, that is focused directly on mental retardation and developmental disabilities, while others—Pregnancy, Birth, and the Infant; Birth Defects; Nutrition; and Human Learning and Behavior—are closely related (TABLE 1).

Expenditures for technology assessment and transfer are found in most programs of the Institute and totalled $11,233,327 in 1984 (TABLE 2).

Although the dollar amount represents a small percentage of the total NICHD budget, the public policy implications of the research programs supported (or potentially supported) by the NICHD are profound (e.g., fetal research, antenatal diagnosis, treatment of handicapped infants—"Baby Doe,"—*in vitro* fertilization, and other alternative means of reproduction.) The NICHD activity that provides the data for technology assessment is clinical trials. They provide the data for formal activities in technology assessment, such as Consensus Development Conferences. Examples of clinical trials include:

- Stereotyped behavior and thioridiazine pharmacodynamics
- Cooperative clinical comparison of chorion villi sampling and amniocentesis
- Effects of maternal PKU on pregnancy outcome
- National collaborative cysteamine study

A total of 52 Consensus Development Conferences have been held at the NIH since 1977. Of this total, six have been related to mental retardation and developmental disabilities. The NICHD has developed and conducted three of these conferences, and assisted in two others (TABLE 3).

It is the purpose of this paper to describe the methodology used at the NICHD conferences and discuss its unique advantages. Before we go on, however, it should be noted that NICHD formal efforts in technology assessment are not confined to consensus conferences. In April 1985, a two-year effort sponsored by the NICHD and NINCDS to illuminate the "Prenatal and Perinatal Factors Associated with Brain Disorders" resulted in the publication referred to by Dr. Brann earlier in this conference. In addition, the institute sponsors, and cosponsors,

TABLE 1. The Programs of the National Institute of Child Health and Human Development (NICHD)

- Pregnancy, Birth, and the Infant
- Birth Defects
- Sudden Infant Death Syndrome
- Mental Retardation
- Nutrition
- Physical Growth
- Human Learning and Behavior
- Fertility, Infertility
- Contraceptive Development
- Contraceptive Evaluation
- Population Dynamics

TABLE 2. NICHD Technology Assessment/Transfer, FY 1984

Clinical Trials	$10,975,650
Technology Transfer (extramural)	257,677
NICHD 1984 Actual Obligations	$226,360,245

TABLE 3. NIH Consensus Development Conferences Related to Mental Retardation and Developmental Disabilities

Sponsor	Title	Date Held
NICHD	Antenatal Diagnosis	March 1979
NINCDS/NICHD	Febrile Seizures	May 1980
NICHD	Cesarean Childbirth	September 1980
NINCDS/NCI	Computed Tomographic Scanning of the Brain	November 1981
NIAID/NICHD	Defined Diets and Childhood Hyperactivity	January 1982
NICHD/FDA	Use of Diagnostic Ultrasound Imaging in Pregnancy	February 1984

public forums such as one held in November 1983 on "Gene Therapy," and an upcoming forum to be held in July 1985 on "Malpractice Issues in Childbirth" (with the International Childbirth Education Association and the Division of Maternal and Child Health, Health Resources and Services Administration, PHS).

The NIH consensus development conferences are developed under guidelines[2] promulgated by the Office of Medical Applications of Research (OMAR), a part of the Office of the Director, NIH. Operating within the broad purview of these guidelines, the NICHD has developed its own unique methodology to handle the difficult topics that it selects, the strong public policy implications that derive therefrom, and the needs of transfer and dissemination. The three Consensus Development Conferences, developed and conducted by the NICHD, are "Antenatal Diagnosis,"[3] "Cesarean Childbirth,"[4] and the "Use of Diagnostic Ultrasound Imaging in Pregnancy."[5] The reports of these meetings have been widely disseminated to the public, practioners, and policymakers. The reports have been used extensively in residency training programs, as well as in other teaching settings. We are currently conducting an evaluation study to determine the impact of the "Cesarean Childbirth" report on repeat sections and hospital policy regarding the father in the delivery room.

The process begins with the selection of a topic. During this process, the NIH-OMAR guidelines are particularly useful. They state that a topic selected for a conference must meet the following criteria:

1. The subject under consideration should be medically important.
2. There should be a scientific controversy that would be clarified by the consensus approach or a gap between current knowledge and practice that a Consensus Development Conference might help to narrow.
3. The topic must have an adequately defined and available base of scientific information to answer the previously posed questions.
4. The topic should be amenable to clarification on technical grounds and the outcome should not depend mainly on the impressions or value judgments of panelists.
5. The timing of the conference should be such that it is likely to have a meaningful impact; that is, it should neither be so early in the developmental course of a new technology that data are insufficient nor so late that the conference merely reiterates a consensus already arrived at by the profession.

In addition, the following desirable elements are identified for the topics:

1. Public health importance. The topic should affect a significant number of people.
2. Health care cost impact. The topic may have implications for reimbursement by agencies such as the Health Care Financing Administration.
3. Preventive impact.
4. Public interest.

The topics selected by the NICHD have met all the primary and secondary criteria. Once a topic is selected, a planning meeting is held. Three tasks are undertaken at the meeting:

1. Selection of a chairperson;
2. Specification of the types of individuals to serve on the panel (e.g., physicians, basic scientists, epidemiologists, etc.);
3. Identification of the questions to be answered by the panel during the course of the consensus process (e.g., "For what purposes is ultrasound now used in pregnancy? For each way, what is the evidence that ultrasound improves patient management and/or outcome of pregnancy?").

Selection of a chairperson is a critical step in the process. The individual must be respected, knowledgeable, and possess credibility in the various communities involved in the topic without being an advocate for one point of view. He or she must also be skillful in dealing with controversial subject matter in a public setting. The composition of the panel is critical as well. In addition to physician members, both academic and practicing, others are included, such as, allied health personnel, basic scientists, behavioral and social scientists, and epidemiologists. In addition, nonscientific members of the panel are important and include (when appropriate) lawyers, ethicists, and professional managers. Lastly, patients and patient advocates are included. NICHD has had singular success in locating superb patient advocates who have made major contributions to the three panels that we have convened.

Once a chairperson is chosen, a panel selected, and questions identified, a major literature search is undertaken with the assistance of the National Library of Medicine to identify the world literature with an emphasis on the availability of data. The chairperson calls the first meeting of the panel about a year or 18 months before the consensus conference. At the first meeting, the questions, the results of the literature search, and a formal "charge" delivered by the Associate Director of NIH for Medical Applications of Research are presented to the panel. The panel uses the time between its initial meeting and a month before the conference to produce a draft report that includes a draft consensus statement. The panel may take steps to fill gaps in the literature by holding a hearing to receive unpublished data (which was done in the case of "Diagnostic Ultrasound") or create new data sets (which was done in the case of "Cesarean Childbirth"). All of the panel's activity is carefully documented in the draft report: The draft is widely circulated to all interested organizations and individuals. They are encouraged to submit written comments and/or to give oral testimony at the consensus meeting.

The consensus conference is an open public meeting. It consists of a summary presentation of the draft report by the members of the panel, invited comments on the draft report by senior figures in fields related to the topic, the presentation of additional data not considered by the panel, testimony from individuals and organizations on the draft report, and a closed executive session for the panel to resolve and include the additional information received with the draft report to produce the final consensus statement. The meeting ends in a final open session with the delivery of the final consensus statement by the panel chairperson. Subsequent to the meeting, the panel makes final changes in the draft report to bring it in alignment with the final consensus statement. The statement and the report are widely disseminated. An important part of the dissemination process is accomplished through cooperation with various professional organizations such as the American College of Obstetricians and Gynecologists, the American College of Nurse Midwives, the American Institute of Ultrasound in Medicine, the Society of Diagnostic Medical Sonographers, and so forth.

There are many advantages to using this methodology:

1. Thorough documentation and entry into the literature of sources of information and references through publication by the Government Printing Office.
2. Attendees are provided a thorough background document to prepare for the consensus conference.
3. All members of the panel (especially the lay members) are brought up to the same level of knowledge.
4. The report is available for the use of practitioners and teaching programs.
5. Consensus statements can be more detailed in response to the proposed questions.
6. Information sources are more complete because of the inclusion of:
   a. MEDLARS searches,
   b. Unpublished data.
7. Organizational participation broadens the information sources and enhances dissemination.
8. The format provides more organized and complete presentation of background information.
9. The draft report provides a focus for comments and suggestions for change.

Thus, the methodology it has developed allows the NICHD to handle the delicate public policy implications of the topics that it has selected for technology assessment in an open, thoroughly documented manner. Through the publication of a complete final report, the public, health-care practitioners, the research community, and policy makers are provided with information needed to balance the "burden and benefit"[6] of high technology.

REFERENCES

1. 1978. Assessing the Efficacy and Safety of Medical Technologies: 3–4. Office of Technology Assessment. Washington, DC.
2. 1983. Guidelines for the Selection and Management of Consensus Development Conferences. Office of Medical Applications of Research. National Institutes of Health. Bethesda, MD.
3. 1979. Antenatal Diagnosis. National Institutes of Health. NIH Publication No. 80-1973. Bethesda, MD.
4. 1981. Cesarean Childbirth. National Institutes of Health. NIH Publication No. 82-2067. Bethesda, MD.
5. 1984. Diagnostic Ultrasound. National Institutes of Health. NIH Publication No. 84-667. Bethesda, MD.
6. JENNETT, B. 1985. High technology medicine: How defined and how regarded. Milbank Mem. Fund Q. Winter: 141–173.

DISCUSSION

MARCEL KINSBOURNE: One point that you didn't make explicitly, but that I am sure you are implying, is that at these meetings it is intended to have informed scientific opinion on both sides of the issue, rather than scientists ranged against the public, and that opposing scientific positions be represented on the panel. That did not happen in the meeting that you mentioned, "Defined Diets and Childhood Hyperactivity."

Let me develop this because it was perhaps a unique situation and not representative. One session of the two-day meeting had to do with diets (and that included the issue of sugar). The investigators who were on the panel were all ones who saw no merit in the additive-free diet.

However, there were two papers published in *Science* in 1980 supportive of the diet. None of the authors of those papers was invited to take part. Had they been, they might even have sided with their colleagues on the other side in minimizing the public health importance of the diet, although perhaps disagreeing that the diet has no effect. In my opinion, the fact that both sides were not represented weakens the chance that consensus will in fact be attained on the basis of such a meeting. It really defeats its purpose, which is an excellent one.

JAMES G. HILL: I completely agree with you. We had tried in that meeting to get the National Institute of Allergy and Infectious Disease to use a methodology closer to what we use; for various reasons, they didn't. Your comment really supports the approach we take. If indeed the NICHD had been running that meeting, then we would have formed the panel the way we usually do it and would have made sure that both sides were represented and that during the course of the development of the report the kind of interaction that you suggested would indeed take place. You are giving us an illustration of why the approach that we take is more valid.

Knowledge to Practice: A Researchable Issue

JAMES J. GALLAGHER

Kenan Professor of Education
Frank Porter Graham Child Development Center
The University of North Carolina at Chapel Hill
Chapel Hill, North Carolina

Technology transfer is often thought of as the effective movement and use of hardware from one place to another. However, technology has a much broader scope than that, as pointed out in a recent report of the Office of Technology Assessment[1]: "Technology in its broadest sense is the application of an organized body of knowledge to practical purposes. This definition encompasses physical objects such as wheelchairs or subway elevators and also processes, such as, vocational rehabilitation or reimbursement systems."

The transfer of valid research findings to various target groups in the society is a question of substantial impact to the society, but one that has received little research attention in its own right. The individual scientist is likely to consider his/her job well done if a significant problem is successfully attacked through the design of experiments, the construction of testable hypotheses, and the building of theories and models that explain or predict the phenomenon under study. The generation of new knowledge is the goal. Once a scientist has found significant information, the communication targets are colleagues in the scientific field and the knowledge-diffusion strategy may be to place a reprint in the mail.

RESPONSIBILITIES OF THE SCIENTIFIC COMMUNITY

Few scientists concern themselves with how their discoveries become distributed, applied, transformed, or used in the larger society, contenting themselves with the knowledge that "good science" would eventually be found useful. Their interests actually are often in their continued work, and their intellectual adventures in model development and hypothesis testing. But if the individual scientist can justify a lack of interest or responsibility in dissemination to the larger public, can the same be said of the larger scientific community? Are there not several major responsibilities that the scientific community needs to consider?[2]

1. To explain to the public the role that science plays in the reorganization of knowledge, and the expected time lag between *discovery* and *implementation*.
2. To cooperate enthusiastically with policies guarding against the potential of abuse of children in research studies.
3. To organize systems that maximize translation and delivery of valid knowledge to interested consumers.

How knowledge becomes used is a problem worthy of study in its own right, however, and there have been increasing attempts to study and to enhance such delivery of new knowledge.[3-7]

It is not unknown for significant knowledge to be delayed in its application by the unavailability or inability of other scientists to perceive the potential applications, or by an unfavorable political climate discouraging implementation. The toughest double play to make this season or any season is from the *laboratories* to *field tests* to *implementation*.

The insistence of the society, as represented by the Congress, that we should pay more attention to pathological conditions such as stroke, mental retardation, or cancer, rather than the more basic scientific problems of physiology or human development, does not imply lack of confidence in the scientific community, or in the scientists' ability to *eventually* discover causes and cures for these conditions. It is a reflection of impatience! To tell a White House staff person who has cancer that we will have this thing licked in half a century, or to tell a parent who has a child with mental retardation that in two or three generations we will be able to provide constructive answers, is not likely to be either encouraging or consoling.

The legislative focus on conditions such as mental retardation is made with the clear intent that scientific discovery, and the application on issues directly relating to that condition, would be accelerated. Part of that acceleration would come from systematically and promptly moving valid ideas and practices through major stages in the transfer of research to application. The question is, how does knowledge become applied appropriately at the direct service level and what are the means for minimizing the time necessary for such transport?

KNOWLEDGE TO PRACTICE

Levien[8] and Glaser[9] traced past scientific discoveries and have estimated that anywhere from thirty to fifty years is required under current circumstances to travel from the basic discovery to the useful application (See TABLE 1). In a previous paper[2] I had identified five stages through which ideas must pass before useful implementation takes place in education. The five stages listed in TABLE 2 involve the discovery of new knowledge, the application of that knowledge to a particular target group, the generation of systematic programing using this knowledge, the field test of programs involved in the use of the knowledge, and finally, the general applications in the field itself. The challenge is how to move the

TABLE 1. Time Lapse between Discovery and Implementation[a]

Innovation	Year of First Conception	Year of First Realization	Duration in Years
Heart pacemaker	1928	1960	32
Input–output economic analysis	1936	1964	28
Hybrid corn	1908	1933	25
Electrophotography	1937	1959	22
Magnetic ferrites	1933	1955	22
Hybrid small grains	1937	1956	19
Green revolution: wheat	1950	1966	16
Organophosphorus insecticides	1934	1947	13
Oral contraceptive	1951	1960	9
Video tape recorder	1950	1956	6
Average duration			19.2

[a] After Glaser.[9]

TABLE 2. Stages of Applying Results from Basic Research to Programs for Children with Disabilities

Stages	Characteristics
I. Discovery of knowledge	The generation of new facts, ideas, and concepts about children or social systems or instructional strategies.
II. Knowledge applied to target group	The knowledge available through Stage I must be applied to some groups of exceptional children and families by scientists interested in their special problems and familiar with the basic knowledge fields.
III. Generation of systematic programing for exceptional children	Knowledge gained in Stages I and II must be organized and specially designed to meet specific needs, problems, developmental levels, etc., of target groups of exceptional children. This material needs to be developed and applied in special environments where *control* is possible.
IV. Field test of programs	The merits of the special programs or procedures developed in III have to be proved in realistic situations. These trials demonstrate utility to local educators. Adequate evaluation is often missing from casual field testing and must be incorporated in effective programs.
V. General applications to the field	After sufficient field testing or trials of the procedures or programs have been conducted, there is general acceptance and adoption in the field. At this point, the new knowledge, now transformed completely becomes known as "common sense," or normal operations.

knowledge, and the practices built upon that knowledge, through these stages more efficiently, and to remove the barriers that stand in its way.

Education is one of those fields in which we aspire to bring to bear the maximum use of our knowledge (the health care delivery systems are another), and the enormity of the task is impressive.[10,11] There are currently over 40 million children ages five to seventeen in public school programs in the United States and 2-½ million teachers and support personnel in the public schools.[12] More than one out of every one hundred persons in the United States is directly involved in the reception of, or the providing of, educational instruction at the elementary, secondary, or postsecondary level. Furthermore, the educational enterprise is not arranged like a corporation where the power is at the top and orders flow down through the various levels of the organization. Instead, there are approximately 12,000 school districts making independent decisions, often with little more than a nod to state departments of education and only a passing glance at the United States Department of Education. How can we distribute good practice? How can we use the best of what we know how to do?

It is tempting to draw an analogy to the distribution of knowledge about the efficacy of new drugs, but that is not a good analogy. There is no profit motive in education to spur the delivery of new ideas as is the case with a profitable drug, nor is there a regiment of detail men out in the field to encourage their use. Even more important, however, is that new drugs are applied in a doctor/patient relationship in a one-on-one personal relationship, whereas educational or psychological ideas applied to the schools will have an impact on a wide spectrum of students and staff, to say nothing of the community or society as a whole.

The basic idea that is embodied in the P.L. 94-142, The Education for All Handicapped Children Act, that children with handicaps are better off in an educational setting with nonhandicapped children, has set large parts of the educational establishment on its ear for the past decade trying to adapt to this legislative mandate. One can postulate a general principle from that experience and many similar ones that *the more people that will have to change their own behavior because of a new idea, the greater the resistance there will be to the idea's acceptance.*

Still, history shows us that adaptations are made, discoveries are passed on, so some investment into the systematic study of the transfer process itself would be useful. We can trace one such process as an illustration. B. F. Skinner[13,14] developed a set of learning principles that he hoped to apply to all living organisms. His work is clearly one of the major events in the social sciences in this century. Much of his basic research work was done with animals in laboratory settings.

Skinner's major contribution to learning theory was to identify reinforcement following behavior, positive and negative, as a key concept to the shaping of subsequent behavior. But what does his work with pigeons have to do with mental retardation? Such a finding on reinforcement and operant conditioning can be taken in many directions. For example, Skinner extended his own work by initiating an era of programed learning with the instruction itemized into small steps in which each of the correct responses could receive positive reinforcement. His interest in mental retardation was casual to say the least.

STEP I TO STEP II—DISCOVERY TO APPLICATION WITH TARGET GROUP

It is the application of these principles to children with mental retardation that would mark Step II in the transfer of knowledge to practice. It took a number of years to bring together scientists who were particularly interested in the application of these principles to children and adults with mental retardation and developmental disabilities. The work of scientists at the Mental Retardation Centers, particularly at Washington and Kansas, were noteworthy in their application of Skinnerian principles to a variety of problems related to retarded children.[15,16]

The focus of the application of the principles of operant conditioning was to remove obnoxious behavior that interfered with learning and to encourage constructive behavior that enhanced the student's willingness to pay attention, persist through difficulties, and so forth. Yet, it is one thing to use the same principles in a research or laboratory setting, with merely a different target group of subjects, and it is quite another to organize the principles into specific protocols or curriculum that can be applied in less controlled environments, such as the school.

Implementation

The movement from Stage I to Stage II can be enhanced by earmarking research monies to specific targets, that is, appropriating research funds that would be spent on MR/DD for example. The creation of organizations such as the Mental Retardation Centers can provide a multidisciplinary base for colleagues interested in executing research on this problem.

STEP II TO STEP III—APPLICATION WITH TARGET GROUP TO GENERATION OF SPECIFIC PROGRAMS

This process of incorporating Skinner's principles into instructional programs beyond the laboratory probably took place through journal articles, convention programs, and university seminars to a mix of educational personnel in graduate university courses, professional school instruction, and at research centers that stressed a multidisciplinary approach to problems.

Implementation

The movement from research to program development is a more difficult transition. The researcher and the program developer are often physically separated, one in the university, the other in the service-delivery setting. One must rely at present on the uncertain bridges of college classroom, professional conventions, and journals to interest the practitioner in the potential of the new idea. The substantial money expended on the graduate training of leadership personnel in programs for disabled children through Public Law 88-164 introduced many key persons to the Skinnerian model, and many were able to see the applications and apply them in a practical setting.

STEP III TO STEP IV—GENERATION OF SPECIFIC PROGRAMING TO FIELD TEST

The fourth step introduces systematic field tests of the programs and is designed to demonstrate utility to educators, providing the professional credibility necessary for a full-scale acceptance of what often has to be a major shift in procedures in teacher training and in organization of school programs. Special educational research and demonstration funds have helped to create such opportunities in the past.

Implementation

From Step III to Step IV, program development to field test, requires access to a service-delivery system, a set of classrooms, or a school system where these ideas or protocols could be tested. Having people in educational administrative leadership and administrative positions who had knowledge and enthusiasm for these operant-conditioning principles helped to legitimize these practices educationally. Money for systematic demonstration coming from the Office of Special Education Programs in the United States Office of Education also gave a stimulus for the test of these ideas.[17]

STEP IV TO STEP V—FIELD TEST TO IMPLEMENTATION

Implementation is a process still going on today with the practices built upon Skinnerian principles. This stage may be marked by practitioners talking to one another about such practices without researchers taking part.

Implementation

Finally, one can tell when the implementation has been substantially made in the service-delivery field whenever the protocol or procedures are no longer even identified as research. Currently teachers are providing more *positive reinforcement* to their disabled students, and many of them believe that they do so out of common sense or historical or traditional educational practice. They have learned it through modeling their behavior after other master teachers using these methods. If they were asked if these procedures can be clearly linked to research, they would likely say, "No, it is just common sense."

RESEARCH ON THE PROCESS

Some researchable questions on the transfer process of knowledge to practice through this sequence might be the following:

(1)   A retrospective study of the progress through the system of a concept such as *metacognition*. Where and at what times did such an idea first begin to appear in each of the stages, and what were the communication channels?

(2)   Another approach would be to introduce a new term deliberately and then follow its progress. A term like *Type III error,* picking the wrong problem to attack, and then seeing how long and what avenues that idea takes to get into common usage would be a researchable issue.

(3)   The transition between Steps II and III bridging the gap between the scientific community and service providers could be traced by documenting the friendship networks of scientists influential in translating ideas to see if they have accomplished this transfer through direct and personal contact.

Three facilitators that would seem to be worthy of consideration to aid the transfer process would be the allocation of funds for: (1) research on this topic of transfer, (2) specifically encourage movement through these stages by supporting innovative program development and demonstration, and (3) training personnel in the use of these new procedures, so that practitioners can become comfortable with them and not resist them as new and strange approaches.

REFERENCES

1. —. 1982. Technology and Handicapped People. Office of Technology Assessment. Washington, DC. p. 51.
2. GALLAGHER, J. J. 1978. Organizational needs for quality special education. *In* M. Reynolds, Ed. Futures of Education for Exceptional Students: Emerging Structures. Council for Exceptional Children. Reston, VA.
3. ROGERS, E. 1962. Diffusion of Innovations. The Free Press of Glencoe. New York.
4. ROGERS, E. & F. SHOEMAKER. 1971. Communication of Innovations: A Cross-Cultural Approach. Free Press. New York.
5. HAVELOCK, R. G. 1973. The Change Agents' Guide to Innovation in Education. Prentice-Hall. Englewood Cliffs, NJ.
6. HOUSE, E., T. KERINS & J. STEELE. 1972. A test of the research and development model of change. Educ. Adminis. Q. 8(1).
7. TROHANIS, P. 1982. Planning for dissemination. *In* P. Trohanis, Ed. Strategies for Change.: 95–110. University of North Carolina at Chapel Hill. Technical Assistance Development System (TADS)—Frank Porter Graham Child Development Center. Chapel Hill, NC.
8. LEVIEN, R. 1971. National Institute of Education: Preliminary Plan for the Proposed Institute. Rand Corporation. Santa Monica, CA.
9. GLASER, E., ED. 1976. Putting Knowledge to Use. Human Interaction Research Institute. Los Angeles.
10. BRICKELL, H. M. 1961. Organizing New York State for Educational Change. State Education Department. Albany, NY.
11. KIESLER, S. & S. TURNER. EDS. 1977. Fundamental research and the process of education. (Final report to the National Institute of Education.) National Academy of Sciences. Washington, DC.
12. DEARMAN, N. & V. PLISKO, 1981. The Condition of Education. National Center for Educational Statistics. U.S. Department of Education. Washington, DC.
13. SKINNER, B. 1938. The Behavior of Organisms. Appleton-Century-Crofts. New York.
14. SKINNER, B. 1953. Science and Human Behavior. MacMillan. New York.
15. SACKETT. G., ED. 1978. Observing Behavior. Volume II: Data Collection and Analysis Methods. University Park Press. Baltimore, MD.
16. BROOKS, P., R. SPERBER &. C. MCCAULEY, EDS. 1984. Learning and Cognition in the Mentally Retarded. Lawrence Erlbaum Associates. Hillsdale, NJ.
17. GALLAGHER, J. J. 1975. Technical assistance and the nonsystem of American education. *In* National Technical Assistance Systems in Special Education. M. C. Reynolds, Ed.: 1–11. Leadership Training Institute/Special Education, University of Minnesota. Minneapolis, MN.

Technology Transfer: Lessons from Experience—The Communication Process

DONALD A. SNIDER

Institute for Basic Research in Developmental Disabilities
New York State Office of Mental Retardation &
Developmental Disabilities
Staten Island, New York 10314

Technology transfer is the process whereby research knowledge is disseminated from scientists to and among practitioners in order to fulfill actual or potential human needs. Technology transfer has occurred very rapidly in acute health care settings and, in fact, is seen as a major factor responsible for improvements in the quality of health care, as well as the upward spiral of health care costs.[1] On the other hand, historical accounts of the mental retardation (MR) field[2-5] would seem to suggest that technology transfer has proceeded generally much more slowly than in acute health care delivery.

Within the field of mental retardation, innovations used for diagnostic and preventive purposes appear to have diffused more rapidly than have treatment technologies per se. Furthermore, innovations in treatment technologies, at least until the introduction of pharmacotherapy and behavior modification, appear to have been the result of what Roberts[6] called "demand pull" (inspired by the practical necessities of coping with problems in the field) rather than "technology push" (arising from new research findings or technological capability).

This paper examines how we might accelerate technology transfer in and to the mental retardation field by applying some of the lessons derived from studies of the diffusion of innovations, a term used here synonomously with technology transfer. Although there are excellent review articles on various technologies in the field of mental retardation (e.g. behavior modification), there is no tradition of studies of diffusion of innovations of which I am aware; therefore, we must extrapolate and draw our lessons from studies in other fields, particularly medicine.

The specific focus of this paper is on communication, for technology transfer or the diffusion of innovations is a communication process. The extent to which we will be successful in bringing state-of-the-art technologies to bear on the prevention, diagnosis, and treatment of mental retardation will depend on our ability to communicate.

Lasswell, writing almost 40 years ago,[7] characterized the elements in the communication process as source, message, channel, receivers, and effects. These elements are shown in TABLE 1 along with their analogue in technology transfer. These elements in the communication process will be used as a framework for examining selected studies of technology transfer.

The classical study of technology transfer or the diffusion of an innovation was by Coleman, Katz, and Menzel, who studied the adoption of a new antibiotic by physicians in an Illinois community.[8] It should be recognized that the Coleman study has been criticized[9] because the adoption of new drugs for clinical use may differ from other technology transfer situations, limiting generalizability. I would argue, however, that the Coleman study provides some significant lessons about

TABLE 1. Elements in the Communication Process and Their Analogues in Technology Transfer

Elements	Counterparts
Sources	Scientists and clinicians who create knowledge and technologies; educators
Messages	Descriptions of the efficacy of the innovation or technology itself (a process, piece of equipment, or product)
Channels	Journals, meetings, seminars, academic courses, informal networks
Receivers	Practitioners, administrators, other scientists
Effects	Adoption, implementation, utilization, acquisition

both the communication process and the strategies needed to introduce new technologies into the mental retardation field. Some of the lessons from the Coleman study are:

1. Informal communication channels among the practitioners within the community, rather than formal channels (e.g., journals, scientific meetings, and drug promotional activity) were the more effective means of disseminating the information.
2. Receiver characteristics played an important role in determining the effects of a message. The first clinicians to adopt the new technology tend to be those with a higher income, a larger practice, a more cosmopolitan scope in their travel and reading, and a history of attending out-of-town specialty meetings.
3. The effects on receivers are not immediate and uniform, but occur over a period of time. A few clinicians will adopt a new technology quickly, followed shortly by the majority of practitioners, and finally the minority of "laggards."

Some criticisms of this model are warranted, notably that innovations diffuse frequently within organizations or among affiliated persons, rather than among persons operating independently.[9,10] For example, the diffusion of infant screening for phenylketonuria (PKU) and hypothyroidism has been as a result of states mandating the tests, which then are performed in hospitals or by reference laboratories. However, studies of burn care,[11] computed tomography scanning,[12] and other technologies indicate that Coleman provided a useful paradigm for examining many crucial aspects of the communication process underlying technology transfer.

The innovation itself is also a critical part of the communication process. Rogers[9] noted properties of the message describing an innovation that impede or facilitate its diffusion. These properties include relative advantage and complexity.

Relative advantage is the degree to which an innovation is perceived by the receivers as better than the practice it replaces. In determining relative advantage one must not only look at the efficacy of the technology, but also at safety and economic factors. The economic factors include its cost compared to that of the technology it replaces, the cost of the illness it prevents or cures, and the administrative overhead. These are all critical aspects of the message that the clinician or administrator seeks and upon which the decision to adopt a new technology is based.

Complexity, the degree to which an innovation is perceived as difficult to understand and use, is another property that affects its adoption and use. Technologies that are perceived to be more complex are less likely to be tried, or they will appeal only to a select audience (receivers), such as clinicians in teaching hospitals. Complexity encompasses both the skill needed to use a technology, and the expertise needed to understand its functioning and nature.[13] These represent different communication problems. While it may not be necessary for the technician to understand the theory behind the EEG, neurologists must have a comprehensive understanding of both theoretical and implementation aspects of the technology if they are to make valid interpretations of the electroencephalograms and understand the limitations of the procedure.

The National Institute of Child Health and Human Development's (NICHD) model for consensus development conferences described by James G. Hill earlier in this workshop[14] represents a systematic attempt to evaluate the relative advantage of competing technologies and communicate those results to the clinician community. The consensus development conferences on antenatal diagnosis, febrile seizures, and computed tomographic scanning are of particular relevance to the mental retardation field. Documentation of complexity issues, as well as the relative advantages of the technologies in question, are important functions of the reports on these conferences.

Above, the importance of informal communications channels in technology transfer was described, and several studies reinforce this lesson from Coleman *et al.*[8] Refereed journals, one important type of formal communication channel, promote knowledge and provide quality control of research through peer review; however, studies by Allen[15] indicated that these journals and other formal communication channels are not as helpful as informal contacts in changing the behavior of clinicians.

There is another problem with formal communication channels, namely, they are overloaded by the large and ever-increasing quantity of biomedical publications.[18] Bernstein noted, for example, that the National Library of Medicine collects 22,000 serials worldwide per year. In Bernstein's view the corollary to the increase in journals is that ". . . a significant fraction of the literature is redundant, less relevant, less correct, or less new, and hence less useful than the rest." The problems of "overloaded" communication channels and "noise" (i.e., less correct and less useful messages) are serious problems indeed.

IMPEDIMENTS AND ISSUES IN TECHNOLOGY TRANSFER

What are some of the lessons from experience about that special communication process known as technology transfer that could help bring improved treatment as well as diagnostic and prevention methods to the mental retardation field?

First, we need to establish a clear message—a consensus—about the relative advantages of various treatment, prevention, and diagnostic technologies for problems faced by the mental retardation field. This includes both biomedical and nonbiomedical technologies. Clearly, there currently are available some highly effective early interventions and other technologies to improve the quality of life of many persons affected by mental retardation. A first step will be to catalog these technologies and the circumstances under which they can improve the quality of life and improve the functioning of mentally retarded people. Although some of these treatment technologies may not meet all of NICHD's criteria for holding a consensus development conference described previously by Hill (e.g., the subject under consideration may not be medically important),[14] that should not deter us from evaluating these technologies and disseminating a clear message about their efficacy, safety, and costs. We also need to say what does not work in order to minimize the use of ineffective technologies or incorrect applications of proven methods. Majkowski,[17] for example, has noted that the effects of anticonvulsants on mentally retarded persons with epilepsy are often different than for nonretarded persons with epilepsy.

Second, while adequate formal channels of communication in the mental retardation field now exist for disseminating information in journals and at conferences, we need to improve our use of informal channels. We need to inform and educate the informal opinion leaders within our systems of the value of the innovations, so that they in turn can facilitate the diffusion process. We must also interact with those informal opinion leaders in the technology assessment and transfer process in other ways. We need them to assist us in the assessment process. We will also need to provide them with opportunities to interact directly with the scientists and clinicians who have created these methods. This will help give the opinion leaders "ownership" of the innovations and also benefit their colleagues, with whom they are likely to share this new knowledge enthusiastically. It will also benefit the scientists by showing them where real needs are, stimulating the further research and development needed by those working in the field of mental retardation.

Third, we need to increase the amount of attention medical schools give to mental retardation. We have heard at this workshop about some of the innovative techniques used in schools with University-Affiliated Programs (UAPs), but I cannot help wondering about what is going on at schools without UAPs. As far as continuing medical education is concerned, I am seeing more and more announcements for programs on mental retardation; however, we must determine if these courses (messages) are reaching the professionals who work with the mentally retarded. If, as I would expect, these messages are not getting through, we need to try other approaches.

REFERENCES

1. ALTMAN, S. A. & ROBERT BLENDON, Eds. Medical Technology: The Culprit Behind Health Care Costs? Proceedings of the 1977 Sun Valley Forum on National Health. DHEW Publication (PHS) 79-3216. Washington, DC.
2. SCHEERENBERGER, R. C. 1983. A History of Mental Retardation. Brookes Publishing Co. Baltimore, MD.
3. ROBACK, A. A. & T. KIERNAN. 1969. Pictorial History of Psychology and Psychiatry. Philosophical Library. New York.
4. CONOLLY, J. 1968. The Construction and Government of Lunatic Asylums. Dawsons of Pall Mall. London. (Originally published in 1847.)
5. CLELAND, C. C. 1978. Mental Retardation: A Developmental Approach. Prentice-Hall. Englewood Cliffs, NJ.
6. ROBERTS, E. B. 1981. Influences on innovation: Extrapolations to biomedical technology. *In* Biomedical Innovation. E. B. Roberts *et al.,* Eds. MIT Press. Cambridge, MA.
7. LASSWELL, HAROLD D. 1948. The structure and function of communication in society. *In* The Communication of Ideas. Lyman Bryson, Ed. Institute of Religious Studies. New York.
8. COLEMAN, J. S., E. KATZ & H. MENZEL. 1966. Medical Innovation—A Diffusion Study. Bobbs-Merrill. Indianapolis, IN.
9. ROGERS, E. M. WITH F. F. SHOEMAKER. 1971. Communication of Innovations: Across-Cultural Approach. Free Press. New York.

10. 1976. Office of Technology Assessment. Development of Medical Technology: Opportunities for Assessment. Publication Number OTA-H-34. Government Printing Office. Washington, DC.
11. JORDAN, H. S. et al. Diffusion of innovations in burn care: Selected findings. Burns 9(4): 271–279.
12. CREDITOR, M. & J. GARRETT. 1977. The information base for diffusion of technology: Computed tomography scanning. N. Engl. J. Med. 297: 49–52.
13. BERNSTEIN, L. M., V. H. BEAVEN, J. R. KIMBERLY & M. K. MOCH. 1975. Attributes of innovations in medical technology and the diffusion process. In The Diffusion of Medical Technology. G. Gordon & G. L. Fisher, Eds. Ballinger. Cambridge, MA.
14. HILL, JAMES G. 1986. Technology assessment at the national institute of child health and human development: An alternative model. N. Y. Acad. Sci. This volume.
15. ALLEN, T. J. The role of person-to-person communications in the transfer of technological knowledge. In Biomedical Innovation. E. B. Roberts et al., Eds. MIT Press. Cambridge, MA.
15. BERNSTEIN, L. M., E. R. SIEGEL & C. M. GOLDSTEIN. 1980. The hepatitis knowledge base, prototype information transfer system. Ann. Intern. Med. 93(Part 2): 169–181.
16. MAJKOWSKI, J. 1982. In Mental Retardation. Rayner, Sture & Lindstrom, Eds. Bertil. Lund, Sweden.

DISCUSSION

ROBERT GUTHRIE: I want to comment on a factor that has been neglected in the discussion of technology transfer, and that is the very effective role parents' organizations can play in the process. In São Paulo, a large state in Brazil, a parents' organization for the mentally retarded created public support more than 10 years ago for a newborn screening program. This group helped to pass a law mandating screening for PKU and hypothyroidism, the first law of its kind outside of the United States. It is ironic that only in the United States and in developing countries have such laws been necessary and that both in the United States and in Brazil it was parents' groups who successfully lobbied for such laws. In getting all newborn infants screened, the parents' groups have been more effective than professional organizations.

DONALD SNIDER: I agree; Ted Brown and I are working with the National Fragile X Syndrome Family Support Group to promote research and prevention programs. From a communication point of view, a support group represents a credible source, and their messages on sensitive topics are likely to carry more weight than those from a government-sponsored research organization.

ALLEN CROCKER: Although there have been long delays of 20–40 years between discovery and screening, in some cases there has been rapid application of information. I refer particularly to O'Brien's discovery in 1969 of hexosaminidase A, the enzyme deficient in individuals with Tay-Sachs disease. By 1971 large-scale screening was already being done in Baltimore for detection of carriers of Tay-Sachs disease. Again, this was largely as a result of sponsorship by parents' organizations.

PHILLIP SWENDER: What are the consequences of random or uncontrolled technology transfer, as, for example, in the use of beta-blocking agents to manage behavior? Some physicians use beta blockers in persons with behavioral disorders, particularly rage attacks, because it was found to blunt the "fight" reaction in persons with superventricular tachycardia. This use of beta blockers has occurred in a very random and uncontrolled way, and has now become "institutionalized" as a form of treatment and perhaps misused in a number of circumstances. What are your thoughts on the control of random technology transfer?

DONALD SNIDER: This type of random technology transfer is inevitable, but your example illustrates the need to assess technology and to then disseminate the results in a very targeted way in our field. Technologies can be applied improperly.

Index of Contributors